PSYCHOPHYSIOLOGY OF THE FRONTAL LOBES

CONTRIBUTORS

E. YU. ARTEMIEVA
A. S. ASLANOV
O. P. BARANOVSKAYA
CARMINE D. CLEMENTE
E. DONCHIN
JOAQUIN M. FUSTER
N. A. GAVRILOVA
M. GERBNER
L. K. GERBRANDT
JANE GRUENINGER
WALTER GRUENINGER

E. D. HOMSKAYA
DONALD B. LINDSLEY
M. N. LIVANOV
A. R. LURIA
D. A. OTTO
K. H. PRIBRAM
STEVEN C. ROSEN
EBERHARDT K. SAUERLAND
E. G. SIMERNITSKAYA
JAMES E. SKINNER
JOHN S. STAMM

W. GREY WALTER

PSYCHOPHYSIOLOGY OF THE FRONTAL LOBES

EDITED BY

K. H. PRIBRAM

Stanford University
Stanford, California

A. R. LURIA

University of Moscow
Moscow, U.S.S.R.

1973

ACADEMIC PRESS   New York and London
A Subsidiary of Harcourt Brace Jovanovich, Publishers

COPYRIGHT © 1973, BY ACADEMIC PRESS, INC.
ALL RIGHTS RESERVED.
NO PART OF THIS PUBLICATION MAY BE REPRODUCED OR
TRANSMITTED IN ANY FORM OR BY ANY MEANS, ELECTRONIC
OR MECHANICAL, INCLUDING PHOTOCOPY, RECORDING, OR ANY
INFORMATION STORAGE AND RETRIEVAL SYSTEM, WITHOUT
PERMISSION IN WRITING FROM THE PUBLISHER.

ACADEMIC PRESS, INC.
111 Fifth Avenue, New York, New York 10003

United Kingdom Edition published by
ACADEMIC PRESS, INC. (LONDON) LTD.
24/28 Oval Road, London NW1

LIBRARY OF CONGRESS CATALOG CARD NUMBER: 73-8251

PRINTED IN THE UNITED STATES OF AMERICA

CONTENTS

List of Contributors   vii
Preface   ix

Part One   Introduction

1. The Frontal Lobes and the Regulation of Behavior
   A. R. Luria   3

Part Two   The Effect of Frontal Lesions on the Electrical Activity of the Brain of Man

2. Application of the Method of Evoked Potentials to the Analysis of Activation Processes in Patients with Lesions of the Frontal Lobes
   E. G. Simernitskaya   29
3. Changes in the Asymmetry of EEG Waves in Different Functional States in Normal Subjects and in Patients with Lesions of the Frontal Lobes
   E. Yu. Artemieva and E. D. Homskaya   53
4. Changes in the Electroencephalogram Frequency Spectrum during the Presentation of Neutral and Meaningful Stimuli to Patients with Lesions of the Frontal Lobes
   O. P. Baranovskaya and E. D. Homskaya   71

Part Three   The Nature of the Electrical Activity of the Frontal Cortex in Man

5. Correlation of Biopotentials in the Frontal Parts of the Human Brain
   M. N. Livanov, N. A. Gavrilova, and A. S. Aslanov   91
6. Human Frontal Lobe Function in Sensory-Motor Association
   W. Grey Walter   109

Part Four The Nature of Electrical Activity in the Frontal Cortex of Nonhuman Primates

7. While a Monkey Waits
   E. Donchin, D. A. Otto, L. K. Gerbrandt, and K. H. Pribram — 125
8. The Locus and Crucial Time of Implication of Prefrontal Cortex in the Delayed Response Task
   John S. Stamm and Steven C. Rosen — 139

Part Five The Relationship between Frontal Cortex and Subcortical Brain Function

9. Transient Memory and Neuronal Activity in the Thalamus
   Joaquin M. Fuster — 157
10. The Role of the Brain Stem in Orbital Cortex Induced Inhibition of Somatic Reflexes
    Eberhardt K. Sauerland and Carmine D. Clemente — 167
11. The Nonspecific Mediothalamic-Frontocortical System: Its Influence on Electrocortical and Behavior
    James E. Skinner and Donald B. Lindsley — 185

Part Six Experimentally Based Models of Frontal Lobe Function

12. Study on the Functional Mechanisms of the Dorsolateral Frontal Lobe Cortex
    M. Gerbner — 237
13. The Primate Frontal Cortex and Allassostasis
    Walter Grueninger and Jane Grueninger — 253

Part Seven Conclusion

14. The Primate Frontal Cortex—Executive of the Brain
    K. H. Pribram — 293

Author Index — 315
Subject Index — 321

LIST OF CONTRIBUTORS

Numbers in parentheses indicate the pages on which the authors' contributions begin.

E. YU. ARTEMIEVA, Department of Psychology, University of Moscow, Moscow, U.S.S.R. (53)

A. S. ASLANOV, Institute of Higher Nervous Activity, Academy of Sciences of the U.S.S.R., Moscow, U.S.S.R. (91)

O. P. BARANOVSKAYA, Department of Psychology, University of Moscow, Moscow, U.S.S.R. (71)

CARMINE D. CLEMENTE, Department of Anatomy and the Brain Research Institute, School of Medicine, University of California at Los Angeles, Los Angeles, California (167)

E. DONCHIN,* Neurobiology Branch, NASA Ames Research Center, Moffett Field, California (125)

JOAQUIN M. FUSTER, Brain Research Institute and Department of Psychiatry, School of Medicine, University of California at Los Angeles, Los Angeles, California (157)

N. A. GAVRILOVA, Institute of Higher Nervous Activity, Academy of Sciences of the U.S.S.R. Moscow, U.S.S.R. (91)

M. GERBNER, Institute of Psychology, Hungarian Academy of Sciences, Budapest, Hungary (237)

L. K. GERBRANDT, Department of Psychology, Stanford University, Stanford, California (125)

JANE GRUENINGER, Department of Psychology, Stanford University, Stanford, California (253)

WALTER GRUENINGER, Department of Psychology, Stanford University, Stanford, California (253)

*Present address: Department of Psychology, University of Illinois, Urbana, Illinois.

LIST OF CONTRIBUTORS

E. D. HOMSKAYA, Department of Neuropsychology, University of Moscow, Moscow, U.S.S.R. (53, 71)

DONALD B. LINDSLEY, Departments of Psychology and Physiology and the Brain Research Institute, University of California at Los Angeles, Los Angeles, California (185)

M. N. LIVANOV, Institute of Higher Nervous Activity, Academy of Sciences of the U.S.S.R., Moscow (91)

A. R. LURIA, University of Moscow, Moscow, U.S.S.R. (3)

D. A. OTTO,* Department of Psychology, Stanford University, Stanford, California (125)

K. H. PRIBRAM, Department of Psychology, Stanford University, Stanford, California (125, 293)

STEVEN C. ROSEN, Department of Psychology, State University of New York, Stony Brook, New York (139)

EBERHARDT K. SAUERLAND,† Department of Anatomy and the Brain Research Institute School of Medicine, University of California at Los Angeles, Los Angeles, California (167)

E. G. SIMERNITSKAYA, Department of Psychology, University of Moscow, Moscow, U.S.S.R. (29)

JAMES E. SKINNER, Physiology Department, Baylor College of Medicine, and Neurophysiology Department, The Methodist Hospital, Houston, Texas (185)

JOHN S. STAMM, Department of Psychology, State University of New York, Stony Brook, New York (139)

W. GREY WALTER, Burden Neurological Institute, Bristol, England (109)

*Present address: Environmental Protection Agency, Clinical Environmental Research Laboratories, University of North Carolina, Chapel Hill, North Carolina

†Present address: Department of Anatomy, The University of Texas, Medical Branch, Galveston, Texas.

PREFACE

This volume on the *Psychophysiology of the Frontal Lobes* was initially conceived as a report of a symposium on frontal lobe function held in 1966 at Moscow during the 18th International Congress of Psychology. In this sense it was meant to be a companion to the report of another symposium held during that congress: the *Biology of Memory* edited by Pribram and Broadbent (1970).

For a variety of reasons, publications become delayed. In the case of the *Biology of Memory* volume, the delay allowed an opportunity to incorporate relevant reports presented at the centennial meeting of the American Psychological Association, and thus both enlarge the scope and the currency of that publication.

With regard to the manuscript of the *Psychophysiology of the Frontal Lobes*, the problem was compounded. Initially, we were faced with a then relatively recent and comprehensive book, *The Frontal Granular Cortex and Behavior* (Warren and Akert, 1964), which reported a symposium held at Pennsylvania State University. We therefore did not want to duplicate the contents of that publication, nor did we want our volume to become a repository for whatever trivial additions to knowledge might have accrued from slight modifications of experimental design, replication with larger n of earlier studies, etc. We therefore decided to all but eliminate from our report the anatomical and brain-lesion–behavioral studies that had made up the bulk of the Pennsylvania State symposium. This left us primarily with the electrophysiological–behavioral contributions, which had come of age and accrued since the earlier symposium. As in the case of the *Biology of Memory* manuscript, additional papers were recruited because of their relevancy to the topic.

The wisdom of our choice became progressively clearer as the delay in publication became extended. The Pennsylvania State symposiasts (including the editors of the present volume) felt the urge to reconvene in a workshop to gain understanding of some of the issues raised in the earlier symposium. Jerzy Konorski found the opportunity to make such a convention possible in conjunction with the 25th International Congress of Physiological Sciences, and it was held in August, 1971 at Jablona near Warsaw, Poland. The papers were published a little over a year later as a supplement to the *Acta Neurobiologiae Experimentalis*, a publication of the Polish Academy of Sciences (Konorski, Teubner, and

Zernicki, 1972). For the student interested in frontal lobe function, our volume thus becomes the third in a decade of intensive research on the problem. He should know, therefore, what distinguishes it from the other two. This can best be accomplished by first presenting a brief review of the highlights of the Jablona reports.

Hans-Lukas Teuber ably summarizes the Jablona conference under the captions "where," "when," "what," and "how," and we shall follow his suggestion here. With respect to "where," the conferees have fairly well identified two gradients of localization of disturbance of behavior by lesions: a dorsal—ventral gradient that relates to the spatial versus nonspatial dimensions of tasks and a posterior—anterior gradient that relates to a delay factor. The data for this insight came largely from the laboratory of Mishkin and Rosvold at the National Institutes of Health at Bethesda (Washington, D.C.), though amply confirmed by studies performed by Konorski's group in Warsaw and our own at Stanford and Moscow.

Another aspect of the question "where" has received extensive exploration—again by Rosvold and his associates. This question asks what neural systems other than the frontal are especially involved in the behavior that is characteristically disturbed by frontal lesions. Early experiments by Rosvold (Rosvold and Delgado, 1953) and Pribram (Migler, 1958) based on still earlier work by Mettler (1949) had implicated the head of the caudate nucleus in delayed alternation performance. Rosvold has shown that other parts of the basal ganglia (globus pallidus) are involved as are the parts of the dorsal thalamus especially related to the basal ganglia (n. centrum medianum). Divac and Potegal in two additional papers spell out the frontal lobe—basal ganglia relationship in detail.

What then of the frontolimbic relationship that one of the editors of the present volume has been pursuing since 1948 (Kaada, Pribram, and Epstein, 1949; Pribram, 1958, 1960, 1971)? Rosvold and Nauta both address this problem and confirm the relationship behaviorally and anatomically, and Butter adds interesting detail in his manuscript. The connection between basal ganglia and limbic systems also becomes clear in Rosvold's work: anterior, ventral, and medial parts of the frontal lobe are related to a ventromedial sector of the caudate nucleus, and it is this system that relates to the limbic formations. Recall that it is this system that is at the "nonspatial" and "delay" ends of the gradients that have been identified behaviorally.

The question "when" is addressed in papers by Warren (and one by Lukaszewska) on phylogeny, and a much needed study by Patricia Goldman on ontogeny. These papers resolve many of the discrepancies that have plagued the literature because different species and animals of different ages have been indiscriminately studied. Teuber brings the comparison of species home to man. He discusses the continuities and discontinuities that develop when, for ethical reasons, animal models must be employed to study the human condition.

The question "what" is answered by the Jablona conferees almost to a man (and woman) by recourse to physiology rather than psychology. Teuber, in his summary chapter, throws out a caution and quotes Milner to the same effect; but for the most part, the conference is addressed to the physiological basis rather than to the psychological effects of frontal lobe function. Thus, the failure on spatial dimensions of tasks is attributed to kinesthetic impairments, the failure on nonspatial dimensions, to connections with hypothalamic physiological drive mechanisms. The effects of frontal lesions on delayed performances is attributed to a memory trace deficit despite Stamm's and Niki's neurophysiological contributions that show frontal activity to occur primarily at the time of stimulus presentation. More of this in a moment.

The participants imply, though they do not state this explicitly, that the cerebral cortex as a whole is organized according to the three primitive embryological layers: ectoderm, mesoderm, and entoderm. Thus, for them, the posterior parts of the cortex deal with exteroception, the middle parts, with the control of muscle, and the mediobasal–limbic parts with visceral function. It is in the frontal cortex that mesodermal and entodermal functions come together. Whether, in fact, this is an accurate picture, or whether the participants are so imbued with this trichotomous view of brain function that they impose it on their data, needs careful examination. (They are, of course, not alone in these views: see for example, Paul MacLean's papers on "The Visceral Brain," 1949, and "The Triune Brain," in press.) In any case, little attempt is made by the authors to show a connection between these physiological conceptions and the related cognitive, conative, and affective psychological processes—again, the exception is Teuber who invokes the concept of feed-forward control by way of corollary discharge in an attempt to explain "how" the frontal lobe does its job.

It is to the "how" of frontal lobe function that the volume on the *Psychophysiology of the Frontal Lobes* is addressed. In short, our set of manuscripts begins pretty much where the Jablona conference leaves off, even though the work reported here often antedates or has been done coincident with the others. The point of contact is Stamm's contribution mentioned above, which, in somewhat different form, appears in both volumes.

In essence, this volume addresses questions left largely unexplored in the Pennsylvania State and Jablona conferences. As in the earlier publications, firm answers are not achieved, but directions are indicated and, what is perhaps even more important, whole sets of new techniques, hardly mentioned in the other publications, are brought to bear on elucidating the frontal lobe enigma. These applications are in their infancy but promise to provide a much richer vista since they fill the chasm between the straight anatomical and the behavioral-assessment-of-brain-lesion studies.

Most of the studies reported in this volume were performed in Moscow and on the American Pacific Coast, just as most of the studies reported at Jablona were performed in Warsaw and on the Eastern Seaboard of the United

States. Again, circumstance, not intent, caused this separation of efforts. Thus, the almost simultaneous publication of the two volumes takes on the additional role of bringing together, for a wider audience, complementary studies and views of a single community of scholars. We feel, therefore, that our efforts and those of our associates in the laboratory and at Academic Press have not been in vain, and we thank all of the contributors, publishers, editors, typists, translators, and readers for patiently accomplishing what was not always an easy task.

Karl H. Pribram

REFERENCES

Kaada, B. R., Pribram, K. H., and Epstein, J. A. (1949). Respiratory and vascular responses in monkeys from temporal pole, insula, orbital surface and cingulate gyrus. A preliminary report. *J. Neurophysiol.* **12**, 347-456.
Konorski, J., Teuber, H. L., and Zernicki, B. (Eds.) (1971). The frontal granular cortex and behavior. *Acta Neurobiologiae Experimentalis.* International Congress of Physiological Sciences, Jablona, Poland: 32(2).
MacLean, P. D. (1949). Psychosomatic disease and the "visceral brain"; recent developments bearing on the Papez theory of emotion. *Psychosom. Med.* **11**, 338-353.
MacLean, P. D. (1972). The triune brain. Paper presented at the Sesquicentennial Celebration of the Institute of Living. Hartford, In Press.
Mettler, F. A. (Ed.) (1949). "Selective Partial Ablation of the Frontal Cortex: A Correlative Study of its Effects on Human Psychotic Subjects." Hoeber, New York.
Migler, B. (1958). The effect of lesions to the caudate nuclei and corpus callosum on delayed alternation in the monkey. Master's thesis, University of Pittsburgh, Pittsburgh.
Pribram, K. H. (1958). Comparative neurology and the evolution of behavior. *In* "Behavior and Evolution" (A. Roe and G. G. Simpson, Eds.), pp. 140-164. Yale University Press, New Haven.
Pribram, K. H. (1960). The intrinsic systems of the forebrain. *In* "Handbook of Physiology, Neurophysiology II" (J. Field, H. W. Magoun, and V. E. Hall, Eds.), pp. 1323-1344. American Physiological Society, Washington, D.C.
Pribram, K. H. (1971). "Languages of the Brain." Prentice-Hall, Englewood Cliffs, New Jersey.
Pribram, K. H., and Broadbent, D. E. (Eds.) (1970). "Biology of Memory." Academic Press, New York.
Rosvold, H. E., and Delgado, J. M. R. (1953). The effect on the behavior of monkeys of electrically stimulating or destroying small areas within the frontal lobes. *Amer. Psychologist* **8**, 425-426.
Warren, J. M., and Akert, K. (Eds.) (1964). "The Frontal Granular Cortex and Behavior." McGraw-Hill, New York.

Part One

INTRODUCTION

Chapter 1

THE FRONTAL LOBES AND THE REGULATION OF BEHAVIOR

A. R. LURIA

*University of Moscow
Moscow, U.S.S.R.*

New attainments in the field of neuropsychology in the last decades have resulted in a radical revision of our views concerning one of the most intricate problems of natural science—the localization of functions in the brain cortex.

Classical concepts advocating *direct localization* of complex psychological functions within limited parts of the cortex, which once played essential roles in the development of neurology, have now become serious obstacles to further research in this branch of knowledge and have been discarded. Similarly, the naïve concept of *antilocalizationism*, in which the brain was regarded as a single unit and in which the character of functional disturbances was determined by the mass action of the affected brain, have now become things of the past. In their place, concepts of "system" localization of functions have become established; according to these concepts, each mode of psychological activity is a functional system based upon a complex interaction of jointly functioning parts of the brain, each of which contributes to the mode's expression (Luria, 1966a, b).

Radical changes have also taken place in the basic physiological concepts concerning the main principles governing nervous system activity; the classical concept of a reflex arc has given way to the concept of a servomechanism, a view that allows an approach to the analysis of the brain as a self-regulating system. A new theory of complex interaction between the specific and nonspecific systems in the neural process has come into being, and interaction between the activating apparatuses of the brain stem and the differentiated parts of the brain cortex appears now in a new light. A theory of the principal modes of activity of single neurons has been developed in which one part of the neuron is considered to show a highly specific character, whereas the other part is now regarded as being composed of extremely complex units that react sensitively to any changes in the incoming signal, compare the actual influences with past experience, and regulate the state of general brain activity.

These recent hypotheses have made possible a number of new approaches to the analysis of the functions of one of the most complex neural structures—the frontal lobes of the human brain.

I shall try to present herein a review of modern concepts relating to this problem, perhaps the most difficult problem confronting neural science today.

I. THE PROBLEM OF THE FRONTAL LOBES

Theories about the functions of the frontal lobes have always been contradictory. Neurologists who had observed patients with lesions of the frontal lobes but who did not detect any disturbances either in sensitivity or within the motor and reflex spheres often concluded that patients with such lesions remained symptomless and that this "youngest and least differentiated" part of the brain, in the words of Hughlings Jackson, had no strictly defined functions. However, psychiatrists who studied the distinctive features of behavior of patients with massive frontal lobe lesions and who described the "disturbances of motives (*Mangel an Antrieb*) and the absence of criticism" observed in such patients were inclined to regard the frontal lobes as one of the most important apparatuses of the human brain. Unfortunately, they too were unable to characterize the essence of the disorders observed in distinct physiological and psychological terms.

It is noteworthy that even leading physiologists experienced similar difficulties. I. P. Pavlov, for example, who observed dogs deprived of their frontal lobes, stated that the system of conditioned (salivary) reflexes did not undergo any substantial change, whereas general expedient behavior proved to be affected to such a degree that a lobectomized dog could be regarded as a "completely mutilated animal with few manifestations of expedient behavior left ... a profound invalid and helpless idiot ..." (Pavlov, 1949). Bekhterev also pointed out the great changes that take place in the behavior of such animals. He considered the frontal lobes of the dogs' brain to insure the "evaluation of the results of their actions and direct the movements to conform with this evaluation": thus, they perform a "psychoregulating function" Bekhterev, 1907, pp. 1964-1968). However, even in these cases the observers failed to analyze more thoroughly the mechanisms responsible for the disturbances of these higher forms of behavior, and science is still faced with the task of explaining frontal lobe function in clear terms accessible to further analysis. One is left with the impression that this is due to the fact that the concepts of classical physiology, especially the reflex arc scheme, are not adequate for disclosing the nature of frontal lobe activity and that explanation will come from considering the brain as a highly organized self-regulating system.

Some more successful attempts to elucidate frontal lobe function in such terms have recently been made by a number of scientists, who have investigated the behavior of *animals* (e.g., Pribram and his co-workers such as Rosvold and Mishkin, and by Konorsky and his group) and an analysis based on these principles of *human* frontal lobe function has been made in my laboratory over the last three decades.

The rest of this chapter presents a summary of the basic data obtained from these investigations.

II. THE FRONTAL LOBES AND REGULATION OF THE ACTIVATION PROCESSES

Each human activity starts from definite intention, directed at a definite goal, and is regulated by a definite program which demands that a constant cortical tone be maintained. It is known that when this tone declines, for example, when the cortex is in an inhibitory phasic state, selective responses of the higher neural processes tend to disappear: the normal finding that strong stimuli (or their traces) cause pronounced reactions and weak stimuli (or their traces) provoke less pronounced reactions, cease, and the mutilated cortex either begins to respond to strong and weak stimuli with more or less equal reactions (equalization phase) or the weak stimuli (or their traces) begin to provoke even more intense reactions than the strong ones did (paradoxical phase). The laws governing these phasic states have been thoroughly studied by the Pavlovian school. Their manifestations can readily be observed in a human when he is asleep or half awake. Such observations suggest that no organized thought is possible in these states and that selective connections are replaced by nonselective associations deprived of their purposeful character. It is possible that much of the peculiar logic of dreams can be explained by these physiological facts.

Anatomical, psychological, and clinical data obtained over the past years lead to the assumption that an essential role is played by the interrelations of the brain stem apparatuses and the frontal (especially mediofrontal) cortex in the maintenance of the latter's active state.

As is known, the nonspecific activating system of the brain stem is closely bound to the cerebral cortex, especially the cortex of the limbic region and medial parts of the frontal lobe. It is likewise known that descending impulses, which regulate the activating formation by modifying the state of cortical activity to conform to the subject's arising intentions, come predominantly from formations in the frontal parts of the brain.

These morphological and morphophysiological facts, the fundamentals of which were discovered by a brilliant group of researchers (Magoun, Moruzzi, Jasper, and others), have been more clearly elucidated in research by Grey Walter and M. N. Livanov, as well as in a number of investigations carried out over the past years by Homskaya and her co-workers (Luria and Homskaya, Eds., 1966; Homskaya, 1972).

According to Grey Walter, any expectation calls forth distinctive slow waves or "expectancy waves" that emerge first in the frontal cortex and then spread to other regions. His investigations showed that if the probability of emergence of an expected signal diminishes, a decrease of the expectancy waves occurs; when the instruction that has provoked the state of heightened expectation is countermanded, these waves fully disappear.

Almost similar observations were obtained by M. N. Livanov (see Chapter 5 in this volume). His investigations established that intense intellectual activity, for example, in the course of solving a complicated arithmetical problem, leads to the emergence of synchronous waveforms recorded from the frontal cortex. When the problem is solved and the intellectual activity ends, the number of synchronous waveforms decreases. The number shows a particular increase in states of permanent intellectual strain (for example, in the acute paranoid state) and decreased under the action of pharmacological sedatives.

All these facts show that the frontal parts of the brain cortex play an essential role in the regulation of the state of activation that arises as a result of some task given to the subject. Investigation of this function of the frontal lobes in normal subjects and of disturbances in the higher forms of regulation of the active states in subjects with frontal lobe lesions constitutes one of the principal objectives of our laboratory.

It is well known that the appearance of any new or significant stimulus provokes an orienting reaction, which, as shown by Sokolov and his collaborators (Sokolov, 1958, 1960), manifests itself in a number of vegetative and electrophysiological symptoms (constriction of the vessels of the arm and dilation of the vessels of the head, galvanic skin reactions, depression of the alpha rhythm, etc.). With gradual habituation to these stimuli, the vegetative components of the orienting reaction disappear. They reappear every time any change is made in the stimuli reaching the subject.

Investigations by Vinogradova (1959) and Homskaya (1960, 1961, 1965, 1966, 1972) showed, however, that it is possible to increase markedly the stability of such orienting manifestations and to make the response practically inextinguishable over a long period of time. This possibility emerges when a verbal instruction that imparts meaning to the stimulus is given to the subject.

Such increased stability of the vegetative or electrophysiological components of the orienting reaction occurs, for example, when the subject is given the instruction to watch whether certain changes will appear in the stimuli (such as in their force, duration, or quality), when the subject must count the number of presented stimuli, or when a given stimulus turns into a signal to perform some action (for example, when the subject must press a key in response to the presentation of each signal).

As established by the investigations of Homskaya, in such experiments the vegetative components of the orienting reaction, arising in response to the presentation of a signaling stimulus, persisted for a long time and became practically inextinguishable in the course of 15 to 20 presentations of the signals. With lesions of the extrafrontal parts of the hemispheres of the brain, it remains possible to stabilize the state of activation in the cortex by means of a verbal instruction that imparts meaning to the stimulus; however, this stability is greatly decreased when lesions of the frontal lobe are present.

Clinical observations have shown that patients with lesions of the postcentral, temporal, or parietooccipital parts of the brain may exhibit appreciable kinesthetic, auditory, or visual defects. In these cases, brain lesions may lead to serious disturbances of gnosis, praxis, and speech. These patients may experience considerable difficulty in spatial orientation, as well as in intellectual operations. Their attention, however, remains quite sustained; they fulfill their instructions with concentration, and intent firmly directs their activity. If we take into account the truly gigantic tasks that these patients are sometimes compelled to perform in order to compensate for their defects in the course of long rehabilitative training, the fact that their active state is stable is evident (Luria, 1972).

These phenomena were observed by Homskaya and her co-workers in a series of special investigations carried out on patients with lesions (tumors, wounds, hemorrhages) localized in the posterior parts of the hemispheres.

Their observations indicated that the vegetative or electroencephalographic components of the orienting reaction in these patients could become rapidly extinguished or sometimes (with massive foci) could exhibit a distorted, and in some cases paradoxical, character. However, when meaning was imparted to the stimuli by means of a verbal instruction, or, in other words, when the stimuli were imparted to the patient's conscious activity, these pathological phenomena were obliterated, and the vegetative or electrophysiological components of the orienting reflex assumed a more stable character. The verbal instruction stabilized the orienting reaction in these patients, and the symptoms of activation that were observed in them became steadier. The possibility of evoking stable activation of the cortex with the help of a verbal instruction, i.e., the possibility of regulating the state of the cortex by means of the second signaling system, caused these patients to approximate normal subjects.

Quite a different picture was observed in patients with pronounced lesions of the frontal lobes, and especially, of the polar, medial, and mediobasal parts of the frontal cortex.

Observations made by Homskaya showed that the vegetative (vascular and skin-galvanic) components of the orienting reflex, which, as in the above-described patients, could bear a pathologically distorted character, differed in one essential respect: the verbal instruction that included these stimulations in some kind of purposeful activity and imparted meaning to them did not lead here to a stabilization of the vegetative components of the orienting reflex. Here the corresponding reactions either were activated, although for a very short period of time, becoming extinguished at once, or did not appear at all, and the patient continued (sometimes successfully, and sometimes not) to complete the given task in the absence of the usual background of a stable increase in the tone of the brain cortex.

It was characteristic that such impediments to stable activation of the cortex as verbal instructions produced less derangement of response with lesions of the posterofrontal parts of the cortex and considerably more pronounced derangement with lesions of the polar, medial and mediobasal parts. These data agree with clinical observations of the general disturbance in sustained attention and active state in these patients, who easily lost the given problem as well as their intention, and who manifested some other forms of derangement of purposeful activity (see, for example, Homskaya, 1966, 1972).

Similar information was obtained from investigations of the electrophysiological components of the orienting reflex. Such investigations were carried out by Baranovskaya and Homskaya (1966). They showed that indifferent (i.e., neutral) stimuli usually led to a depression of the alpha rhythm and to a shift of the frequency spectrum to the zone of higher frequencies. It is characteristic that after the verbal instruction, which included the signals presented to the subject in his conscious activity (in other words, which imparted a meaning to the stimuli), the alpha rhythm depression assumed a considerably more pronounced character and became more stable. Similar phenomena were observed in patients with lesions localized outside the frontal parts of the cortex: sometimes these phenomena were observed against a background of considerable deviation in the general characteristics of the electroencephalogram. In patients with lesions localized within the frontal lobes (especially in their polar, medial, and mediobasal parts), the action of the verbal instruction, which imparts meaning to the stimuli, did not lead to such results: either there were no symptoms of cortical activation at all, or these symptoms appeared only for a very short time.

Similar data were obtained as a result of analyzing another electrophysiological indicator of the activation processes, suggested by Genkin (1963). Studying the dynamics of asymmetry of the ascending and descending phases of the alpha waves, he established that subjects who are in a state of rest exhibit a certain rhythm of oscillation of the asymmetry index that changes every 6–8 sec (Homskaya, 1972) during active attention to intellectual work, these periodic oscillations break down, and the magnitude of this breakdown may be regarded as symptomatic of the degree of activation (Artemieva and Homskaya, 1966).

Observations disclosed that, in patients with lesions localized outside the frontal parts of the brain, this periodic oscillation of the ascending and descending fronts of the alpha waves may be of a less uniform character; however, when the patients were involved in intellectual activity that strained their attention (for example, mental computation), the derangements in the periodicity of oscillations, the asymmetry index, became more distinct and showed the same law-governed character as in normal subjects. No such regular changes of the periodicity of the oscillations were observed in patients with lesions of the frontal lobes (and especially of their mediobasal parts); the character of periodic oscillations of this index did not show any substantial changes (Artemieva and Homskaya, 1966; Artemieva, 1965).

Consequently, in patients with massive lesions of the frontal lobes it proves impossible to evoke a stable activation of the brain cortex with the help of a verbal instruction and these patients lack a very essential condition for fulfilling a complex task.

Also of great interest are data, that were obtained by E. G. Simernitskaya. Examining potentials evoked during the presentation of various stimuli, she was able to establish their specific relation to separate points of the brain cortex. During cutaneous stimulation, potentials appeared in the sensory-motor cortex, and during visual stimulation, they appeared in the occipital cortex. The experiments showed that an active expectation of a signal called forth by a verbal instruction brought about a marked intensification of the potentials, which was most pronounced in the early stages of the experiments.

Similar phenomena were observed in patients with focal lesions localized outside the frontal lobes of the brain. In contrast to these results, in patients with massive lesions of the frontal parts of the brain, the verbal instruction did not result in intensification of the potentials evoked; in some cases, it led to a paradoxical emergence of symptoms of transmarginal inhibition, which indicated the inability in such cases to provoke a stable process of activation by means of a verbal instruction (Simernitskaya and Homskaya, 1966; Simernitskaya, 1970).

All of these data, obtained with the help of various methods but aimed at solving the same problem, convincingly show that lesions of the frontal lobes, which may not affect the elementary (direct) modes of the orienting reaction, and which sometimes even manifest themselves in its pathological revival, make it impossible to evoke a state of activation in the cortex by means of a verbal instruction. In other words, frontal lobe lesions violate the physiological basis underlying the regulation of the higher, specifically human, forms of attention.

The above conclusion is particularly well supported by data from lesions of the polar, medial, and probably mediobasal, parts of the frontal cortex, which, according to available data, are components of neural systems that insure the descending influences on the reticular formation and thus directly participate in the most complex forms of regulation of the states of neural activity. Similar results were obtained in a series of studies conducted by Filippycheva (1966) and Filippycheva and Fatler (1970) using a polygraphic registration of vegetative and electrophysiological symptoms.

III. FRONTAL LOBES AND THE EXECUTION OF VERBALLY PROGRAMMED BEHAVIORAL PROCESSES

Derangements in the regulation of the states of neural activity arising from lesions of the frontal cortex, the objective symptoms of which have been described in Section II, must inevitably lead to marked disturbances of complex forms of behavior. Naturally, these disturbances are manifest not as much in the elementary receptor processes and motor acts as in the complex forms of self-regulating activity that proceed from a definite intention and develop according to a definite program.

Let us consider these disturbances in detail, trying to induce them under experimental conditions and to investigate the mechanisms by which they are governed. We shall begin with symptoms that any observant investigator can discern in patients with massive (more often, bilateral) lesions of the frontal lobes.

If the patient's hands are lying on top of his blanket, he can respond to the command, "Lift the hand" without difficulty. However, during the performance of the second or third trial, symptoms of inactivity may appear; the movements slow down, the hands are not lifted as high, and after several repetitions of the orders the movements may be fully discontinued. Now let us repeat the same experiment under different conditions. This time, the hands of the patient (with a massive bilateral lesion of the frontal lobes) are under the blanket; in order to execute the instruction, "Lift the hand," he must perform a complex series of movements. First, he must free his hand from under the blanket, and only then can he lift it. It should be pointed out that the first part of this program was not mentioned in the instructions, and this intermediate intention must be formulated by the patient himself. As a rule, patients with massive bilateral lesions (tumors) of the frontal lobes do not perform such an action and soon replace the required movements with an echolalic repetition of the instruction: "Yes, lift the hand."

This observation supports the conclusion that patients with massive frontal lobe lesions are able to execute a direct order (which remains possible for them only within certain limits), but are unable to execute a complex program of actions if some links of it have not been formulated in the instruction.

Further observations make it possible to specify the conditions under which activity required of patients of this group becomes impossible and to ascertain the behavioral elements that replace the execution of a program inaccessible to these patients.

If we ask a patient with a massive lesion of the frontal lobes to repeat echopraxically the movements of the experimenter (for example, to lift his finger or fist if the experimenter's finger or fist is lifted), he will execute this task without difficulty, especially if the movements are separated from one another by considerable intervals.

Let us now ask the patient to perform movements that conflict with those performed by the experimenter, for example, to lift the fist when the experimenter lifts his finger, and vice versa. Observations will show at once that this conflict task is very difficult for the patients and sometimes even impossible.

A patient with massive lesions of the frontal lobes, as a rule, can easily repeat such an instruction to the experimenter (at first he repeats it in parts and subsequently, in full); he can memorize it and reproduce it over a rather long period of time (sometimes after one and even several days). However, his attempts to execute this instruction show considerable difficulty; his first attempt is successful but after two or three trials he becomes unable to execute the task and, when perceiving the experimenter's movement, he no longer regards it as a conditioned signal. He ceases to recode this signal in conformation with the instruction and at once begins to replace the conditioned movement, which is required by the instruction, with a simple imitative stimulus-bound repetition of the experimenter's movements. At first, such a slip from the given program is accompanied by attempts to correct the error: when reproducing the instruction aloud, the patient corrects the admitted error. Subsequently, however, the corrections disappear and the required movement is replaced by a simple imitative reproduction of the experimenter's movement.

Numerous variants of such experiments (a request to knock twice in response to one knock or vice versa, a request to perform a slight movement in response to a strong signal or vice versa, etc.) produce the same results. We can therefore conclude that, although patients with massive lesions of the frontal lobes preserve the ability directly to obey a signal, they are unable to inhibit this imitative tendency and to subordinate their behavior to an internal scheme by recoding the signal.

Similar observations may be obtained from patients with massive lesions of the frontal lobes placed in conditions that require constant switch-overs from one program to another.

Let us give such a patient the task of responding to two alternating signals with conditioned movements, without requesting him to perform movements that are in conflict with the signals applied. For example, let us ask him to lift the right arm in response to one knock and the left arm in response to two knocks. This instruction is easily repeated by our patients (first in part and subsequently in full) and is easily retained by them. Even in this case, however, patients with massive lesions of the frontal lobes soon begin to experience marked difficulties in accomplishing the task. These difficulties are connected with the need for making constant switchovers from one movement to another. They must overcome the pathological inertness of the established stereotype, particularly pronounced in patients with lesions of the convexital (motor) parts of the frontal lobes. This complication is responsible for the fact that a patient with massive lesions of these parts of the brain very quickly shows a tendency to inertly reproduce the same movement (for example, lifting the same arm in response to both signals), or, as it often happens, to alternate, also inertly, both reactions (by lifting the right and left arms), irrespective of the signal that is presented to him. The phenomena of pathological inertness of the established stereotype and the substitution of an inert reproduction of the motor stereotype for the execution of the required program are in these cases particularly distinct, if we first strengthen this stereotype by means of a regular alternation of both signals (R–L, R–L, R–L) and then interrupt it by presenting the signals irregularly (R–L, R–L, R–R, L–L). As a result of such a breakdown of the stereotype, patients with a pronounced frontal syndrome as a rule, cease to adapt their behavior to the given program and begin to replace the necessary coding of the movements by an inert reproduction of the elaborated motor stereotype.

In these cases, too, the errors committed may be corrected by matching the given movements to the instruction that is repeated by the patient aloud. This correction rapidly disappears, however, and the program reactions of the patient become an inert motor stereotype that as the experiment continues may (in cases of particularly massive lesions of the frontal lobe) even lead to a deformation of the previously retained verbal instruction.

Disturbances in the regulation of behavior elicited by complex tasks are particularly obvious in experiments where patients with massive lesions of the frontal lobes are asked to execute asymmetric programs easily conflicting with the tendency to execute symmetric programs. A typical example of such an asymmetric program is the task of laying out rows consisting of one white and two black checkers or of drawing rows consisting of two crosses and one circle. In these cases, patients with massive lesions of the frontal lobes, who easily retain the verbal instruction and as easily correct errors committed by other persons, very soon begin to deform their own actions and replace the asymmetric program by a symmetric one, inertly forming rows consisting of two black and two white checkers or drawing rows of two crosses and two circles. These observations were considered in detail in some other works (Lebedinsky, 1966) and there is no need to repeat here the analyses contained in these publications. As in the previously described experiments, such deformation of an action, which easily loses its program, may be observed only in patients with massive lesions of the frontal lobes, whereas patients with other focal lesions do not exhibit such instability in their intentions, once formed.

These difficulties in the execution of complex programs and the inability to suppress involuntarily emerging inert stereotypes may be observed with particu-

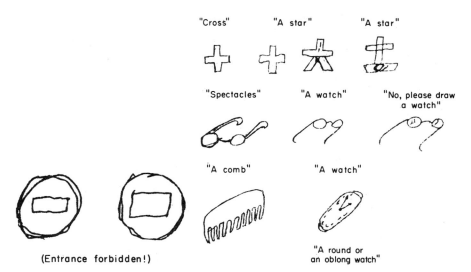

FIG. 1. Drawing by a patient with a left frontal tumor.

FIG. 2. Drawing by a patient with aneurism of the anterior cerebri communicans anterior and lesion of both frontal lobes.

lar clarity in experiments where patients with massive frontal lobe lesions are asked to draw pictures suggested in the verbal instructions. In these cases, the act of graphical presentation of objects constitutes by itself a complex motor program, and the tendency to replace it by an inert stereotype may here assume a particularly distinct character. Two examples that may help to analyze this disintegration of behavior in patients with massive lesions of the frontal lobes will be cited here. In the first case (Figure 1) a patient with a massive tumor of the frontal lobes was asked to draw "a circle and a minus"; the patient first drew a circle and then, without displacing his hand, drew a rectangle inside the circle and wrote the following words: "Entrance strictly forbidden." The firm stereotypes of the patient's former profession (he was a chauffeur) were revived by the first part of the program set and turned the execution of the given instruction into a reproduction of a prohibitive road sign. In other words, such patients replace the fulfillment of the given program by an inert stereotype. No less illustrative is the second example, which relates to another patient with a massive bilateral tumor of the frontal lobes. Having drawn spectacles without difficulty, this patient was unable to switch over to the execution of another task, to draw a watch. He made several attempts to comply with the instruction, but continued inertly to draw spectacles, including in his drawing figures and hands (Figure 2). In another publication (Luria, 1966a, b) we cited some other examples, so we shall not analyze these disturbances in greater detail here.

Numerous observations (Luria and Homskaya, eds., 1966) showed that the above-described disturbance of behavior that is regulated by a complex verbal program arises not only in patients with massive (bilateral) lesions of the frontal lobes; symptoms of this disturbance may be observed also in patients with less pronounced (mostly, left-side) lesions of the frontal lobes. They may be expressed as a replacement of the execution of complex programs by impulsively emerging fragments or inert stereotypes that at first are corrected by the patients by matching the given action to the initial task and then very quickly become uncorrectable.

Such disturbances may be observed not only in the execution of complex motor tasks but also, and even more distinctly, in the execution of verbal programs that require constant switch-overs from one action to another and the constant necessity of internal coding of the performed act.

Probably the best example in this respect is the accomplishment by such patients of a task that is well known in the clinic, making successive subtractions of 7 from 100. In these cases, the patients are constantly compelled to effect a switch over by converting the differences into the minuend and to skip over tens, dividing the subtrahend 7 into its component parts, and accomplishing the subtraction by means of two successive operations. Because patients with a lesion of the frontal lobes tend to replace the execution of flexible programs by inert stereotypes, the investigator may very often observe here a replacement of complex flexible operations by an inert repetition of a solution that has once arisen (for example, 100, 93, 83, 73, 63), by an uncontrolled repeating stereotype of skipping over tens (for example, 100, 93, 86, 69), or by a reduction of the operation performed and mechanical addition of a part of the subtrahend 7 to the next ten (100, 93, 84); the error may be decoded here as $93-3=90$ with a replacement of the further operation $(90-4)$ by an addition of the last subtrahend to the next ten, which gives 84. A thorough analysis of the rules by which the execution of flexible verbal programs by patients with massive frontal lobe lesions becomes deranged makes it possible to ascertain and describe some important laws that may provide highly valuable information concerning the role of the human frontal lobes in the regulation of complexly programmed forms of behavior.

IV. FRONTAL LOBES AND PROBLEM-SOLVING BEHAVIOR

Disturbances in the execution of complex programs constitute only one aspect of the behavioral defects that arise in massive lesions of the frontal lobes. Another, still more pronounced, aspect of the frontal syndrome is a disturbance of the independent choice of programs of action. This type of disturbance may be observed even in cases of relatively mild lesions of the frontal lobes, and primarily, lesions of the prefrontal parts of the cortex. Like the previously described disturbances, they may be observed in different forms of activity.

During our investigations, which lasted several years, we more often observed such derangements in patients given tasks of constructive activity, where the correct solution depended upon a preliminary systematic orientation in the conditions (orienting basis of the activity), that makes it possible to form an adequate hypothesis and to choose a proper system of operations from among numerous alternatives.

The accomplishment of visual constructive tasks in the so-called Kohs block and Link cube present a typical example of such activity.

In the first task, the subject must lay out a definite pattern with the help of blocks, some sides of which are divided diagonally into two parts of different colors. Whereas the task of laying out simple patterns (for example, consisting of one yellow and one blue square) does not require any preliminary complex analysis of the given conditions, more complex patterns (for example, a big blue triangle against a yellow background formed of two yellow and blue blocks) markedly complicate the task: the subject must analyze the principle of construction, converting three perceptive elements of the figure (background–triangle–background) into two construction elements (yellow and blue blocks, of which one triangle can be formed), and only such preliminary coding allows a successful solution of the given task.

Something analogous takes place in experiments with constructing the Link cube. Here, the subject must form one big yellow cube with the help of 27 small blocks whose sides are of different colors; in some of the blocks three sides are yellow, in others, two sides; in still others, only one side is yellow, whereas one block is colorless. In order to solve the task, the subject must first calculate the required number of elements of each group. This allows him to place the blocks with three yellow sides in the corners of the big cube, the blocks with two yellow sides along its edges, and the remaining blocks in the middle of each surface of the cube and inside it.

Investigations carried out by Tsvetkova (1966a, b) and Gadzhiev (1966) showed that the solution of this task, which is quite accessible to every educated adult, greatly suffers in patients with even a mild frontal syndrome. Patients of this group divide their activity into two stages: initially they orient themselves to the conditions necessary for the execution of the task, in order to elaborate a program of proper actions after the preliminary analysis and calculation of the elements required has been carried out. As a rule, patients with lesions of the frontal lobes, with their distinctive impulsiveness, never start with a preliminary analysis of the task's conditions but immediately attempt to solve. This leads to typical errors in planless solving attempts that usually remain uncorrected. Such disturbances are typical of the behavior of patients with frontal lobe lesions but they are never encountered in patients with other localized lesions (where the orienting basis of activity, as a rule, remains intact and only the ways in which this activity manifests itself may suffer).

FIG. 3. Eye movements of a normal subject during a 3 min observation of the picture *The Unexpected Return*, by I. E. Repin (Yarbuss, 1965). (1) Free observation. (2) Question: Is the family rich or poor? (3) Question: How old are the people in the picture? (4) Question: What was everybody doing before the man returned home? (5) Question: What kind of clothes are they wearing? (6) Question: What pieces of furniture are in the room? (7) Question: How long was the man absent?

FIG. 4. (Opposite page) Eye movements of two patients with lesions of the frontal lobes during a 3 min observation of the same picture as in Fig. 3 (Luria et al., 1966). (1)–(7) as in Fig. 3.

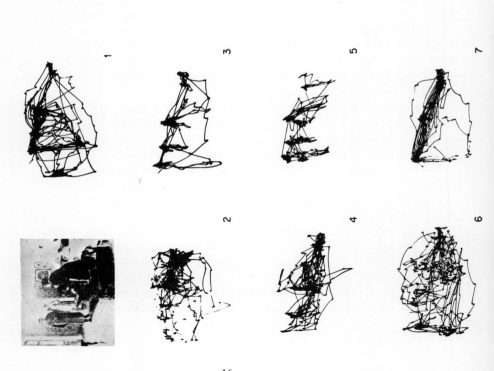

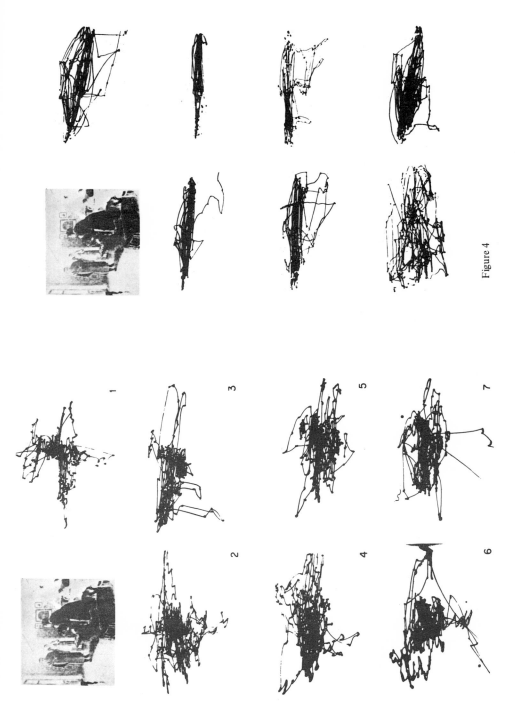

Figure 4

Disturbances of preliminary orientation in complex tasks to be executed by patients with lesions of the frontal lobes assume a particularly distinct character in experiments where the patients are asked to examine a picture having a certain subject and to draw conclusions concerning its content. The method of recording the movements of the eyes, which has now become possible because of a number of special investigations [in the U.S.S.R., the investigations by Yarbuss (1965) who elaborated a method of recording delicate movements of the eyes with the aid of a light beam that is reflected in a mirror fastened to the sclera of the eye], made it possible to reveal some details previously inaccessible to analysis.

The investigations demonstrated that if a normal adult is asked to examine a complex picture for 3 min, for example, Repin's *The Unexpected Return,* his gaze singles out some essential elements of information from this picture and confronts them. As a result, the subject comes to the proper solution after an intensive preliminary orientation in the details of the picture and after singling out the complexes that determine the sense of the picture. If the task that is formulated during the examination of the picture is changed, there occurs a corresponding change of the units of information singled out by the subject, and the movements of the eyes recorded on photographic paper become quite different, reflecting the character of the preliminarily formulated task. In Figure 3 (Yarbuss, 1965) it may be seen that the movements of the eyes during the examination of the aforementioned picture, when the subject had to answer the question, "What is the age of the persons shown in the picture?" proved to differ from those that were observed when the task was to answer another question: "Is the family shown in the picture rich or poor?" or, "How are the persons shown in the picture dressed?" The question, "How many years, in your opinion, did the man who unexpectedly returned home spend in tsarist exile?" evoked intensive activity aimed at comparing the age of the persons portrayed, providing the grounds for the final answer.

Of a quite different character was the examination of the same picture by patients with massive lesions of the frontal lobes. Figure 4, taken from an investigation carried out by Karpov in our laboratory (Luria *et al.*, 1966; Karpov *et al.*, 1968) and showing records of the movements of eyes of two patients with massive lesions of the frontal lobes, indicates that organized visual analysis of the material is here fully absent; patients with massive lesions of the frontal lobes do not perform any preliminary orienting activity. They do not single out the most informative details of the picture and do not confront them. The gaze of such a patient wanders chaotically about the whole picture, without changing its focus under the influence of the task set; sometimes it assumes the character of inertly repeating movements that in no way reflect any thinking activity. It is clear, therefore, that the patient's answers do not result from active analysis but present occasional guesses that come into his mind from passively emerging stereotypes of previous experience. (Answering the question, "How many years, in your opinion, did the man, who unexpectedly returned home, spend in tsarist exile?" the patient says: "Probably 20 or 30 years. In those days the terms of exile were quite long . . .") Thus, the intellectual activity of such a patient is deprived of the orienting basis of action, which, according to modern psychology, constitutes a highly important foundation of any intellectual act.

There is no need to remind the reader here that this type of disturbance of visual thought is characteristic only of lesions of the frontal lobes and markedly differs from disturbances of the operative side of visual perception. The latter may result from lesions of the parietooccipital parts of the cortex when the patient's attempts to orient himself in the content of the picture remain intact and only the technique of the spatial analysis of the elements is violated.

The following essential question arises: What is the structure form of the loss of preliminary orientation in a complex task and what are its consequences?

One such experiment was suggested by Sokolov (1960) and carried out by patients with lesions of the frontal lobe by Tikhomirov (1966.)

In order to answer this question, it is necessary to perform special experiments that could model, possibly in more distinct forms, the entire complexity of the obstacles that are encountered by the cognitive activity of man.

If a subject with closed eyes is asked to feel the contours of two letters with his fingers (for example, of the Russian letters H and П) laid out with the help of checkers and then to say which of these letters is presented by the shape, his cognitive activity will change with the gradual change of his actions. First he will feel all the elements of the given contour in sequence, and the orienting phase of his activity will be prolonged. When exploration begins to decrease, the subject will ignore all the elements not possessing distinctive features (for example, the lateral walls of both letters do not differ from each other). The subject's fingers begin to single out only those checkers that bear some useful information: by singling out a checker that enters the upper crosspiece he can immediately conclude that this is the contour of the letter П, whereas the absence of such a checker or the presence of a crosspiece in the middle of the vertical strictly indicates the contour of the letter H. Consequently, a reduction of the search and the singling out of the most informative points constitute the basic feature of the phase that is called by us the *orienting basis of action.*

Tikhomirov's investigation shows that active attempts to single out the points that bear the greatest amount of distinctive information are preserved in patients with lesions of the posterior, for example, temporal or parietooccipital, parts of the brain, being made difficult in the latter case only by a disturbance of spatial orientation. However, in patients with pronounced lesions of the frontal lobes, the organization of the observed activity greatly changes. Such patients continue to feel the contours of the letters randomly with their fingers; they do not single out the most informative points, never curtail their actions, and simply guess which of the two letters is under their finger. Thus, the preliminary phase of activity, namely, the ascertainment of the most informative attributes, which subsequently form the basis of a hypothesis concerning the required answer, may be completely absent in these patients, and their guess does not depend on any preliminary orienting activity. At the same time this experiment reveals another peculiar feature characteristic of patients with a pronounced frontal syndrome: as a rule, their hypothesis does not turn into a starting point for its subsequent verification; having expressed it, usually they never return to the given contour and do not match it to their supposition. Not only is the initial phase, i.e., the orienting basis of action, absent in them, but also the final phase, the phase of its verification. Thus, the complex apparatus that was designated by P. K. Anokhin as the *acceptor of the consequences of action* is also violated in these patients.

The above-described disturbances of activity in patients with lesions of the frontal lobes manifest themselves with particular clarity also in the patients' *verbal thought processes*. In view of this, analysis of the ways in which patients with a distinct frontal syndrome solve more or less complex intellectual problems is of particular interest to us.

As is known, formal logical operations may remain intact enough in patients with lesions of the frontal lobes. In cases where such operations make use of consolidated logical stereotypes, without the patients having to choose from several equiprobable alternatives, the establishment of proper logical relationships proceeds without any appreciable difficulties. For example, patients with a pronounced frontal syndrome can easily find relations of contrast, relations between the whole and a part, and relations between genus and species, or can even solve problems in analogy by matching to a given notion some other notion that is properly related to it. However, a special investigation demonstrated that when such a patient is given a similar task that includes a choice from several equiprobable alternatives (for example, when he is presented with an elective variant of a test in analogy), the task becomes practically insoluble and the patient begins to operate within any randomly arising relationship, being unable to make the necessary choice (Luria and Lebedinsky, 1968). This means that the basic thought defect characteristic of patients with lesions of the frontal lobes consists not in the absence of logical codes in them but in the ability to make a goal-directed choice from a number of equiprobable alternatives.

This is particularly obvious in patients with a pronounced frontal syndrome when they solve complex arithmetic problems. As shown by a special investigation (Luria and Tsvetkova, 1966, 1967), such patients easily cope with simple problems that have single solutions, i.e., that do not require any choice from several equiprobable alternatives; but they are unable to solve a problem that requires preliminary orientation within its conditions and formation of a hypothesis, which in normal subjects leads to a proper choice from several equiprobable alternatives and to the performance of some intermediate operations insuring the attainment of the final goal.

The disturbance in the organization of the intellectual act that arises in patients with the frontal syndrome proves to be particularly distinct. Such patients do not subject the conditions of the problem to preliminary analysis and do not confront their separate parts. Instead, as a rule, they single out random fragments of the conditions and begin to perform partial logical operations, without attempting to formulate a general strategy and without confronting this operation with other elements of the conditions of the problem; neither do they match the result obtained to the initial conditions.

Let us suppose that such a patient is given the following problem, which is easily solved by children at the age of 12: "18 books were placed on two shelves; on one of the shelves there were twice as many books as on the other. How many books were there on each shelf?" The patient, correctly repeating the condition of the problem, may easily replace the final question by a more accessible one: "How many books were there on both shelves?" At the same time, he will not confront this question with the initial condition and will not take notice of the fact that the answer to it has already been included in the condition of the problem. If we draw the patient's attention to his wrong repetition of the question and make him formulate it correctly, the solution of the problem will still remain unsecured. Without forming any proper strategy, the patient at once begins to perform a fragmentary operation: "twice as many ... means 2 times $18 = 36$"; at the same time he does not match the result obtained to the initial condition and does not notice the contradiction.

Two basic features characteristic of behavioral disturbances in lesions of the frontal lobes (namely, the disintegration of the orienting basis of action accompanied by the inability to effect a preliminary analysis of the given condition or to make a subsequent choice from several equiprobable alternatives, and the inability to match the final results to the initial conditions) are reflected in the process of problem solving by patients with lesions of the frontal lobes more distinctly than in any other activity. The above-described features markedly differ from the character of the disturbances that are displayed in problem solving by patients with lesions of the posterior (parietooccipital) parts of the cortex. Attempts at preliminary orientation in the conditions of the problem remain intact in these patients, but the performance of proper operations encounters certain, typical difficulties that are connected with the disintegration of the spatial and numeric mental processes (schemata). It is natural, therefore, that research into the process of accomplishing complex intellectual activity discloses the most delicate symptoms of frontal lobe lesions and provides one of the most reliable methods for early diagnosis of these lesions.

V. CONCLUSIONS AND PERSPECTIVES

I have now concluded the review of some basic information that has been obtained during the past 30 years in my laboratory from studying the functions of human frontal lobes. Let me summarize these facts and outline the main prospects for further research.

The data obtained made it possible to reveal the basic functions performed by the frontal lobes, the important organ that effects the highest forms of coding of human behavior and the regulation of psychological processes.

These data show that the frontal lobes play an essential part in the higher forms of regulating the states of activity. They control the active state of the cortex, which is necessary for the accomplishment of complex tasks, and play an important role also in the execution of intentions that determine the direction of human activity and impart to the latter an elective and purposive character. Numerous observations have also revealed the role of the frontal lobes in the execution of complex programs of activity, the formation of the orienting basis of action, and the organization of its strategy. Further, their role in the process of matching the effect or consequence of action to the initial intention which is the basis of the highly important function of the modification of action. All this undoubtedly makes possible a further approach to disclosing the nature of the "derangement of initiative" and "defects of criticism" that psychiatrists describe as resulting from lesions of the frontal lobes, although their physiological mechanisms remain obscure.

The result of the investigations described in this chapter represent the first attempts to replace these empirical notions with scientific ones. However, it would be too presumptuous to believe that these investigations provide a basis sufficient to disclose the physiological mechanisms of the functions of the frontal lobes. The research so far done by us in this direction may be rightfully regarded only as a description of the structure of the behavioral disturbances arising in lesions of the frontal lobes. A thorough investigation of the changes in the dynamics of the neural processes that are the causes of these disturbances remains to be made.

The disturbances in regulation of the active states of the brain, set forth in Section I of this chapter, lead to the assumption that the above-described changes are caused by disturbances of the tone of the cortex and by inability of the cortex to retain its heightened state of activation. (Patients with a pronounced frontal syndrome may preserve their heightened orienting reactions to elementary stimuli, but the possibility of evoking a stabilization of the states of activity by means of a verbal instruction disappears.) What could be the physiological mechanisms of this defect?

Some facts indicate that the endurance of the activity of cells of the frontal lobes declines in these cases to such a considerable degree that any attempt to retain their stable active state is frustrated by a phenomenon that is close to parabiotic inhibition. Such an hypotheses, however, requires verification and this can be done only with the help of special neurophysiological methods, the application of which is still a thing of the future.

The second problem that requires further investigation is the pinpointing of the exact loci of the frontal lobe lesions that are responsible for the above-described derangements of behavior.

In the course of our investigation we intentionally spoke of the frontal lobes as a single unit and tried to characterize the principal behavioral disturbances arising as a result of their lesions and to contrast these cases with lesions localized outside the frontal lobes. However, that which was admissible in the first stages of research turns into an obstacle in its later stages.

A considerable number of facts have been accumulated showing that the frontal lobes do not present a single and homogeneous system and that the role played by their separate parts in the organization of behavior is not uniform. It is already quite well known that the polar and *medial* (and perhaps also mediobasal) parts of the frontal lobes participate in the regulation of the state of activation and that their resection results in the inability to protractedly maintain the state of cortical activity by the task offered to the patient and in a derangement of selectivity leading to rapid inhibition of the selective traces, the formation of diffuse associations, and, in the final analysis, the emergence of a confused state of consciousness (Luria *et al.*, 1967). Furthermore, we know that the convexal parts of the frontal lobe (and especially the posterofrontal parts of the cortex) preserve an intimate connection with the cortical parts of the motor analyzer and that lesion in this location as a rule, leads to a motor stereotype and to a derangement in the complex forms of organization of movements that manifests itself against the background of absolutely intact consciousness. Finally, we know that there exists a distinct division of function between the frontal parts of the dominant (left) and subdominant (right) hemisphere, and that unilateral lesions lead to the emergence of different syndromes: in some cases lesions cause a disturbance of the programs of verbal thought and in others, a disturbance of emotions.

All of this information, however, is insufficient for a complete understanding of the functions of the frontal lobe syndrome and further elaborate investigations of separate morphofunctional formations entering it are required.

This concludes the review of an initial approach to the question of the functions of the frontal lobes in the regulation of behavior. I leave the solution of other questions touched upon in this paper to those future investigators who tackle the analysis of this perhaps most intricate field of natural science.

REFERENCES

Artemieva, E. Yu. (1965). Periodical oscillations of asymmetry of the EEG-waves as an indicator of activity. Thesis, Moscow University (in Russian).

Artemieva, E. Yu., and Homskaya, E. D. (1966). Changes in asymmetry of EEG waves in different functional states of activity in normal subjects and in patients with lesions of the frontal lobes. In "Frontal Lobes and Regulation of Psychological Processes" (A. R. Luria and E. D. Homskaya, eds.), pp. 294-313. Moscow Univ. Press, Moscow (in Russian).

Baranovskaya, O. P., and Homskaya, E. D. (1966). Influences of attention on the frequency spectrum of EEG in normal subjects and in patients with lesions of the frontal lobes. In "Frontal Lobes and Regulation of Psychological Processes" (A. R. Luria and E. D. Homskaya, eds.), pp. 254-276. Moscow Univ. Press, Moscow (in Russian).

Bekhterev, V. M. (1905-1907). "Fundamentals of the Theory of Brain Function," Vols. 1-7. St. Petersburg (in Russian).

Filippycheva, N. A. (1966). Neurophysiological mechanisms of disturbances of motor reactions in lesions of the frontal lobes. In "Frontal Lobes and Regulation of the Psychological Processes" (A. R. Luria and E. D. Homskaya, eds.), pp. 398-430. Moscow University Press, Moskow (in Russian).

Filippycheva, N. A., and Fatler, T. O. (1970). Functional states of the human brain associated with gliomas of the mesial structures of the hemispheres. *Journal of Neurology and Psychiatry* **70**, 646-654 (in Russian).

Gadzhiev, S. G. (1966). Disturbances in constructive intellectual activity in patients with lesions of the frontal lobes. In "Frontal Lobes and Regulation of Psychological Processes" (A. R. Luria and E. D. Homskaya, eds.), pp. 618-640, Moscow Univ. Press, Moscow (in Russian).

Genkin, A. A. (1962). Application of statistical description of ascending and descending fronts of EEG waves as an indicator of processes associated with intellectual activity. *Proc. Acad. Pedagog. Sci.* R.S.F.S.R. N.Y., pp. 99-102 (in Russian).

Genkin, A. A. (1963). On the asymmetry of the length of the phases of EEG waves in intellectual activity. *Proc. Acad. Sci.* R.S.F.S.R. **149**, No. 6, pp. 1460-1463 (in Russian).

Homskaya, E. D. (1960). The influence of the verbal instruction on vascular and skin-galvanic components of the orienting reflex in local brain lesions. *Proc. Acad. Pedagog. Sci.* R.S.F.S.R. No. 6, pp. 106-110 (in Russian).

Homskaya, E. D. (1961). The role of verbal instruction in the vegetative components of the orienting reflex in local brain lesions. *Proc. Acad. Pedagog. Sci.* R.S.F.S.R. Pap. No. 1, pp. 117-122, Pap. No. 2, pp. 103-107 (in Russian).

Homskaya, E. D. (1965). Regulations of the vegetative components of the orienting reflex with verbal instructions in patients with local lesions of the brain. *Quest. Psychol.* No. 1 (in Russian).

Homskaya, E. D. (1966). Vegetative components of orienting reflex to indifferent and significant stimuli in patients with lesions of the frontal lobes. In "Frontal Lobes and Regulation of Psychological Processes" (A. R. Luria and E. D. Homskaya, eds.), pp. 190-253. Moscow Univ. Press, Moscow (in Russian).

Homskaya, E. D. (1972). "Brain and Activation." Moscow University Press, Moscow (in Russian).

Karpov, B. A., Luria, A. R., and Yarbuss, A. L. (1968). Disturbances of the structure of active perception in lesions of the posterior and anterior regions of the brain. *Neuropsychologia* **6** pp. 157-166.

Lebedinsky, V. V. (1966). Performance of symmetrical and asymmetrical programs in patients with lesions of the frontal lobes. *In* "Frontal Lobes and Regulation of Psychological Processes" (A. R. Luria and E. D. Hosmkaya, eds.), pp. 567-603. Moscow Univ. Press, Moscow (in Russian).

Luria, A. R. (1966a). "Higher Cortical Functions in Man." Basic Books, New York.

Luria, A. R. (1966b). "Human Brain and Psychological Processes." Harper, New York.

Luria, A. R. (1969). The frontal lobe syndrome. *In* "Handbook of Clinical Neurology" (P. J. Vinken and G. W. Bruyn, eds.), Vol. 2, pp. 725-768. North-Holland Publ., Amsterdam.

Luria, A. R. (1972). "The Man with a Shattered World." Basic Books, New York.

Luria, A. R. (1973). "The Working Brain." Penguin.

Luria, A. R., and Homskaya, E. D. (1964). Disturbances in the regulative role of speech with frontal lobe lesions. *In* "The Frontal Granular Cortex and Behavior" (J. M. Warren and K. Akert, eds.), pp. 353-371. McGraw-Hill, New York.

Luria, A. R., and Homskaya, E. D., eds. (1966). "Frontal Lobes and Regulation of Psychological Processes." Moscow Univ. Press, Moscow (in Russian).

Luria, A. R., and Lebedinsky, V. V. (1968). Disturbances of the logical operations in patients with lesions of the frontal lobes. *In* "Psychological Studies." Moscow Univ. Press, Moscow.

Luria, A. R., and Tsvetkova, L. S. (1966). "Neuropsychological Analysis of the Solution of Arithmetical Problems." "Prosveschenije," Moscow (in Russian).

Luria, A. R., and Tsvetkova, L. S. (1967). "Les troubles de la resolution des problèmes." Gauthier-Villars, Paris.

Luria, A. R., Karpov, B. A., and Yarbuss, A. L. (1966). Disturbances of active visual perception with lesions of the frontal lobes. *Cortex* **2**, 202-212.

Luria, A. R., Homskaya, E. D., Blinkov, S. M., and Critchley, M. (1967). Impaired selectivity of mental processes in association with a lesion of the frontal lobe. *Neuropsychologia* **5**, 105-117.

Pavlov, I. P. (1949). "Complete Works," Vol. III. Acad. Sci. Publ. House, Moscow-Leningrad (in Russian).

Simernitskaya, E. G. (1970). Evoked potentials as an indicator of the active process. Moscow Univ. Press, Moscow (in Russian).

Simernitskaya, E. G., and Homskaya, E. D. (1966). Changes in evoked potentials to significant stimuli in normal subjects and in lesions of the frontal lobes. *In* "Frontal Lobes and Regulation of Psychological Processes" (A. R. Luria and E. D. Homskaya, eds.), pp. 277-293. Moscow Univ. Press, Moscow (in Russian).

Sokolov, E. N. (1958). "Perception and Conditional Reflex." Moscow Univ. Press, Moscow (in Russian).

Sokolov, E. N. (1960). Probability model of perception. *Quest. Psychol.* No. 2 (in Russian).

Tikhomirov, O. K. (1966). Disturbances in programs of active search behavior in patients with lesions of the frontal lobes. *In* "Frontal Lobes and Regulation of Psychological Processes" (A. R. Luria and E. D. Homskaya, eds.), pp. 604-617. Moscow Univ. Press, Moscow (in Russian).

Tsvetkova, L. S. (1966a). Disturbances in constructive intellectual activity in lesions of occipitoparietal and of frontal regions of the brain. *In* "Frontal Lobes and Regulation of Psychological Processes" (A. R. Luria and E. D. Homskaya, eds.), pp. 641-663. Moscow Univ. Press, Moscow (in Russian).

Tsvetkova, L. S. (1966b). Disturbances of arithmetical problem solving in patients with lesions of the frontal lobes. *In* "Frontal Lobes and Regulation of Psychological Processes" (A. R. Luria and E. D. Homskaya, eds.), pp. 677-705. Moscow Univ. Press, Moscow (in Russian).

Vinogradova, O. S. (1959). The role of the orienting reflex information of the conditional connections. *In* "Orienting Reflex and Problems of Higher Nervous Activity," pp. 86-160. Acad. Pedagog. Sci., Moscow (in Russian).

Yarbuss, A. L. (1965). The Role of Eye Movements in the Visual Processes." "Nauka," Moscow (in Russian).

Part Two

THE EFFECT OF FRONTAL LESIONS ON THE ELECTRICAL ACTIVITY OF THE BRAIN OF MAN

Chapter 2

APPLICATION OF THE METHOD OF EVOKED POTENTIALS TO THE ANALYSIS OF ACTIVATION PROCESSES IN PATIENTS WITH LESIONS OF THE FRONTAL LOBES

E. G. SIMERNITSKAYA

Department of Psychology
University of Moscow
Moscow, U.S.S.R.

A characteristic feature of modern neurophysiology is its trend toward correlating the electrical phenomena of various brain formations with behavioral reactions of the organism. Of particular interest from this point of view are the characteristics of the changes in the evoked potential (EP) of the cerebral cortex, which are thought to be the results of numerous excitations coming from various formations to the synaptic structures of the cortex (Anokhin, 1962). The application of this method in neuropsychology may help us comprehend some of the neurophysiological mechanisms that underlie selective psychological processes.

The purpose of the present chapter is to investigate the correlation of EP's with specific features of the subject's activity organized with the help of a verbal system. Another special purpose is to study the possibility of regulating the EP in patients with lesions of the frontal lobes.

As is known, the frontal lobes play an essential part in the regulation of activation states that arise under the influence of a task put before the subject (Walter, 1966; Livanov *et al.*, 1966). It is also known that in lesions of the frontal lobes, the highly differentiated forms of activation, particularly the physical form of the orienting reaction, suffer the most (Homskaya, 1966). It may be assumed that an analysis of the EP's which undergo substantial changes within fractions of a second, may give a clue to the workings of the high-speed system regulating the activation processes in patients with lesions of the anterior parts of the brain. Understanding the functioning of this system, which insures a rapid interchange of states in various parts of the brain during a specific activity, may facilitate the ascertainment of the neurophysiological mechanisms that are responsible for violations of the purposive character and electivity of the mental processes when lesions of the anterior parts of the cerebral hemispheres are present.

I. METHODS

The method of eliciting the EP by neutral and meaningful stimuli (flashes of light from an Alvar Photostimulator) was applied in the present research. The investigation consisted of seven series of single flashes, each series including ten flashes.

In the first series, no special verbal instruction preceded the optic stimulations. There was only a general instruction of the following type: "Sit quietly. You mustn't do anything." Then a meaning was imparted to the flashes with the help of the following three verbal instructions.

(1) "The eighth to tenth flashes will be followed by electrical shock in the left arm." (In all cases no painful reinforcement was applied.)

(2) "Attend to the duration of the flashes, and after each second flash state which of them was shorter." (In all of the investigations, the duration of the flashes was always about 0.5 msec.)

(3) "When you see a flash, press the button with the left hand as quickly as possible."

The choice among the aforementioned instructions was determined by the fact that each of them was aimed at activating a definite cortical zone constituting the central makeup of a corresponding neural analyzer. Thus, electrocutaneous stimulation was regarded as adequate to activate the sensory-motor zone of the right hemisphere; the task of differentiating the durations of the flashes was considered adequate to activate the visual areas of the cortex; and the sensory–motor component of the motor response was considered adequate to activate the visual areas and motor zone of the left (working) hemisphere.

In accordance with this scheme, electroencephalograms (EEG's) were recorded from the parietooccipital and central–frontal regions of both hemispheres. The bipolar method of recording was used.

After each instruction, the task was cancelled; thus, it was possible to discover to what extent the resulting modifications were really caused by the instruction.

Evoked potentials were recorded on an Alvar electroencephalograph; subsequently, sections of the curve having durations of 500 msec (100 msec prior to the action of the stimulus and 400 msec after it) were magnified by means of an epidiascope. Single curves obtained in this way in normal subjects and in patients with local brain lesions were averaged within an interval of 10 msec, separately in each of the four analyzed regions (parietooccipital and central-frontal regions of the right and left hemispheres) in all seven series of investigations.

In order to obtain objective indices distinctive of normal people, the results of investigations carried out on ten normal subjects were processed by group averaging. We used the method of statistical analysis of the EP with an evaluation of the confidence interval limits and variances and applied statistical tests aimed at determining the significant difference of the EP curves (after Student) obtained in different series of the investigation. The application of these methods made it possible to establish statistically significant characteristics of rest EP's (i.e., of states when the stimulations were not preceded by special instructions) as well as to establish their dynamics when the meaning of the stimulus changed. The processing of the EEG was accomplished on an electronic computer of the M-20 type according to a specially designed program.

We investigated ten normal subjects, 15 patients with focal lesions of the anterior parts of the brain, and 12 patients with lesions of its posterior parts. Here we shall confine ourselves to describing individual causes of lesions of the anterior parts of the cerebral hemispheres, which allow description of variants of possible derangements in the regulation of the states of activity in lesions of the frontal lobes.

II. INFLUENCE OF A VERBAL INSTRUCTION ON THE EP IN NORMAL SUBJECTS

The results of investigations carried out on normal subjects show that the EP may reflect an elective activation that is connected with the content of the actions performed by the subject. Depending upon the signaled meaning of the stimulus, which introduces the expectation of a painful stimulation and of a motor or sensory reaction, the dynamics of the time and spatial changes of the EP prove to be different.

Figure 1 presents the EP of subject E during various forms of activity. Background EP's, i.e., EP's to indifferent stimulation, are represented in the parietooccipital regions by a positive—negative oscillation (Figure 1A, column 1). A positive oscillation arising with a latency of 50 msec represents a part of the primary response. The secondary response, in the shape of a stretched out slow negative wave, emerges with a latency of 100 msec and reaches its maximum in 300 msec. In the anterior parts of the frontal lobes, the EP has the form of a two-humped negative wave, where the first peak is observed after 100 msec and the second after 250 msec.

Figure 1A, columns 2 and 3, demonstrates two types of EP modifications recorded in this subject when he expected an electric shock.

In Figure 1A, column 2, the modification of the EP bears a generalized character and manifests itself as a considerable reduction of the EP time parameters in both the anterior and posterior parts of the cerebral hemispheres. No positive oscillation of the primary response is observed in the parietooccipital regions; the peak of the negative oscillation arises at 150 msec and not at 300 msec and the general duration of the EP declines.

In the central–frontal regions the first component of the negative oscillation greatly increases and the maximum is reached earlier. The symmetric response turns into a markedly asymmetric one, with a predominance of the first peak. Subsequently, as the eighth to tenth meaningful flashes gradually approach, the zone where the EP manifests certain distinctions in comparison with the background becomes narrowed. From Figure 1A, column 3, it may be seen that in response to the eighth meaningful flash, the parameters and form of the EP in the parietooccipital regions are similar to those of the rest EP, whereas in the central–frontal regions, the degree of asymmetry of the EP increases, with a depression of the second peak of the negative wave that is strongly pronounced in the background.

When the instruction has been countermanded in response to the second stimulation, the asymmetry of the EP, which now does not differ essentially from the background EP, disappears from the anterior parts of the hemispheres (Figure 1, column 4).

Thus, the expectation of a painful stimulation as a result of a verbal instruction evokes the same changes in the character of the bioelectrical activity that are observed in experiments with a real application of electrocutaneous stimulation (Kornmüller, 1932; Livanov, 1934; Shumilina, 1949). These changes are predominantly localized in the central–frontal regions and consist of a shortening of the cycle of the EP.

During the task of discriminating between flashes of light according to their duration (the duration of the flashes remained constant throughout the investigation), analogous modifications of the EP were recorded in the parietooccipital parts of the cortex. They did not emerge immediately after the introduction of the verbal instruction but only with the second presentation of the stimulus (Figure 1B, column 2).

A comparison of Figures 1A and 1B shows that the shortening of the interval between the flash and the peak of the negative wave (from 300 to 150 msec) is determined not by a change of the negative oscillation recorded in the background but by the emergence of a new and additional component of the EP. This conclusion is confirmed by the possible simultaneous manifestation of both components of the negative oscillation (Figure 1B, column 2); the time parameters of the late component remain within the same limits (the interval between the flash and the peak is 300 msec). Gradually, during subsequent stimulations, the amplitude of the late wave decreases to its full depression (Figure 1B columns 3–5), leading to a reduction in the total duration of the EP.

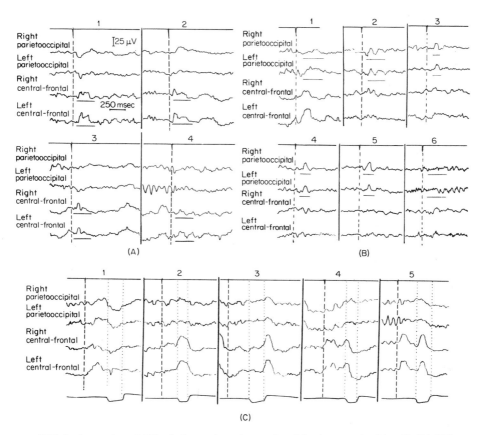

FIG. 1. Dynamics of EP's during various forms of activity in normal subject E. (A) EP's during the expectation of an electric shock; (B) EP's during the task of discriminating between flashes of light according to their durations; (C) EP's during the accomplishment of a motor task. Dashed vertical lines mark the presentation of the stimulus.

In the central–frontal regions the first presentations of the stimulus do not evoke any substantial changes in the parameters and form of the EP in comparison with those called forth by indifferent stimuli. However, in response to the fourth flash (Figure 1B, column 4), a depression of the EP is evident in the anterior parts of the hemispheres and persists up to the end of this series.

The cancellation of the instruction eliminates the effect of activation and brings the parameters of the EP back to their initial values (Figure 1B, column 6).

Quite different was the character of the EP in this subject in response to flashes of light that were signals for motor reactions (Figure 1C). The first presentation of the stimulus did not result in any appreciable modifications of the EP, the form and parameters of which did not differ essentially from background traces (Figure 1A, column 1). In response to the second flash of light a splitting of the slow negative wave into two components was recorded from the posterior parts of the hemispheres, just as it was observed during the task of discriminating between the flashes according to their duration (Figure 1C, column 2). In contradistinction to the task of differentiating between visual stimulations, in this case the splitting of the negative wave was not accompanied by a depression of its late component, the amplitude of which not only remained within the previous limits (Figure 1C, column 2) but exceeded the early component of the EP (Figure 1C, column 4).

In the central–frontal regions the form and parameters of the EP also substantially differed from analogous traces in the previous series of the investigation. A comparison of the time parameters of EP's recorded in these experimental conditions with the latency of voluntary motor reactions shows that the EP may exhibit different relationships with the motor response, either preceding it (Figure 1C, column 1) or coinciding with it (Figure 1C, column 2); EP's may also appear after the termination of the motor response (Figure 1C, column 3). When the motor habits become automatic, two high-amplitude waves begin to be recorded in the central–frontal regions. One of them precedes the motor response, apparently reflecting the subject's intention to produce a rapid response to the stimulus, and the other is accompanied by a motor response (Figure 1C, column 5).

Thus, depending upon the nature of the activity accomplished by the subject, the modifications of the EP's in the parietooccipital and central–frontal regions have a specific character; this testifies to an elective participation of the cortical structures in the formation of different functional systems.

The results of visual analysis were confirmed by the statistical processing of the EP's that had been obtained in different series of the investigation. This processing showed that different signaled meanings of the stimulus led to significant modifications that were recorded in various complexes of jointly functioning zones of the cortex. When, in different experimental situations, statistically significant modifications of the EP showed up in the same cortical zone, they emerged at different stages of the evoked response (see Table 1). In particular, both the expectation of an electrical shock and the task of discriminating between flashes according to their duration were accompanied by statistically significant modifications in the EP of the central–frontal parts of the right hemisphere; however, in the first case, they emerged during the 70th msec and in the second case, during the 240th msec. i.e., they reflected the dynamics of different components of the evoked response.

An analysis of the correlation of the EP with the specific features of the subject's response shows that it is precisely the content of the verbal instructions that determines the elective manifestation of individual components of the EP. Thus, the introduction of verbal instructions, which are adequate to activate a specific analyzer, calls forth an intensification (or emergence) of a negative potential with a latency of 70-100 msec, whereas in the case of nonsignaling stimuli, or when instructions are addressed to another analyzer, a later negative oscillation with an interval of 180-240 msec between the flash and the peak appears with particular clarity. The data obtained are in accord with the widely held hypothesis that the excitatory processes are accompanied by an intensification of high-frequency electrical oscillations, whereas a decline of the functional state leads to a change of the harmonic composition of the EEG toward a predominance of the low-frequency components over the high-frequency ones (Golikov, 1956; Smirnov, 1957; Rusinov, 1957; Maiorchik, 1960; Sokolov, 1960).

III. INFLUENCE OF A VERBAL INSTRUCTION ON THE EP IN PATIENTS WITH LESIONS OF THE FRONTAL LOBES

An analogous investigation carried out on patients with lesions of the frontal lobes produced quite different results. An analysis of the data obtained from this investigation at once reveals the distinctive features of the EP in comparison with the normal subject's traces. Whereas in the normal subjects there was observed an intensification of specific components of the EP, depending upon the kind of activity accomplished by the subjects, in some cases of frontal lobe lesions such reactions were recorded merely in response to indifferent flashes of light. These reactions reflected the initial states of specific cortical areas that arose under the influence of the pathological process. Thus, in some patients, only an early negative potential with a latency of 70-100 msec was recorded in the zone of focus, whereas no significant late component of the response was evident; this, apparently, reflected cortical irritation provoked by the influence of the pathological process.

In contrast, in other cases, the electrical activity, including the EP, slowed down. From the point of view of modern concepts of the EEG, this may be regarded as a decrease in the lability of cortical cells in the pathological hemisphere compared to the normal hemisphere.

Finally, a marked reduction of all components of the EP in response to nonmeaningful stimuli was observed in some patients with lesions of the frontal lobes. It may be assumed that the absence of an EP, the presence of which requires a definite level of cortical excitability (Puchinskaya, 1963), indicates a greatly weakened reactivity of the cortical cells.

Thus, the study of EP to neutral flashes of light in patients with lesions of the anterior parts of the hemispheres makes possible a differentiated approach to the

TABLE 1

Diagram of Significant Modifications of the Average Values of the EP during Various Forms of Activity[a]

1. Locations of modifications during the expectation of electric shock

msec:	20	40	60	80	100	120	140	160	180	200	220	240	260	280	300	320	340	360
Right parieto-occipital																		
Left parieto-occipital																		
Right central-frontal				■	■													
Left central-frontal																		

2. Spatial distribution of modifications during light flash duration discrimination task

msec:	20	40	60	80	100	120	140	160	180	200	220	240	260	280	300	320	340	360
Right parieto-occipital								■	■									
Left parieto-occipital									■	■	■	■	■	■				
Right central-frontal													■	■				
Left central-frontal													■	■				

3. Spatial distribution of modifications during a motor task

msec:	20	40	60	80	100	120	140	160	180	200	220	240	260	280	300	320	340	360
Right parieto-occipital				▓	▓	▓	▓											
Left parieto-occipital					▓	▓	▓											
Right central-frontal																		
Left central-frontal													▓	▓	▓	▓	▓	

[a]The shaded portions indicate the sections of time in which the modifications of the average values become significant after presentation of the stimulus.

analysis of the character of the pathologically modified neurodynamics of the cortical elements arising under the influence of a brain tumor.

A still more diverse picture was observed in patients with lesions of the frontal lobes when they were given verbal instructions. The most essential characteristic of these patients was that in most of them the verbal instruction not only resulted in a stabilization of the activation state, but quite often showed a decompensatory factor and produced an inhibitory effect. However, in some of these patients, the regulation of the EP with the help of a verbal instruction remained relatively intact.

Let us now analyze some individual cases of lesions of the anterior parts of the brain, which will allow us to describe certain variants of possible modifications of the EP to signaling stimuli. The purpose of this analysis is to contrast the clinical and pathoanatomical pictures of the brain lesion with modifications of the EP arising as a result of the introduction of verbal instructions. We shall begin with cases that revealed the most pronounced differences of the EP in comparisons with those in normal subjects.

Observation 1: Patient U (diagnosis: sarcoma of the frontoparasagittal region). In the clinical presentation of this patient, symptoms of a lesion of the right frontal lobe were observed against the background of moderate intracranial hypertension that was manifested in mild headaches, indistinct and effaced borders of the optic disks, and increase of the liquid pressure up to 250 mm of water, symptoms of a lesion of the right frontal lobe were observed. The patient exhibited pronounced psychological changes, a slight left-side hemiparesis with a revival of the tendon reflexes and with muscular distonia on the same side, as well as insufficiency of nerve VII of central origin. A neuropsychological investigation revealed in the patient unsustained attention, inability to fulfill complex programs, distinct motor perseverance, and absence of a critical attitude towards his own defects.

Against the background of moderately pronounced general cerebral changes—the EEG showed a focus of pathological electrical activity in the right frontal region.

During the surgical operation, a pale yellow section of the cortex was observed in the superior frontal gyrus of the right hemisphere. In this sector of the brain, the gyri were not differentiable. The brain was dissected in this region in the sagittal direction. At the depth of 2–3 cm a tumor of $4 \times 5 \times 4$ cm was found; it was surrounded by a wide zone of a perifocal reaction. A few days after the operation, the patient died. The pathoanatomical diagnosis was as follows: a sarcoma of the frontoparasagittal region with a big sickle-shaped appendage connected with the dura mater of the convexal surface. There was observed a gross compression by the tumor of the first and, partially, the second frontal gyri on the right, of the first frontal convolution on the left, and of the

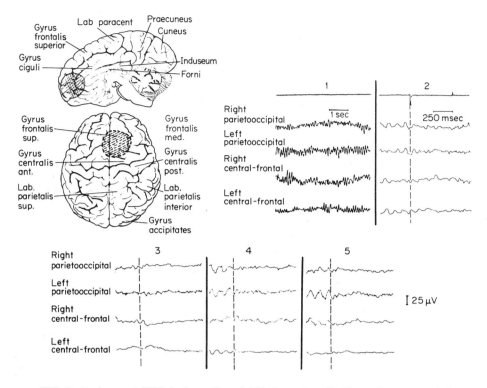

FIG. 2. Background EEG (column 1) and EP's in patient U (diagnosis: tumor of the frontoparasagittal region). (2) EP's to indifferent stimulus; (3–5) EP's after the subject received verbal instructions. Dashed vertical line as in Fig. 1.

geniculum of the corpus callosum, with a deep hollow formed in them, the walls of which were in some places infiltrated by the tumor.

An investigation carried out prior to the operation disclosed a gross pathology of the EP, in spite of the fact that the general cerebral changes of the bioelectrical activity in this patient were feebly pronounced and that a distinct alpha rhythm was recorded in the EEG (Figure 2, column 1). This was reflected in a considerable reduction of the amplitude of the EP, especially of its secondary and later components, which did not exceed the level of spontaneous oscillations of bioelectrical activity (Figure 2, column 2). The primary response was recorded inconsistently, but sometimes it was of a quite pronounced character.

In some cases, this syndrome could be observed also in normal subjects when they were suffering fatigue, or when verbal instructions were addressed to another analyzer (Figure 1C, column 5, lower curves); however, the accomplishment of tasks that were adequate for the given analyzer led to the emergence in the EEG of high-amplitude EP's (Figure 1A, columns 2 and 3; Figure 1C, columns 3–5). In this case, however, the introduction of verbal instructions, aimed at activating various cortical zones, produced no change in the character of the bioelectrical activity: the EP to signaling stimuli, as well as to indifferent stimuli, did not exceed the level of spontaneous oscillations and was not significant (Figure 2, columns 3–5). The absence of a reaction against the background of a regular alpha rhythm correlates, according to data published in the literature (Golikov, 1956; Adamovich, 1956) and, with a considerably weakened reactivity of the cortex and the emergence of phasic states.

Thus, the investigation revealed a considerable decline in the amplitude of all EP components in patients with extensive damage to the cortex of the right frontal lobe, especially of its medial parts, which may testify to a pathological intensification of the inhibitory influences on the ascending activating system. This is confirmed by the impossibility of activating the EP with the help of verbal instructions.

However, the state of heightened activity of the cortex, with a predominance of high-frequency oscillations of the potential and with greatly intensified EP's to indifferent flashes of light, may also be characterized by weakened reactions of the cortex and even by their absence in response to signaling stimuli. The following observation illustrates this fact.

Observation 2: Patient Z (diagnosis: astrocytoma of the left frontal lobe). At the time of hospitalization, the patient complained of a weakness in the right extremities, headaches, and nausea, which were accompanied by a sensation of coldness in the extremities and fainting. The disease developed over 7 months. Clinically, symptoms of increased intracranial pressure (initial phenomena of a choked fundus and a rise of liquid pressure to 340 mm) were observed. Against this background, the syndrome of a lesion of the premotor zone of the left frontal lobe was manifest (a moderate right-side hemiparesis, paresis of a central origin of nerves VII and XII on the right, etc.). A number of symptoms indicated that the pathological process was influencing the diencephalic structures (violation of sleep, fits accompanied by a loss of consciousness, etc.). A neuropsychological investigation revealed in the patient symptoms of a disturbance in the deep parts of the left frontal lobe that was expressed in constrained movements of the right arm, slow elaboration of motor habits in the left arm, unsustained attention, specific intellectual disorders, and lack of self-criticism with respect to her problem. Epileptoid impulses of a diffuse character were observed in the EEG against the background of an intact alpha rhythm; they predominated in the left hemisphere, exhibiting no distinct focal origin.

During the surgical operation, a compact tumor of a considerable size was detected in the depth of the frontal lobe, in the zone of the superior and medial gyri; the tumor was removed. After the operation the patient's state remained very grave, and 1 day later she died as a result of paralysis of respiration and cardiac activity. Autopsy revealed in the zone of the surgical intervention a cavity of 4 × 5 × 4 cm that replaced the white matter of the second and third frontal gyri as well as the adjoining parts of the centrum semiovali, the medial part of the Island of Reil, and the anterior parts of the lenticular nucleus on the left.

An investigation carried out prior to the operation revealed distinct EP's to indifferent flashes of light in both the anterior and posterior parts of the cerebral hemispheres (Figure 3A). The peculiar feature of these responses was that the form and parameters of the EP to nonmeaningful stimuli corresponded to analogous traces recorded only in normal subjects when under conditions of the special activation of attention; this correlated with the data of the EEG, which indicated an irritation of the cortex. The amplitude of the negative components of the EP in the anterior parts of the hemispheres markedly increased, and their latency decreased in comparison with the normal responses. Another distinctive feature of the patient's responses was their pronounced inertness, as a result of which the EP did not change during subsequent stimulations and was not extinguished; they did not show any essential differences in response to the first and eighteenth stimuli (Figure 3A, columns 1 and 2). A considerable intensification of the reaction to such a typical stimulus as flashes of light in combination with a pronounced inertness of the EP, apparently reflects the dominant properties of the focus of the stationary excitation that arose in the frontal lobes under the influence of a protractedly acting tumor (Rusinov, 1958).

Against this background, neither the introduction of a painful reinforcement nor its cancellation (Figure 3B, columns 1 and 2), nor the introduction and cancellation of the visual task (Figure 3B, columns 3 and 4), was accompanied by appreciable activity of the EP. The introduction of a motor task aimed at increasing the excitability of the motor zones of the cortex, which were in a state of heightened excitability, did not lead to intensified excitation. On the contrary, the motor task created conditions for a conversion of the excitation into inhibition, reflected in a sharp increase in the duration of EP's recorded in the central–frontal regions (Figure 3C). Each successive stimulation of this series intensified even more the inhibitory state, which was expressed in a gradual growth of the amplitude and duration of the EP from one stimulation to another (Figure 3C, columns 1–3).

It may be assumed that the pathologically protracted character of the components of EP's in the anterior parts of the hemispheres during the accomplishment of a motor task accounts for the pathology of the patient's motor activity revealed in the course of a neuropsychological investigation. The cancellation of the task returned the parameters of the EP to their initial values (Figure 3C, column 4).

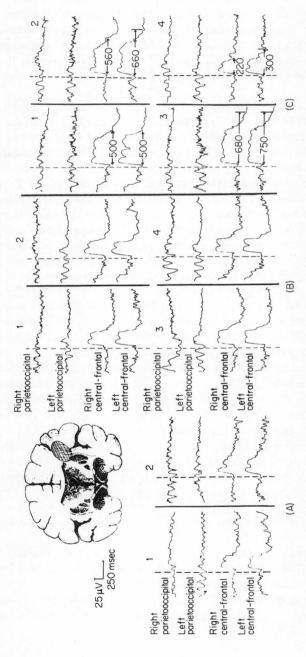

FIG. 3. Patient Z (diagnosis: astrocytoma of the deep parts of the left frontal lobe). (A) EP's to the 1st (column 1) and 18th (2) stimulus in a series of indifferent flashes of light; (B) EP's during the expectation of an electric shock (1), after cancellation of the instruction (2), and during the accomplishment (3) and cancellation (4) of the visual task; (C) EP's during the accomplishment of a motor task.

APPLICATION OF EVOKED POTENTIALS TO ACTIVATION PROCESSES   43

Thus, in patient Z, a deep tumor of the left frontal lobe pressure on the subcortical ganglia and on the formations of the diencephalic area. The investigation established a derangement in the regulation of the EP with the help of verbal instructions, which not only failed to stabilize the activation state but led to distorted reactions in response to a task adequate to the affected analyzer. The emergence of a paradoxical reaction against the background of irritation is explained by the specific features of the focus of dominant excitation, which presents a state of distinctive excitation foreshadowing parabiosis (Ukhtomski, 1925).

The interconnection of the basic neural process and the possibility of a conversion of the excitatory process into inhibition were particularly distinct in the case of patient K.

Observation 3: Patient K (Diagnosis: arachnoidendothelioma of the middle third of the superior longitudinal sinus). Six years prior to this investigation, the patient underwent an operation for the removal of an arachnoid endothelioma of the frontopremotor area, which originated from the superior longitudinal sinus. The operation was accompanied by a regression of the general cerebral and local symptoms. Soon after the operation, however, the patient suffered epileptic attacks, and he was hospitalized for the purpose of medical investigation.

A distinct alpha rhythm was recorded in the EEG of this patient; the general cerebral modifications of bioelectrical activity assumed, to a certain degree, the form of irritation and diffuse epileptoid changes, with a predominance of these phenomena in the frontal parts of the right hemisphere (Figure 4A, column 1). As in the previous case, the EP to neutral flashes differed from analogous normal responses: in response to nonsignaling flashes, the form and parameters of the EP in the central–frontal region of the right hemisphere coincided with corresponding traces recorded in normal subjects only in the state of activation, in particular, during the expectation of an electrical shock (Figure 4A, column 2).

Thus, the character of the EP to nonsignaling stimuli, in combination with the frequency–amplitude characteristics of the EEG, testified to a heightened functional state of the cortical structures with a predominant localization of irritative changes in the anterior parts of the right hemisphere.

The introduction of the verbal instruction, "The eighth to tenth flashes will be followed by an electric shock in the left arm," greatly changed the character of the background EEG: the alpha rhythm assumed a more pointed form, there appeared diffuse slow waves, and a theta rhythm was recorded in the anterior parts (Figure 4B, column 1). Against this background, the presentation of signaling flashes was accompanied by a rapid decline of the amplitude of the EP, and by a practically complete disappearance of EP from the central–frontal regions (Figure 4B, column 2); this indicated a reexcitation of these cortical zones and

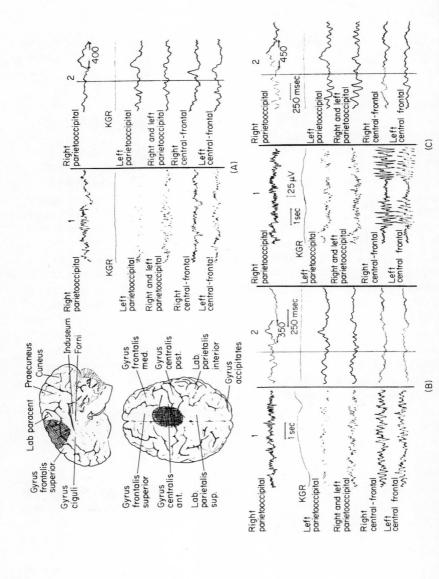

FIG. 4. Patient K (diagnosis: tumor of the middle third of the superior longitudinal sinus). (A) Background EEG (column 1) and EP's to the indifferent stimulus (2); (B) background EEG (1) and EP's during the expectation of an electric shock (2); (C) background EEG (1) and EP's after cancellation of the electric current (2).

their transition to the state of transmarginal inhibition. Of specific interest is the patient's subjective account. When asked whether he had been afraid of the electric current, the patient answered that he had not, although he claimed that he had felt it after the eighth flash (in reality, no electric current was applied at all).

The cancellation of the instruction resulted in a still greater decline of lability (in comparison with the introduction of the verbal instruction); it was accompanied by a marked intensification and greater regularity of bilateral outbursts of the theta activity (Figure 4C, column 1), which proved to be depressed during stimulations (Figure 4C, column 2).

The introduction of a task concerning differentiation of visual stimuli, aimed at activating the relatively intact cortical zones, evoked an activating effect in the shape of a depression of the outbursts of slow activity (Figure 5A, column 1) and shortening of the cycle of EP's in the parietooccipital region of the right hemisphere (Figure 5A, columns 2–5; compare with Figure 4A, column 2; Figure 4B, column 2; and Figure 4C, column 2). Sometimes an earlier negative potential was observed in the posterior parts of the hemispheres. it corresponded to the response recorded in similar conditions in normal subjects (Figure 5A, column 5). The unstable character of this potential indicated the inability of the cortex to retain the state of excitation for a protracted period of time.

The cancellation of the task again led to an increase of the duration of the EP in the posterior parts of the right hemisphere (Figure 5B, column 1) and to a release of the outbursts of bilateral theta activity (Figure 5B, column 2) which, in contrast to series III, were no longer depressed during stimulations (Figure 5B, column 3).

The introduction of a motor task aimed at activating the motor zones of the cortex, which exhibited considerably modified excitability, at first inhibited the slow waves (Figure 6, column 1). Subsequently, beginning with the fifth stimulus, it intensified the inhibitory process still more and led to the emergence of an ultraparadoxical phase, when synchronous slow waves were absent in the background but appeared in response to each stimulus. At the same time, the number of slow waves during successive stimulations showed a gradual increase (Figure 6, columns 2–7).

The cancellation of the task was accompanied by a more intensive functional state, reflected in a depression of the slow waves (Figure 6, column 8).

Thus, a study of the dynamics of EP's in response to different verbal instructions in patient K revealed a sharp decline in the working capacity of the cortical elements of the central–frontal region, causing the introduction of such ordinary tasks as pressing a key, to become powerful obstacles and resulting in the development of transmarginal inhibition. It is noteworthy that during the accomplishment of a visual task addressed to an intact analyzer (in accordance

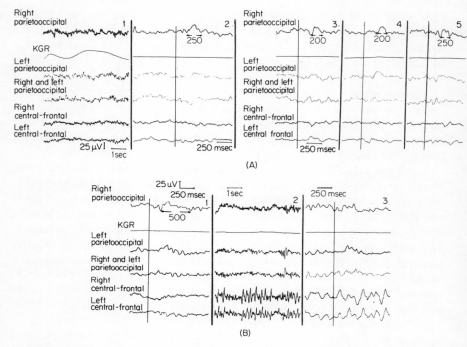

FIG. 5. Patient K (diagnosis: tumor of the middle third of the superior longitudinal sinus). (A) Background EEG (column 1) and EP's during light flash duration discrimination task (2–5); (B) background EEG (2) and EP's after cancellation of the instruction (1 and 3).

with the instruction to differentiate the duration of flashes of light), no transmarginal inhibition was observed.

A peculiar feature of this case was that at the time of the investigation no symptoms of a space-occupying lesion were evident in the deep parts of the brain tumor and subsequent surgical intervention was localized on the medial surface of the posterofrontal and central regions of both hemispheres.

In some patients with lesions of the frontal lobes, the regulation of the EP with the help of verbal instructions remained relatively intact.

Observation 4: Patient A (diagnosis: a tumor of the inferior parts of the posterofrontal region in the left hemisphere). The clinical picture of this patient was as follows: absence of headaches, normal spinal pressure, symptoms of increased intracranial pressure appearing as a slightly effaced nasal portion of the right optic disk and varicose veins in the left fundus. A paresis of nerve VII on the right was clearly manifest, as well as a predominance of tendon reflexes on the right, without any changes in the muscular tone and sensitivity. These

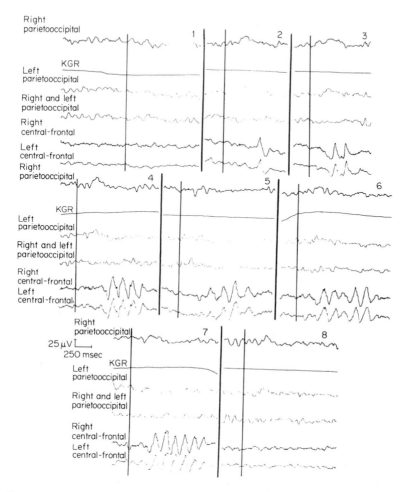

FIG. 6. Patient K (diagnosis: tumor of the middle third of the superior longitudinal sinus). (1–7) EP's during the accomplishment of a motor task; (8) EP after cancellation of the verbal instruction.

data led to the assumption that the patient suffered from a glial tumor that had been growing for some time and that was situated in the inferior part of the left posterofrontal region.

The clinical data concerning the localization of the pathological process were confirmed by pneumoencephalography, which revealed a sharp dislocation to the right of the anterior division of the third ventricle; the latter proved to be narrowed and had the form of an arc convexed to the right. The top of the left inferior horn was deflected backward and downward.

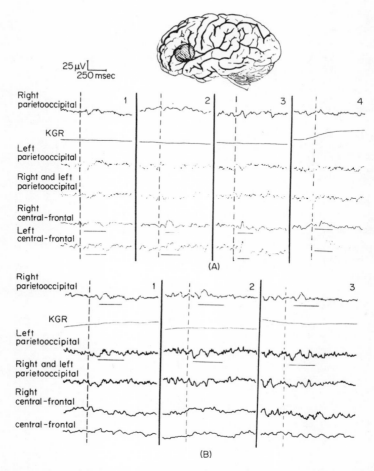

FIG. 7. Patient A (diagnosis: tumor of the inferior parts of the posterofrontal region of the left hemisphere). (A) EP's during the expectation of an electric shock; (B) EP's during a light flash duration discrimination task.

In the EEG, the alpha rhythm was recorded only as separate groups of oscillations. On the right, the alpha rhythm was of a less pronounced character. Beta oscillations and single epileptoid impulses were diffuse, showing a predominance in the central parts of the hemispheres. Separate delta waves were recorded in all regions of the cortex, slightly predominating in the frontal parts of the brain.

Against the background of these pronounced modifications of electrical activity, the expectation of an electric shock called forth by a verbal instruction was accompanied by distinct dynamics of the EP, which indicated an elective reaction of the anterior parts of the hemispheres. Modifications of the EP in the central–frontal regions were not revealed at once (Figure 7A, column 1). With the gradual approach of the signaling stimuli, the EP assumed a more and more pointed form, and the cycle of the EP became shorter (Figure 7A, column 2); as a result, when the tenth signaling flash was presented, the parameters of the EP fully corresponded to the response that had been recorded under similar conditions in normal subjects (Figure 7A, column 3). Cancellation of the verbal instruction immediately eliminated the effect of activation and returned the parameters of the EP to their initial values (Figure 7A, column 4).

A task connected with the differentiation of visual stimuli led in this patient to elective modifications of the EP in the parietooccipital regions. These modifications consisted of the emergence of an additional negative wave, the amplitude of which increased with successive presentations of stimuli (Figure 7B). In the intact hemisphere, the rise of the additional negative wave was accompanied by a depression of the late negative oscillation of the EP, just as was observed in normal subjects, whereas on the side of the lesion both negative waves were recorded simultaneously.

The EP asymmetry recorded from the parietooccipital regions of the cortex of both hemispheres during the accomplishment of a task, which requires a more or less high level of excitation of the visual analyzer, reveals the latent functional insufficiency of the pathological hemisphere in comparison with the normal one; it shows that although the instruction increases the physiological lability of this hemisphere, it does not bring it to the normal level.

IV. SUMMARY

The results of investigations carried out on patients with lesions of the frontal lobes are not uniform, indicating that the relationship of various formations in the frontal lobes to the regulation of activation states differs in each case. Derangements in the regulation of the EP with the help of a system of verbal instructions were particularly distinct in lesions of the mediobasal parts of the hemispheres; this is also true of the case where a subcortical localization of the pathological process exerted an influence on the formations in the diencephalic region. These influences were expressed not only by the absence of an activating effect from the verbal instructions, but also by pathological reactions that indicated the presence of phasic states of altered intensity, which constituted the main background of neural activity in the cerebral cortex of some of these frontal lobe lesioned patients.

It is noteworthy that not all functional loads are equally effective for revealing functional insufficiency in the frontal cortex. The most intensive modifications were observed during the action of verbal instructions addressed to the anterior parts of the hemispheres, in the cortical cells of which the limit of working capacity sharply decreased when the pathological process was localized there. In only one patient, with a tumor of the posterior parietooccipital region of the right hemisphere, that caused a profound state of diffuse inhibition in the cortex, did the instruction connected with the expectation of an electrical shock lead to the emergence of a paradoxical reaction; the latter, however, was rapidly normalized. All the subsequent tasks, including one that was related to the focus of the lesion (discrimination between flashes of light according to their duration), were accompanied by normal reactions testifying to an increase in the lability of the cortical elements when verbal instructions were given.

Thus, an analysis of the possibilities of regulating the EP with the help of a system of verbal instructions showed that not all parts of the cortex have equal effects on its elective activation and that an important role in these mechanisms is played by the frontal lobes.

The fact that derangements in the regulation of the EP manifested themselves differently depending upon the localization of the pathological process within the frontal lobes makes possible a differentiated approach to the functions performed by the frontal lobes in the regulation of the activation state; this is particularly important in the local elective activity that accompanies complex forms of activity effected with the participation of a verbal system of instructions.

REFERENCES

Adamovich, V. A. (1956). On evaluation of the functional state of the cerebral cortex by means of EEG-reactions to closing and opening the eye. *In* "Problems of Theory and Practice of EEG," pp. 109-167. Leningrad (in Russian).

Anokhin, P. K. (1962). New data on specificity characteristics of afferent activations. *Zh. Vyssh. Nerv. Deyatel. im I.P. Pavlova* **12**, No. 3 (in Russian).

Golikov, N. V. (1956). Physiological foundations of EEG theory. *In* "Problems of Theory and Practice of EEG," pp. 3-31. Leningrad (in Russian).

Homskaya, E. D. (1966). Vegetative components of orienting reflex to indifferent and significant stimuli in patients with lesions of the the frontal lobes. *In* "Frontal Lobes and Regulation of Psychological Processes" (A. R. Luria and E. D. Homskaya, eds.), pp. 176-190. Moscow Univ. Press, Moscow (in Russian).

Kornmüller, A. E. (1932). Architektonische Lokalization bioelektrischer Erscheinungen auf der Grosshirarinde. Untersuchungen am Kaninchen bei Augenbelichtung. *J. Psychol. Neurol.* **44**, 447-459.

Livanov, M. N. (1934). Analysis of bioelectrical waves in the cerebral cortex of rabbits. *Sov. Nevropatol. Psikhiat. Psikholog.* **3**, 11-12, 98-116 (in Russian).

Livanov, M. N., Gavrilova, N. A., and Aslanov, A. S. (1966). Correlation of biopotentials in human frontal cortex. *In* "Frontal Lobes and Regulation of Physiological Processes" (A. R. Luria and E. D. Homskaya, eds.), pp. 176-190. Moscow Univ. Press, Moscow (in Russian).

Mayorchik, V. E. (1960). Various forms of synchronization of human cortical rhythm depending on the level of cerebral stem irritation. *Conf. Probl. Electrophysiol. Nerv. Syst., 3rd, 1960* pp. 251-252 (in Russian).

Puchinskaya, L. M. (1963). On the zone of manifestations of the non-specific response in human EEG. *In* "Electrophysiology of the Nervous System," p. 306. Rostov (in Russian).

Rusinov, V. S. (1957). Electrophysiological investigations of the higher nervous activity. *Zh. Vyssh. Nerv. Deyatel. im I.P. Pavlova* 7, 855-867 (in Russian).

Rusinov, V. S. (1958). "General and Local Changes in EEG in Development of Conditioned Reflexes," Colloq. Moscow (in Russian).

Shumilina, A. I. (1949). Functional importance of frontal parts of the brain in conditioned-reflex activity. *In* "Problems of Higher Nervous Activity," pp. 299-305. Moscow.

Smirnov, G. D. (1957). Electrical phenomena in the central nervous system and their changes under some pharmaco-chemical effect on tissue metabolism. Doctoral Thesis, University of Moscow (in Russian).

Sokolov, E. N. (1960). Neuronal model of the stimulus and orienting reflex. *Vopr. Psychol.* 4, pp. 61-72 (in Russian).

Ukhtomski, A. A. (1925). "New Approaches in Reflexology and Physiology of the Nervous System," p. 60 (in Russian).

Walter, W. G. (1966). Human frontal lobe function in regulation of active states. *In* "Frontal Lobes and Regulation of Psychological Processes" (A. R. Luria and E. D. Homskaya, eds.), pp. 156-176. Moscow Univ. Press, Moscow (in Russian).

Chapter 3

CHANGES IN THE ASYMMETRY OF EEG WAVES IN DIFFERENT FUNCTIONAL STATES IN NORMAL SUBJECTS AND IN PATIENTS WITH LESIONS OF THE FRONTAL LOBES

E. YU. ARTEMIEVA   and   E. D. HOMSKAYA

Department of Psychology
University of Moscow
Moscow, U.S.S.R.

Department of Neuropsychology
University of Moscow
Moscow, U.S.S.R.

Any changes of the functional state of the brain, i.e., any increase or decrease of the waking state, is reflected in some change in the EEG. It is known that the activation of attention during intellectual work or emotional stress is accompanied by a depression of the alpha rhythm and by an increase of beta activity (Berger, 1930; Walter, 1950; Mundy-Castle, 1957; Werre, 1957). However, there are data showing that at moments of attention activation (for example, during expectation of a stimulus) an increase of alpha activity may also be observed (Barlow, 1957). In some cases, no appreciable changes in the frequency spectrum of the EEG, including the alpha band, are apparent in states of active attention. Some authors have pointed out that in transition from wakefulness to sleep a regular alpha rhythm predominates in the EEG during the initial and transitional states; this rhythm cannot be visually distinguished from the alpha rhythm in the state of physiological rest, and only in the last stages of a deep sleep do irregular flat slow waves appear in the EEG (Loomis *et al.*, 1936).

Thus, there exist some functional changes, excitatory and inhibitory, that are indistinguishable by visual inspection of the EEG and that are not reflected in any shift of its frequency spectrum. These visually indistinguishable functional states become differentiable if some parameters characterizing the shape of the EEG wave are used, for example, parameters for estimating the asymmetry of slope of the ascending and descending fronts of alpha waves. The use of these parameters is based upon the assumption, corroborated by some experimental data (Bishop, 1962; Chang, 1951), that the processes of oscillation consist of consecutive cycles of activity and that during the time of the alpha oscillation some shifts of the functional state are observed; these shifts predetermine the latencies of the motor and similar behavioral measures.

The present report is based on a study comparing patients with frontal lesions to normal adult subjects. Recordings made from the parietooccipital region in a state of rest have been observed to show a relatively stable average level of asymmetry of the slopes of ascending and descending fronts of the alpha-wave phases.* However, the absolute magnitude of this parameter varies in different subjects (Genkin, 1963). Correlation between the amount of asymmetry and the dominant frequency in the alpha band could not be obtained (Genkin, 1964).

The amount of asymmetry changes during the change of the functional state, e.g., falling asleep, arousal, fatigue, as well as during intellectual work (mental arithmetic, etc.). A high correlation was revealed between this parameter and the speed and success of accomplishment of tasks that require a definite stability of voluntary attention (proof-reading tests, etc.). No uniform changes of this parameter during intellectual work were observed: some subjects are characterized by an increase in the asymmetry of the duration of the EEG phases, others, by a decrease of the same index. In both cases, the average amplitude and the frequency of the EEG oscillations remained stable. The changes of the amount of asymmetry in the state of falling asleep are more homogeneous: as a rule, the amount markedly decreases before the changes of the EEG become visually observable. It was established that the changes of the given parameter during the active state depend upon the initial background: the higher the amount of asymmetry is during the state of rest, the higher it is during the active state (Genkin, 1963). The sign of the reaction likewise depends upon the initial amount of asymmetry. This parameter can be helpful for detecting focal changes in the EEG (Moisseyeva and Genkin, 1963).

Thus, the amount of asymmetry of alpha oscillations apparently reflects some important process of brain activity that is connected with the level of the waking state, i.e., with the current functional state of the brain.

We used the above mentioned parameter Δ_T and found that not only do the average values of asymmetry over long periods of time manifest law-governed changes depending upon different states of the subject, the dynamics of the asymmetry level, too, can provide important information concerning the process under investigation.

I. METHOD AND RESULTS WITH NORMAL SUBJECTS

The present report therefore deals with the behavior of the parameter Δ_1 in time, i.e., the average of asymmetry during 1 sec. The EEG was recorded with

---

*A special parameter, Δ_T, was introduced for measuring the asymmetry. This parameter presents an estimate of the average asymmetry of the slope of the fronts during T sec.

an inkwriter (the speed of the paper being 60 mm/sec) and was divided into periods of 1 sec each. To estimate the value of Δ_1, in each interval [t to ($t + 1$) sec] we used the discrete analog of the integral

$$\int_t^{t+1} \text{sign } X'(t)\, dt$$

where $X(t)$ is the curve of the EEG, $X'(t)$ is its derivative, and

$$\text{sign } u = \begin{cases} +1, & u \geq 0 \\ -1, & u < 0 \end{cases}$$

Separate measurements were made every 25 sec. They were symbolized by small dots if they reflected the phase of increase of the amplitude, and by vertical lines if they reflected the phase of its decrease (Figure 1). In case of coincidence with the coordinate min or max, the symbol of the preceding measurement was applied (see Figure 1). The value of Δ_1 was equal to the ratio of the differences between the number of dots and vertical lines and the average number of oscillations during 1 sec (in our further exposition the parameter Δ_1 will imply precisely this value).

We found that in normal subjects a strict preservation of the periods of rhythmical oscillations of Δ_1 (called by us G waves) was observed (Artemieva et al., 1965). This can be seen in Figure 2 (A-D), which shows the results of analysis of EEG's recorded from the parietooccipital region of the left hemisphere (by the bipolar method) in four normal subjects, 25-35 years of age; all of them had clear alpha rhythms. The values of the period were different in different subjects (6-8 sec; most were 7 sec).

The periods of oscillation are independent of the average amount of the asymmetry. From Figure 2 it may be seen that in Case A the asymmetry is obviously positive, whereas in Case B it is negative; the period of the G waves remains unchanged.

During a subject's active attention to intellectual work, a break in the regularity of the G waves appears: the oscillations of the values of Δ_1 assume a less regular character (see Figure 3A, B, and C). The onset of counting is reflected in a violation of the initial periodicity of oscillations, whereas the termination of counting is reflected in a restitution of this periodicity (see Figure 3A). It must be mentioned that in these subjects the EEG records during rest and during intellectual activity (counting) were visually indistinguishable (see Figure 3A, B, C).

In cases when intellectual activity was associated with a depression of the alpha rhythm, a marked violation of the regularity of oscillations of the alpha-wave asymmetry was observed to precede the depression of the alpha rhythm (see Figure 3D, E). The same breakdown of periodic oscillations, sometimes of a more protracted character, was associated with the end of the depression of the alpha rhythm.

The experiments revealed a close connection between the behavior of the oscillations of Δ_1 (recorded in the above-described manner) and the general state

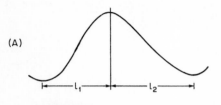

FIG. 1. Scheme for estimating asymmetry.
(A) Asymmetry of a single oscillation:
$$\Delta = \frac{l_1 + l_2}{l_1 + l_2}$$

(B) Coded EEG section with a duration of 2 sec:
$$\Delta = \frac{23 - 17}{10} = 0.6$$

$$\Delta = \frac{24 - 16}{10} = 0.8$$

Detailed explanations are given in the text.

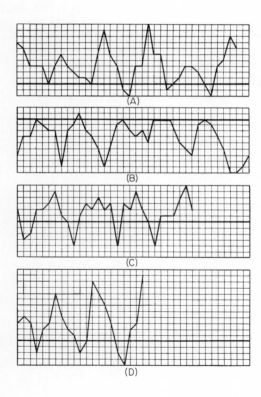

FIG. 2. Slow periodic oscillations of asymmetry (G waves) in the background. (A)–(D) Four different normal subjects. Each segment of the abscissa is 1 sec; the ordinate is values of Δ_1.

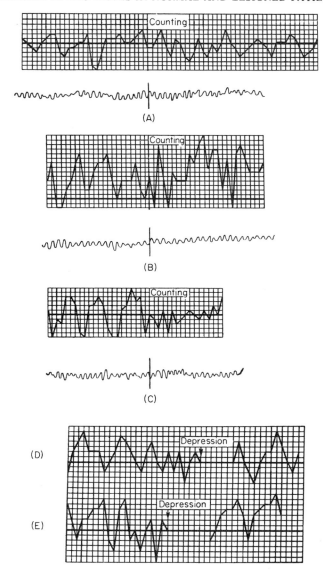

FIG. 3. Changes in G waves during intellectual activity. (A)–(C) Disturbances of rhythmical oscillations of asymmetry during mental computation (multiplication of two-digit numbers) in three different subjects. Under each graph are records of the EEG over 3 sec. Vertical line shows the onset of counting. No visual differences in either period of EEG records are seen. (D) and (E) Disturbances of regular oscillations of asymmetry preceding the depression of the alpha rhythm; in two different normal subjects.

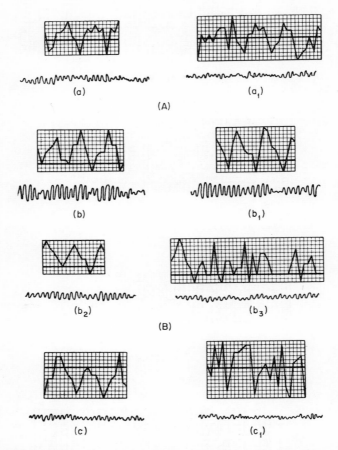

FIG. 4. Slow background periodical oscillations in asymmetry of alpha waves and corresponding EEG records of three normal subjects in states of alertness, excitement and fatigue. (A) Different parts of the background of the same patient; in a state of alertness (a); and in a state of slight fatigue (a_1). (B) In a state of alertness in different (b, b_1) days; at the beginning (b_2) and at the end (b_3) of a long experiment (b_3 in a state of slight excitement). (C) In states of wakefulness (c) and drowsiness (c_1).

of the subject. When the latter was undisturbed and alert, the G waves proved to be stable both during one experiment (Figure 4A, a and a_1) and during different experiments (Figure 4B, b, b_2, and b_3).

When the subject was somewhat excited (Figure 4B, b_3), drowsy (Figure 4C, c_1), or tired (Figure 4A, a_1), no regular periodical oscillations of Δ_1 could be detected, although at other periods G waves were recorded in the same subjects (Figure 4, a, b, b_1, and c). Apparently, this is why we were unable to record regular rhythmical oscillations of Δ_1 in some of the subjects at the beginning of the experiment.

Thus, when the level of wakefulness increases (at moments of strained attention or general excitation) or when it decreases (in fatigue or in a drowsy state), the periodicity of the G waves becomes broken. It is obvious that the preservation of the periodical oscillations of the Δ_1 values depends only upon a definite level of wakefulness or upon a definite functional state. This is also testified to by the fact that breaks in the periodic oscillations of the asymmetry of alpha waves take place just before the depression of the alpha rhythm.

It must be pointed out that sometimes it is quite difficult to detect G waves. This is because of the insufficient statistical reliability of the Δ_1 index (the average from only ten oscillations, which varies considerably) and also to some failures of the parameter Δ_1 itself (Artemieva and Meshalkin, 1965). Therefore, in cases when we fail to find G waves, we are not sure whether this is because of the actual absence of regular oscillations or simply because of the inadequateness of our technique in the given conditions.

II. RELATION TO THE ORIENTING REACTION

We shall now examine what can be determined about the nature of the changes in asymmetry of the alpha waves.

It may be supposed that Δ_1 is related to the nonspecific forms of activity that appear in orienting reactions. The proof of this proposition would be that the dynamics of the oscillations of Δ_1 and the breakdowns in the periodicity of the G waves should obey the laws of the orienting reactions, i.e., they should disappear when the stimuli are repeated and reappear when the stimuli are given meaning. A special study was carried out to verify these data. In eight normal subjects (20–35 years old) with stable alpha rhythms, the behavior of Δ_1 was studied during repeated presentations of intermittent acoustic signals of a medium intensity (60 dB) with a duration of 0.5 sec and with 6–12 sounds in each pattern (first series). In the second series, the verbal instruction, "Count the sounds," was given; in the third series this instruction was countermanded.

The EEG was recorded from the parietooccipital region of the left hemisphere (by the bipolar method), the speed of the paper being 60 mm/sec. The subject was placed in a lightproof chamber. An Alvar electroencephalograph was used. For experimental purposes, we selected subjects with a stable alpha rhythm that exhibited no depression at all or an insignificant depression during the presentation of stimuli. No visual differences in the EEG were seen in the background, during the presentation of neutral stimuli, and during the presentation of meaningful stimuli. In cases when the acoustic stimuli provoked a distinct depression of the alpha rhythm, one or two preliminary trials were carried out, and only then did the basic experiment take place.

Equal sections of the EEG records were analyzed; they related to the time of presentation of the acoustic signals and to the background, which directly preceded the acoustic stimuli. For the purpose of estimating Δ_1, the records were coded in the same way as the 1 sec sections (Figure 1). The ratio of the difference between dots and vertical bars to the average number of oscillations measured during these sections was used as an index of the asymmetry.

Experiments showed that the average value of the asymmetry of the ascending the descending fronts of alpha waves changed in a different way during the first and during subsequent presentations of sounds. During the presentation of the first through twelfth sounds, all subjects exhibited law-governed changes of the asymmetry in comparison with the background (decrease or increase); during subsequent acoustic stimulations (thirteenth through twenty-fifth) the asymmetry of the alpha waves did not change significantly and the ratio between the values of asymmetry in the background and during the presentation of the acoustic stimuli assumed a random character.

The statistical significance of the above-mentioned results was established with the help of the F test (in the simplest form of dispersion analysis).

When the meanings were introduced (by means of the verbal instruction, "Count the sounds"), there appeared again a law-governed change of the asymmetry of the alpha rhythms during the presentation of the acoustic stimuli.

All these differences may easily be seen from a summarized graph (Figure 5A, a-d) where the asymmetry relations both in the background and during the presentation of the sounds are shown. If we evaluate a substantial excess of the asymmetry of the background over the asymmetry during stimulation as +2, a moderate excess as +1, reverse relations as −1 and −2, and absence of any changes as 0, then the ordinate of any point on the graph will be represented by a number equal to the sum of the values in both channels. When the value was expressed by one channel, it was doubled (this doubling could only mask the differences but it did not contribute to the emergence of illusory differences where no real differences were observed). The graph shows that during the presentation of the first through twelfth sounds a significant change of the asymmetry of the alpha waves took place (nearly all points are above the abscissa line)(Figure 5A, a); when the next series of sounds (thirteenth through twenty-fifth) was presented, the points changed their places, reflecting more occasional changes of the asymmetry (Figure 5A, b). The instruction to count the sounds caused, in all eight subjects, marked changes of the asymmetry during the presentation of the acoustic stimuli (Figure 5A, c). These changes were statistically significant ($p = 0.05$). The elimination of the instruction resulted in a restitution of the occasional relations between the asymmetry in the background and during the presentation of the acoustic stimuli (Figure 5A, d).

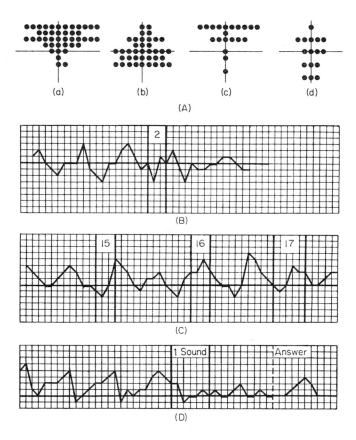

FIG. 5. Changes of asymmetry during the presentation of indifferent and signaling stimuli in normal subjects. (A) Changes of average asymmetry of the duration of alpha rhythm phases (summary data relating to eight subjects). Explanation in the text. (B–D) Changes of G waves in a normal subject. (B) Presentation of the second indifferent sound: the periodicity of the asymmetry oscillations is broken. (C) Presentation of the 15th, 16th, and 17th sounds: the periodicity of the asymmetry oscillations is conserved. (D) Presentation of first sounds having a signaling meaning again result in a breakdown of the G waves.

Changes in the asymmetry during the presentation of irrelevant (neutral) sounds and during their counting are reflected also in the G waves (Figure 5B-D). The presentation of the second intermittent acoustic stimulus in accompanied by a breakdown of the periodic oscillations of asymmetry of the alpha waves (Figure 5B). However, during the action of the fifteenth through seventeenth acoustic stimuli, the G waves are of a relatively regular character (the period of oscillation is 6–7 sec, with a deviation of 0.7; see Figure 5C).

When the subjects start to count the series of sounds (Figure 5D), the periodical oscillations of the asymmetry become again deranged.

Thus, the behavior of the average values of asymmetry of the alpha waves, as well as the dynamics of the periodic oscillations of the Δ_1 (G waves) in the background and during the action of both irrelevant and meaningful acoustic signals, are similar to the behavior of various components of the orienting reaction within the system of unconditional and conditional reflexes: they disappear when the stimulus is frequently repeated (similar to the habituation of orienting reactions to neutral stimuli) and reappear when the stimuli acquire a meaning (similar to the rehabilitation of orienting reactions during the formation of a conditioned connection). These data prove that the changes in the asymmetry of the alpha waves and in the dynamics of the periodic oscillations of Δ_1 reflect the nonspecific form of activity along with the other components of the orienting reaction. Therefore, the correlation between the oscillations in the asymmetry of the alpha waves and the behavior of other components of the orienting reaction (vegetative, sensory, and motor) is of considerable interest.

III. SUBJECTS WITH FRONTAL LESIONS

As a result of a number of investigations (Homskaya, 1960, 1961, 1964; Luria and Homskaya, 1964), it was established that lesions of the frontal lobes of the brain lead to pathological changes in the orienting reaction. As a rule the effects of lesions are expressed as weakness and instability of the orienting reactions to neutral, and especially, to meaningful stimuli. The vascular, galvanic skin reactions and the reaction of the alpha rhythm depression, which are components of the orienting reaction, manifested in such patients during the presentation of neutral stimuli (light or sound) especially when they are of great intensity, are feebler, become easily extinguished, and are not reinstated (or the reinstatement is of a transitory character) when the stimuli receive meaning (i.e., when the subjects are asked to evaluate the signals according to their number, duration, or pitch or to press a key in response to a signal).

These data give reason to assume that some other indices of the state of activity in patients with lesions of the frontal lobes, in particular, the average level of asymmetry of the EEG waves and its periodic oscillations, essentially differ from those in normal subjects.

We investigated nine patients with different lesions of the frontal lobes. In six of these patients, the focus of the lesion was in the right frontal lobe; in one patient it was in the left lobe; and in two patients it spread to both hemispheres, the left frontal lobe being more affected then the right one.

Four patients had extracerebral tumors of the frontal lobes and five had intracerebral tumors. These tumors were located in different parts of the frontal lobes. In two patients, they were located in the interhemispheral fissure in the region of the poles and medial parts of the lobes (more on the left). In two other patients, they were located in the depth of the posterior right frontal lobe; another two patients had tumors in the same parts, with an infiltration into the lateral ventricle or in the region of the septum pellucidum. In one patient, the tumor was located in the medial parts and in the pole of the right frontal lobe, adjoining the premotor and temporal regions on the right, and in another, in the convexal part of the left frontal lobe. In the last patient, a tumor existed in the frontotemporal region, in the depth of the Sylvian fissure, with an infiltration inward to the mediobasal parts of the frontal lobe. In all patients but one, the location of the tumors was established through surgical operation.

The hypertensive—hydrocephalic syndrome was manifested differently in the patients investigated. One patient had no symptoms of hypertension whatsoever; five patients exhibited weak and moderately pronounced intracranial hypertension, e.g., headaches, slight stagnant phenomena in the ocular fundus, decline of vision, high pressure of the spinal fluid (240–250 mm of water). All other patients had a strongly pronounced syndrome of increased intracranial pressure, e.g., acute headaches, sometimes accompanied by nausea and vomiting, papilloedema, decline of vision, increased intracranial pressure (350 mm of water), and changes in the bones of the cella turcica on the roentgenogram.

The neurological local symptoms included: hyposmia or anosmia on one side (in four patients) or on both sides (in two patients); epileptic attacks of the adversive type (in two patients) or general (in three patients), exophthalmia on the side of the lesion (in one patient), bilateral or unilateral pathological reflexes (oral automatism, the grasping reflex, Babinski's reflex); paresis of the facial nerve on one or both sides; aminia; asymmetry of the tendon reflexes; slightly pronounced pyramidal insufficiency (mostly in the arm); modification of the tone in the same extremity; violation of the statics and gait of the astasia—abasia types (one case). Three patients manifested symptoms of the influence of the pathological process on the brain stem structures (paresis of the upward look, spontaneous nystagmus, hyperreflexia of caloric nystagmus).

In all patients the electroencephalogram showed foci of pathological activity in the corresponding parts of one or both frontal lobes.

Investigations of the higher cortical functions revealed in six patients only slight symptoms of a functional disturbance of the frontal lobes; under conditions of intact orientation in the external environment and absence of any derangements in the praxis of posture, gnosis, and speech, the following phenomena were observed in the patients: slightly insufficient self-criticism, inadequacy, emotional instability, slight motor perseverance without timely correction, impulsiveness of movements, nonsustained attention, slight but specific derangements of the dynamics of the mental processes (impulsiveness, thought fragmentation, inclination to stereotypy, delayed corrections). In a lesion of the frontotemporal systems, mild defects of the verbal–auditory memory were in evidence.

The remaining three patients exhibited more pronounced symptoms of lesions of the frontal lobes, although their orientation in space and to some degree in time proved to be relatively intact. These symptoms included: general sluggishness; lack of spontaneity; absence of any complaints; echolalia of verbal answers; the mirror symptom during trials for the praxis of posture; sometimes echopraxia; marked disturbances in serial movements, which were expressed in impeded switching of the persistent motor perseverances without clear realization of the errors committed; unsustained attention; inactivity of the intellectual processes (difficulties in serial counting and narration of texts, poor presentation of the anamnesis, etc.); general exhaustibility of all psychological processes.

Thus, in the main, we investigated patients with a relatively mild frontal syndrome, appearing against the background of weak, moderate or even considerable intracranial hypertension. Only a few of the patients manifested a distinct frontal syndrome with moderate or well-expressed symptoms of increased intracranial pressure. All the patients (except one) were investigated prior to the operation.

The analysis of electroencephalograms (EEG) recorded from the parieto-occipital parts of the normal hemisphere (by the bipolar method) revealed in these patients periodic changes in the asymmetry of the ascending and descending phases of the alpha and theta waves,* just as they were observed in normal subjects. In the patients, however, the period of the asymmetry oscillations was shorter and equalled 3.5 sec, or more often, 4.0 sec.

In contrast to normal subjects, who in a state of rest usually showed a strict preservation of the time periods (during 30–50 min of recording), in the patients the values of the periods rapidly changed, the periodicity itself remaining invariable.

For example, in patient US (with an intracerebral tumor of the right posterofrontal region of the brain) series of oscillations with periods of 3, 4, 5, and 6 sec were observed in the EEG background during one and the same experiment (Figure 6).

In the electroencephalogram of this patient an alpha rhythm with a frequency of 9 Hz was manifest in the posterior parts of the brain. It was somewhat disorganized in the right hemisphere:, the beta rhythm was slightly pronounced; delta waves of considerable amplitude and duration were observed in the right hemisphere, predominantly in the frontal region of the brain.

The average level of asymmetry is likewise unstable in patients with lesions of the frontal lobes; it may markedly shift within relatively short periods of time (15–30 sec), which is by no means peculiar even for normal subjects (Figure 6).

*The coding and subsequent estimation of the asymmetry of the theta waves were based on the same principle as in the case of alpha oscillations.

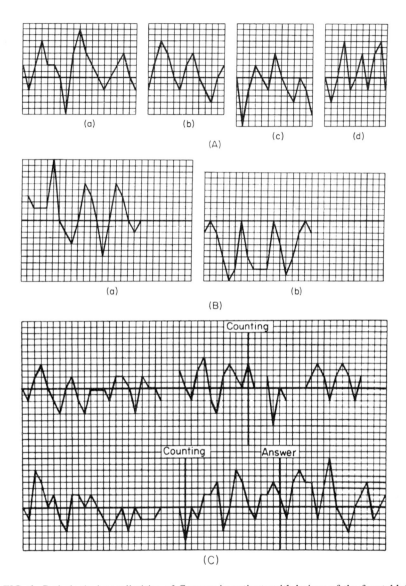

FIG. 6. Pathological peculiarities of G waves in patients with lesions of the frontal lobes in the EEG background (in the course of one experiment) and in the state of activity. (A) Oscillations of asymmetry in the EEG background in patient US. (a)–(d) Series of oscillations with different periods (3, 4, 5, and 6 sec). (B) The same oscillations in patient UST. A marked change of the average level of asymmetry is seen. (C) Dynamics of periodical oscillations of the asymmetry of EEG waves during activity in patient US (intracerebral tumor of the right posterofrontal region of the brain). Explanation in the text.

These facts apparently reflect the insufficient stability of the mechanisms regulating the state of activity, and the ease with which they switch over to new conditions or work.

There is another specific feature of the G waves in our patients: the amplitude of oscillations is greater than the normal level (a statistically significant difference).

A certain interdependence was observed among the decrease of the period, the amplitude of oscillations and the gravity of the patient's state. For example, in patient U (arachnoid endothelioma in the region of the left Sylvian fissure), with mild neurological and neuropsychological symptoms of a lesion of the posterofrontal parts of the brain against the background of slight intracranial hypertension, the period of the G waves was 5 sec and the amplitude was 1.0. In patient S (intracerebral tumor of the posterior parts of the right frontal lobe having the considerable size of 5 times 5 cm) with a pronounced frontal syndrome and absence of any symptoms of intracranial hypertension, the period of the G waves was 4 sec and the amplitude was 1.2. Finally, in patient UR (after a resection of the right frontal lobe) with the syndrome of a profound lesion of both frontal lobes and a pronounced hypertensive–hydrocephalic syndrome, the period of the G waves was 3 sec and the amplitude was 1.8.

The EEG's of patient N of a special character are (intracerebral tumor of the transparent septum with an infiltration into the right frontal lobe) as are those of patient T (intracerebral tumor of the mediobasal part of the right frontal lobe). In these patients, the pathological process affected the deep mediobasal structures of the brain, which adjoin the ventricular system. The neuropsychological picture of the disease in these patients was rather poor: there was observed some impulsiveness of voluntary movements; the patient underestimated the gravity of his disease, could not adequately correct his errors, and manifested unsustained attention on the background of generally intact behavior. The periodicity of the G waves in these patients was of a less pronounced character than in other patients with lesions of the frontal lobes, although the oscillations of the Δ_1 differed from random. The amplitude of these oscillations was small (0.8 and 1.0).

In the EEG of patient T, an irregular alpha rhythm with a frequency of 10–11 oscillations per second was evident in the posterior parts of the cerebral hemispheres; delta waves were observed in all regions of the brain; they markedly predominated in the anterior regions, especially in the right frontal region. Frequent oscillations and epileptoid impulses were also observed in all regions of the brain.

In the EEG of patient N, an alpha rhythm with a frequency of 9–10 oscillations per second and of an irregular amplitude was more pronounced in the posterior parts of the right hemisphere. Stable and flattened delta waves (1–1.5 and 2 oscillations per second) were seen in the anterotemporal regions of the brain; they showed an increase under the presentation of external stimuli.

Thus, preliminary data concerning the asymmetry dynamics of the alpha waves and of the lower frequencies in patients with lesions of the anterior parts of the hemispheres give us reason to assume that derangements in the periodicity of oscillations of the amount of asymmetry of the EEG waves are mostly connected with the mediobasal parts of the frontal lobes. Other parameters (values of the period and of the oscillation amplitude) may undergo certain changes in all patients of the investigated group.

In patient US (intracerebral tumor of the right posterofrontal region of the brain), during mental computation, instead of a protracted disappearance of periodic oscillations of the asymmetry of alpha waves, there was complete absence of any changes in the background periodicity of the asymmetry, a transitory depression with a subsequent rapid reinstatement of periodic oscillations of the asymmetry, or, emergence of relatively regular oscillations of the asymmetry in comparison with the preceding background (Figure 6C).

As is known, a depression of the alpha rhythm is likewise possible in normal subjects during intellectual activity; but this depression is preceded and followed by protracted changes in the regularity of the periodic oscillations of the alpha wave asymmetry. In patient US the period of depression was short and not accompanied by visible changes of the background periodicity; this indicates a different character of this depression.

Of particular interest is the emergence in patient US of relatively regular oscillations in the asymmetry of alpha waves at moments of intellectual strain, a phenomenon that is diametrically opposed to that observed in normal subjects. This fact indicates a certain change of the patient's functional state, but the direction of this change differs from the normal one.

A study of the dynamics of asymmetry of the alpha waves and of slower oscillations during the action of neutral and meaningful acoustic stimuli in patients with lesions of the frontal lobes also revealed an essentially different picture as compared to that in normal subjects.*

Absence of law-governed changes in the average values of asymmetry of the alpha and theta waves under the action of neutral and meaningful stimuli in patients with lesions of the anterior parts of the cerebral hemispheres may be seen in Figure 7, which presents summary data relating to all the patients investigated. The value of the asymmetry during the presentation of sounds 1–12 and 13–25 (i.e., at different stages of habituation of the orienting reaction), after the verbal instruction, "Count the sounds," and after the countermand of the task are approximately equal (compare with Figure 5, which presents analogous data relating to normal subjects).

Absence of any changes of the asymmetry of the EEG waves in patients with lesions of the frontal lobes during the presentation of meaningful stimuli, which in the case of normal subjects exert particularly effective influence on the

---

*Six patients were investigated in accordance with the given program.

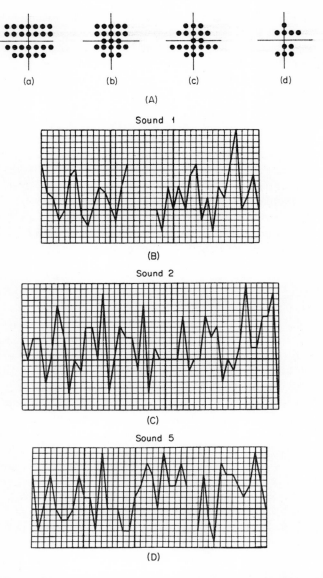

FIG. 7. Changes of asymmetry during presentation of indifferent and signaling stimuli in patients with lesions of the frontal lobes. (A) Changes of average asymmetry of the duration of EEG phases (summary data). Explanation in the text. (B)–(D) Dynamics of G waves during presentation of nonsignaling (B) and signaling [(G) and (D)] sounds in patient UR (after resection of the right frontal lobe) with a grave frontal syndrome. It may be seen that in this patient there are no law-governed changes of the G waves during the presentation of indifferent stimuli (sound 1) and during the counting of the acoustic signals (sounds 2 and 5).

parameter Δ_1, may also be seen in Figure 7; this figure domonstrates the G waves recorded in patient UR (after resection of the right frontal lobe) in the course of counting the acoustic signals. Neither the first neutral sound, nor the second or fifth meaningful sound (Figure 7B and D) could produce any substantial change in the G waves.

IV. SUMMARY

Thus, the peculiarities of the average amount of asymmetry of the alpha and theta waves, as well as of the periodic oscillations of their values during the action of nonsignaling and signaling stimuli in the investigated patients coincide with the dynamics of orienting reactions in the same patients (Homskaya, 1961, 1964; Homskaya et al., 1961); once more proves that the given indices may be regarded as components of the orienting reaction. The above-described phenomena convincingly show that lesions of the frontal lobes violate the brain structures that are responsible for the regulation of the state of activity, as well as of various functional states.

Changes in the asymmetry of duration of the EEG phases, especially the periodic oscillations of the amount of asymmetry, must, apparently, take a definite place among the various electrophysiological manifestations of the orienting reaction such as depression of the alpha rhythm, secondary responses, modifications of the constant potential, and the K complex. The use of this parameter opens up new paths for research into various functional states, both in normal subjects and in patients with local lesions of the brain.

The phenomenon of periodic oscillations of the asymmetry of the EEG waves obviously bears a direct relation to the extra-slow rhythmic oscillations of the potential (described by Aladzhalova, 1962), which were recorded in different structures of the brain. According to that author, these extra-slow electrical phenomena are connected with protracted changes of the excitability of the nervous elements and relate to slow humoral regulations as distinct from the rapidly acting neural regulations.

The similarity of the temporal parameters described by Aladzhalova, the extra-slow rhythmic oscillations and periodic changes of the amount of asymmetry of the EEG waves, leads to the assumption that these oscillations reflect the activity of one and the same slow controlling system that insures a global and protracted reconstruction of the level of brain activity.

REFERENCES

Aladzhalova, N. A. (1962). "Slow Electrical Processes in the Brain." Acad. Sci. USSR Press, Moscow (in Russian).

Artemieva, E. Yu., and Meshalkin, L. D. (1965). Analysis of Δ parameter in asymmetry of the length of the ascending and descending fronts of alpha-rhythm. *Nov. Issled. Pedagog. Nauk.* **4**, 222-227.

Artemieva, E. Yu., Meshalkin, L. D., and Homskaya, E. D. (1965). The periodical oscillations of the asymmetry of ascending and descending fronts of alpha-waves. *Nov. Issled. Pedagog. Nauk.* **138**, 249-251.

Barlow, J. (1957). Cited by de Lange *et al.* (1962).

Berger, H. (1930). Über das Elektroencephalogramm des Menschen. VI. *J. Physiol. Neurol.* No. 40, 160-179.

Bishop, G. H. (1962). "Slow Electrical Processes in the Brain." Acad. Sci. USSR Press, Moscow (in Russian).

Chang, H. T. (1951). Changes in excitability of cerebral cortex following sing electrica shock applied to cortical surface. *J. Neurophysiol* **14**, 25-113.

Genkin, A. A. (1963). The asymmetry of the phases of EEG in mental activity. *Dokl. Akad. Nauk SSSR* **149** No. 6, 1460-1463 (in Russian).

Genkin, A. A. (1964). The length of the ascending and descending front of EEG as a source of information about neurophysiological processes. Ph.D. Thesis, Leningrad (in Russian).

Homskaya, E. D. (1960). The influence of verbal instructions on vascular and galvanic skin components of the orienting reflexes in local brain lesions. *Proc. Acad. Pedagog. Sci. R.S.F.S.R.* Pap. No. 6, pp. 105-110 (in Russian).

Homskaya, E. D. (1961). *Proc. Acad. Pedagog. Sci. R.S.F.S.R.* Pap. Nos. 1 and 2 (in Russian).

Homskaya, E. D. (1964). Verbal regulation of the vegetative components of the orienting reflex in focal brain lesions. *J. Cortex* **1**.

Homskaya, E. D., Konovalov, Yu. W., and Luria, A. R. (1961). Participation of the verbal system in the regulation of the vegetative components of the orienting reflex in different local lesions of the brain. *Probl. Neurosurg.* No. 4.

Loomis, A., Harvey, E., and Hobart, G. (1936). Electrical potentials of the human brain. *J. Exp. Psychol.* **19**, 562-564.

Luria, A. R. and Homskaya, E. D. (1964). Disturbances in the regulative role of speech with frontal lobe lesions. *In* "The Frontal Granular Cortex and Behavior" (A. R. Luria and E. D. Homskaya, eds.). McGraw-Hill, New York.

Moisseyeva, N. J., and Genkin, A. A. (1963). Essays in application of non-parametrical statistics in analysis of EEG in vascular diseases of the brain. *Zh. Neuropatol. Psikhiat. S. S. Korsakova* **63**, 1147-1152 (in Russian).

Mundy-Castle, A. C. (1957). EEG and mental activity. *Electroencephalogr. Clin. Neurophysiol.* **9**, No. 4, 643.

Walter, W. Grey (1950). Normal rhythms, their developments, distribution and significance. *In* "Electroencephalography" (D. Hill and G. Parr, eds.), pp. 203-228. London.

Werre, P. E. (1957). "Interrelation between Psychological and EEG Data in Normal Adults." Leiden.

Chapter 4

CHANGES IN THE ELECTROENCEPHALOGRAM FREQUENCY SPECTRUM DURING THE PRESENTATION OF NEUTRAL AND MEANINGFUL STIMULI TO PATIENTS WITH LESIONS OF THE FRONTAL LOBES

O. P. BARANOVSKAYA and *E. D. HOMSKAYA*

Department of Psychology
University of Moscow
Moscow, U.S.S.R.

Department of Neuropsychology
University of Moscow
Moscow, U.S.S.R.

Among the various electroencephalogram (EEG) indices for the activation reaction, the suppression of the alpha rhythm has been investigated most extensively. However, when describing this suppression, most authors usually confine themselves to stating the presence or abscence of the reaction on the basis of visual observations of the EEG. A quantitative evaluation of the modifications of the main rhythm, obtained with an automatic frequency analyzer, may provide additional data characterizing the suppressive reaction as a component of the orienting reaction.

Investigations of various components of the orienting reaction when lesions of the frontal lobes were present revealed considerable derangements. The vegetative components (vascular, skin galvanic) were weakened in many cases; they became easily extinguished under the presentation of both neutral and meaningful stimuli, particularly when meaning was imparted to the stimulus by means of a verbal instruction. Such derangements are considerably less frequent in lesions of the posterior parts of the cerebral hemispheres (Homskaya, 1960; 1961; Filippycheva, 1963, 1966).

The pathology of the orienting reaction in patients with lesions of the frontal lobes extends also to the EEG indices, being subordinated to the same laws revealed in the study of the vegetative components (Filippycheva, 1963; Artemieva and Homskaya, 1966; Baranovskaya, 1966).

The aforementioned investigators, however, had not yet determined the types of derangements that take place in the EEG component of the activation reaction in the clinical aspects of local brain lesions. Of great interest, therefore, is a quantitative evaluation of the alpha rhythm modifications, as well as research into the composition of the alpha frequencies and comparison of the given indices with definite clinical forms of frontal lobe lesions. Such an approach to the investigation of the EEG component of the orienting reflex will help in the comprehension of the physiological mechanisms that are responsible for the disturbances of mental processes (especially attention in patients with frontal lobe lesions).

As a rule, when studying the suppressive reaction, investigators consider the alpha rhythm range (8–13 Hz) as a homogeneous whole, without due regard for the heterogeneity of its frequency composition. However, Grey Walter (1957, 1959, 1966) pointed out that the alpha rhythm almost invariably includes several frequency components, each of which has its own functional significance. Of undoubtedly great interest is an analysis of the composition of the alpha rhythm, i.e., of the "behavior" of its separate frequencies, as well as a comparison of these frequencies with the nature of modifications in lower and higher frequencies of the EEG spectrum during the presentation of a stimulus. These data, as mentioned earlier, may be obtained using an automatic analyzer of discrete EEG frequencies.

Dietsch (1932), who utilized the method of Fourier, was the first to carry out a frequency analysis of the EEG; he ascertained the main frequency of the alpha rhythm and a number of its harmonic components. Later, Livanov (1938) began to analyze electrocerebrograms according to the method of Bernstein. Subsequently, EEG frequency spectra were analyzed and studied by a number of investigators by means of mathematical and automatic analyses (Grass and Gibbs, 1938; Gibbs, 1939, 1942; Spilberg, 1941; Walter, 1943, 1950, 1966; Mundy-Castle, 1957; Sokolov and Danilova, 1958; Danilova, 1959; Ilyanok, 1960, 1962, 1965).

Of particular interest is the study of the frequency spectrum in various focal lesions of the brain.

In organic brain lesions, slow delta waves, and sometimes theta waves, are, as a rule, recorded near the pathological focus. Far from the primary focus of the lesion, single slow waves are recorded; they alternate with an irregular alpha rhythm of reduced amplitude (Mayorchik and Rusinov, 1954; Mayorchik, 1956, 1957, 1960; Sokolova, 1957; Bekhtereva, 1959). In surface cortical tumors, delta waves are recorded over the site of the tumor and a theta rhythm in its circumference; in deep subcortical tumors that do not spread to the cortex theta waves often appear in the EEG (Walter and Dovey, 1944; Drift, 1957; Small et al., 1961). In focal lesions of the brain, the most appreciable deviation from the normal spectrum is observed in the zone of the delta and alpha frequencies of EEG frequency spectra. This deviation reaches its maximum in the region of the focus. Shifts in the delta range are more pronounced in intracerebral tumors than in extracerebral tumors (Grindel, 1963, 1964).

The principal purpose of this chapter is to examine the EEG frequency spectrum and its modifications under the influence of neutral and meaningful stimuli during orienting reactions in patients with lesions localized in the frontal parts of the cerebral hemispheres and to compare the results with those of investigations carried out on normal subjects. A control investigation was performed on patients with lesions localized outside the frontal lobes.

I. NORMAL SUBJECTS

A. Methods

The EEG was recorded on a ten-channel Alvar encephalograph. The frequency spectrum was recorded with an electric harmonic Walter analyzer. The latter indicated the averaged energy for 10 sec of each of the 24 EEG frequencies, from 1.5 to 30 Hz.

The background EEG was recorded from six symmetric areas of the hemispheres (parietooccipital, parietoecentral, and central–posterofrontal). The effect of acoustic stimuli was observed in records made from the parietooccipital area of one of the hemispheres.

The method of comparing neutral and meaningful stimuli was applied. In 15 groups of intermittent sound neutral and meaningful stimuli were presented to the subjects. Meanings were imparted to the stimuli by means of the verbal instruction: "Count the sounds."

When processing the experimental data, we expressed the amplitude of each frequency in millimeters. For the purpose of characterizing the background EEG, the spectrum of each area was calculated as an average of 10–12 periods of analysis. Ten seconds—one period of the analysis—were investigated just before and during the presentation of the stimuli. Corresponding to the first 5 and to all 15 acoustic stimuli, the average spectra under our investigation consisted of 5 and 15 periods of analysis. The quantitative evaluation of the magnitude of the alpha frequency modification was performed by comparing their amplitudes in the background just before with that during the presentation of the sounds. The energy of the alpha frequencies during stimulus presentation was expressed in percentage of the background energy, which was taken as 100%. For the purpose of comparing the magnitudes of the suppressions in normal subjects and in patients, a conventional value was introduced, an average suppression for each suppressed frequency, which was obtained by dividing the summed magnitude of the suppressions by the number of suppressed alpha frequencies.

We investigated 15 normal subjects and 33 patients with lesions of various localization: the anterior parts of the brain were affected in 23 patients and the focus of the lesion was situated outside the frontal lobes in the other 10 patients. The experiments were carried out in the laboratory of neuropsychology of Moscow State University, which is located in the Burdenko Scientific Research Institute of Neurosurgery.

B. Results

In most of the normal subjects investigated, the alpha range frequencies predominated in the background frequency spectrum of the EEG, their amplitudes exceeding by several times that of lower and higher frequencies. The spectra of various areas of the cerebral hemispheres had relatively similar general configurations, although the amplitude characteristics of the frequencies (especially of the alpha range) varied to a greater or lesser degree in different areas. A decline of the energy of the alpha frequencies was observed: the farther from the occipital areas and the closer to the frontal areas, the more pronounced was this decline, a fact that had been mentioned in the literature (Gibbs, 1942; Ilyanok, 1965).

In the alpha range of most subjects, all alpha frequencies, from 8 to 13 Hz, were observed, and the amplitude of the higher frequencies (10–13 Hz) exceeded that of the lower alpha frequencies (8–9 Hz).

The application of neutral acoustic stimuli resulted in a more or less pronounced decrease of the energy of the alpha frequencies in all subjects. In 15 subjects, the amplitude of the alpha frequencies decreased on the average by 10–20% during the action of the first five acoustic stimuli (Figure 1). Average suppression for each alpha frequency was 17%; all six frequencies of the alpha range were supressed. Supression was the main type of modification of the alpha frequencies; but in some cases the modifications of the alpha range were expressed in a decline of the amplitude of some frequencies and in an increase (enhancement) of others (reaction of the mixed type). Reactions of this type are less frequently observed during the first presentations of the acoustic stimuli than during subsequent ones.

The suppression of the alpha rhythm more often took place against the background of suppression of lower and higher frequencies of the spectrum, or was accompanied by a shift of the frequency spectrum toward the higher frequencies (Figure 2).

The communication of meaning to the stimuli with the instruction, "Count the sounds," markedly intensified the suppressive reaction in comparison with the action of the neutral sounds (Figure 2A).

The difference between the action of neutral and meaningful sounds proved to be statistically significant for all the alpha frequencies. In 15 subjects, the magnitudes of the suppression in separate alpha rhythms amounted on the average to 30–46% (Figure 1). The average suppression for each alpha frequency was 38%. All six frequencies of the alpha range participated in the reaction.

Occasionally, during the action of the meaningful sounds, just as during the neutral sounds, reactions of the mixed type were observed; however, there were fewer of these reactions in response to the meaningful sounds than in response to the neutral sounds. During the first presentations of the meaningful sounds, only single reactions of this type manifested themselves.

The suppression of the alpha rhythms was accompanied by a decline in the amplitude of slow frequencies and a rise in the high-frequency part of the spectrum or by a decrease of the amplitude of all frequencies (Figure 2).

An analysis of the composition of the suppressive reaction in separate alpha frequencies showed that in response to the neutral sounds all the frequencies of the alpha range, both the lower (8–9 Hz) and the higher (10–13 Hz), were almost equally involved in the reaction, whereas during the presentation of the meaningful sounds, there was observed a certain selectivity in the modifications of the alpha frequencies: the higher alpha frequencies proved to be more reactive (Figure 1).

II. SUBJECTS WITH FRONTAL LESIONS

A. Material

Twenty-three patients with lesions of the frontal lobes were investigated; nine of these suffered from intracerebral tumors, ten from extracerebral tumors, and four from various nontumorous diseases (frontonasoorbital hernia, inflammatory processes in the region of the chiasm, aneurism of the anterior conjunctival artery, posttraumatic states). The pathological process was located in the right hemisphere of five patients and in the left hemisphere of nine patients. Bilateral pathological symptoms were observed in nine patients. As regards their localization, these were lesions of the posterofrontal parts of the brain; massive lesions of the white matter of the frontal lobes, subcortical ganglia, and corpus callosum; lesions of the basal divisions or polar divisions of one or both hemispheres; and lesions of the medial parts of both frontal lobes.

In addition to different local symptoms, these patients manifested general cerebral symptoms, such as moderate increased intracranial pressure (nine patients) and a strongly pronounced increased intracranial pressure (four patients); in eight patients, no symptoms of increased pressure were observed.

An investigation of the higher cortical functions revealed symptoms of deranged functioning of the frontal lobes in five patients; in some patients, along with symptoms of frontal insufficiency, more pronounced motor disturbances

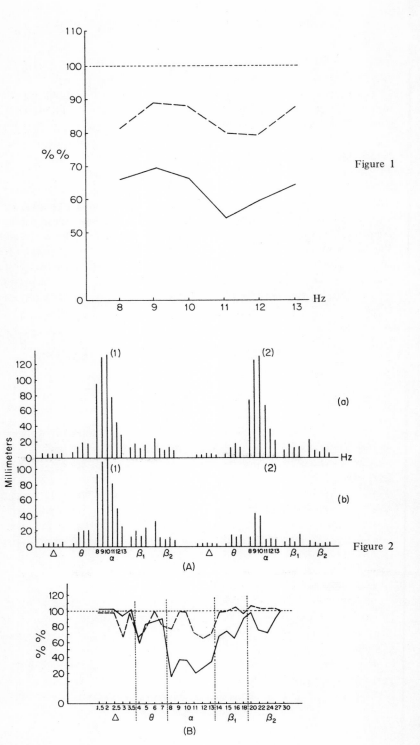

Figure 1

Figure 2

peculiar to lesions of the premotor zone of the brain were found (in 12 patients). In some cases there were symptoms of a lesion in the left temporal area; gross symptoms of a lesion of the frontal parts of the brain were observed in one of the patients. Thus, the clinical as well as the psychopathological symptoms were quite heterogeneous in the group of patients investigated.

B. Results

A characteristic feature of the background EEG in most patients with frontal lesions was an intensification of the slow (delta and theta) part of the spectrum, which most often showed up in the zone of the focus or on the side of the lesion and, less frequently, in both hemispheres. The alpha rhythm did not predominate. A predominance of the alpha range in the EEG spectra was encountered only in four patients and only in limited areas of the brain. In four patients, along with a well-defined alpha rhythm, there was observed an intensification of the ranges of low and high frequencies (spectrum of a diffused type). In the alpha range of the patients, in contrast to that of normal subjects, the amplitudes of the lower frequencies, as a rule, predominated over those of the higher frequencies. An interhemispheric asymmetry, chiefly in the frontal and central divisions, was in evidence.

In addition to the modification of the EEG background in these patients, the activation reaction proceeded differently. Not only did the qualitative aspects of the modifications in the alpha frequencies differ, but also their quantitative aspects. The great variation in the data obtained and the heterogeneity of the clinical picture of the diseases, caused by the different localizations of the pathological processes within the frontal lobes themselves, did not allow us to average the indices of modification of the alpha frequencies in all patients. We were compelled, therefore, to use the method of analyzing similar types of modifications, as well as their intensities, and on this basis to divide the patients into groups so as to utilize individual average values.

All the patients with lesions of the frontal lobes were divided into two groups, each of which was characterized by a similar type, magnitude, and

---

FIG. 1. (Opposite page, top) Modification of the amplitude of alpha frequencies during the action of the first five acoustic indifferent (– – –) and signaling (——) stimuli in normal subjects. Average data from 15 subjects. The abscissa gives the frequencies of the alpha range; the ordinates are the amplitudes of the alpha frequencies expressed in %% to the background, taken as 100%.

FIG. 2. (Opposite page, bottom) Modifications of the EEG frequency spectrum during the action of the first five indifferent and signaling acoustic stimuli in normal subject K. (A) Frequency spectrum in absolute values (mm) in the background (1, a and b), during the action of indifferent stimuli (2, a), and during the action of signaling stimuli (2, b). (B) Modification of the frequency amplitudes during the action of indifferent (– – –) and signaling (——) sounds, in %% of the background.

composition of reaction of the alpha frequencies. Patients of the *first* group manifested considerable disturbances of the orienting reaction in response to neutral, and especially, meaningful stimuli, whereas patients in the *second* group exhibited a relatively normal reactivity of the alpha frequencies under the same conditions. The division of the patients into two groups was based upon the following criteria: (1) average magnitude of suppression falling on one suppressed frequency and (2) average number of suppressing frequencies during the presentation of the first five signaling sounds. In normal subjects, the most intensive decline in the amplitude of the alpha frequencies was observed under precisely these conditions (the average magnitude of the suppression was 38% and the average number of suppressed frequencies was six).

The *first* group included 14 patients with reduced reactivity of the alpha frequencies. The maximum magnitude of the suppression per suppressed frequency did not exceed 27% during the action of the first five signaling sounds and the number of suppressed frequencies did not exceed four.

The *second* group included seven patients with relatively normal reactivity of the alpha frequencies. The average magnitude of the suppression ranged from 32 to 46% and the number of suppressed frequencies ranged from five to six.

The main tendency exhibited in the alpha-frequency modifications in patients of the *first* group during the first five indifferent sounds was an enhancement of the alpha frequencies (reaction of the mixed type and reaction of the pure-enhancement type); it involved one to four frequencies and varied in individual alpha frequencies from 10 to 60%. The modifications in other frequency ranges were expressed either as intensifications of the low and suppression of the high frequencies or as rises of the whole spectrum, and less often as shifts in the spectrum toward higher frequencies. With the subsequent presentation of acoustic stimuli, there sometimes took place a decline of enhancement of the alpha frequencies and its conversion into a suppression (the phenomenon of retarded activation). In other cases, an intensified enhancement of the alpha rhythm or a conversion of the suppression reaction into enhancement was observed.

The introduction of meaning to the stimulus (counting of intermittent sounds in every pattern) resulted in these patients either in a suppression of the alpha frequencies exclusively or in a suppression of some alpha frequencies and an enhancement of others. However, the magnitude of the suppression was small in comparison with that in normal subjects (it did not exceed 27% per suppressed frequency), and in some patients it amounted to only 5–16%, involving not more than four frequencies in the reaction (Figure 3). These modifications were observed against the background of a suppression of all frequencies or a shift in the direction of lower frequencies (Figure 3B). As a result of repeated presentations of the meaningful sounds, the phenomenon of retarded activation, a more intense suppressive reaction, emerged in many of the patients.

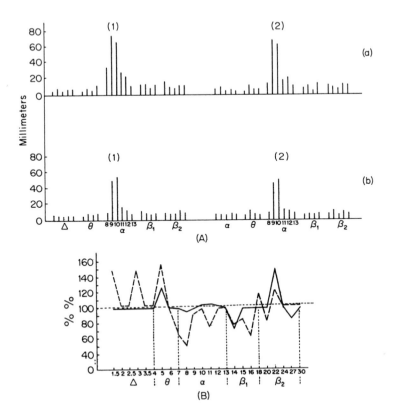

FIG. 3. Modifications in the EEG frequency spectrum in response to the presentation of indifferent and signaling acoustic stimuli in patient K (diagnosis: intracranial tumor spreading to both frontal lobes, corpus callosum, and subcortical ganglia, and affecting the medial parts of the frontal lobes). Designations in the figure are the same as in Fig. 2. Explanation in the text.

An analysis of the behavior of individual alpha frequencies showed that during the presentation of both neutral and meaningful sounds the lower frequencies of the alpha range were more reactive.

In the *second* group of patients, a suppression of the alpha frequencies (varying from 10 to 50% in separate frequencies and involving two to six frequencies, Figure 4) was the predominant type of alpha-range modification in response to neutral sounds. In most patients, this took place against the background of a decline in the energy of low and high frequencies. In the course of subsequent presentations of the sounds, the suppressive reaction was gradually extinguished and an enhancement of separate alpha frequencies emerged.

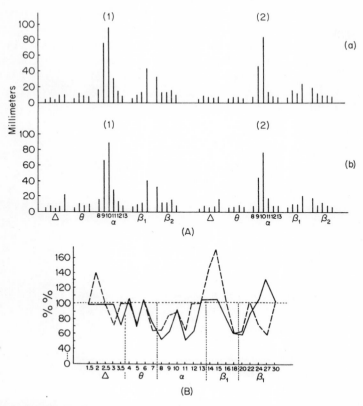

FIG. 4. Modifications in the EEG frequency spectrum in response to the presentation of indifferent and signaling acoustic stimuli in patient D (diagnosis: intercranial tumor, astrcytoma, in the middle parts of the left frontal lobe, with a cyst extending to the pole and base). Designations as in Fig. 2. Explanations in the text.

The introduction of meaning to the stimulus provoked in all patients of the second group a reaction of the pure-suppression type. Its magnitude was within the limits found in normal subjects (Figure 4). In other frequency ranges, a decrease in the energy of all frequencies was often observed (Figure 4B). The phenomenon of retarded activation was absent; both the lower and the higher alpha frequencies participated in the reaction.

A comparison of the experimental results with clinical symptoms in patients with lesions of the frontal lobes did not reveal any correlation among the character of the pathological process, the degree of increased intracranial pressure and the specific features of the EEG reactions to the presentation of acoustic neutral stimuli. On the other hand, a correlation was established between the localization of the pathological process in the frontal lobes and the peculiarities of the reactivity of the alpha frequencies. Within the *first* group, minimum reactivity was manifested by patients with a bilateral localization of the tumor near the midline. In these cases, the pathological process affected the medial cortex of both frontal lobes in the premotor and prefrontal regions or in the region of the pole. In most of these patients, no increased intracranial pressure was observed; only two patients suffered from moderate increased pressure. In the patients of the *second* group, the focus of the lesion was located in other parts of the frontal lobes (e.g., pole, posterior divisions, convexal surface).

III. CONTROL SUBJECTS

A. Material

A control investigation was carried out on ten patients with lesions localized outside the frontal lobes. Some of these patients suffered from lesions of the parietal, parietotemporal and parietooccipital parts of the brain (five patients); in other patients, the focus of the lesion was located in the region of the posterior cranial fossa (three patients) or affected the region of the ventricles (two patients). The pathological process was developing in the right hemisphere (two patients), in the left hemisphere (three patients), and along the middle line (five patients). The character of the pathology was as follows: tumorous lesions of the brain (in six patients); echinococcus with a gross compression of the formations of the striatal syste, internal capsule, and inferior horn of the lateral ventricle (in one patient); chronic basal leucomeningitis and periventricular encephalitis (in one patient); and cystic formation of uncertain etiology (in two patients).

In the clinical picture of the disease, various local and general cerebral symptoms were observed. In most patients these symptoms were expressed as a distinct syndrome of increased intracranial pressure.

A neurophysiological investigation revealed local symptoms that are characteristic of patients with lesions localized in the posterior parts of the cerebral hemispheres (in five patients); in the remaining patients, either no disturbance of the higher cortical functions was revealed, or general disturbances of the mental processes related to intracranial pressure or, in a few instances, phenomena of a secondary frontal syndrome were observed.

B. Results

In most of the control patients, just as in patients with frontal lobe lesions, the background frequency spectra of the EEG had no dominant alpha rhythm in any of the regions investigated and the delta and theta frequency ranges were markedly intensified. A predominance or distinct expression of the alpha rhythm appeared in the spectra of only two patients. An interhemispheral asymmetry (in some or all of the investigated regions in each patient) of the alpha rhythm or of the low frequencies was observed; it exceeded the limits of physiological asymmetry. The focus of the lesion was not uniformly reflected in the frequency spectrum; in some cases it was reflected as an intensification of the slow part of the spectrum, either on the side of the lesion in the zone of the focus or in both hemispheres, and in others it appeared only as a suppression of the alpha rhythm.

However, in most patients with an extrafrontal localization of the lesion, the activation reaction was similar to that in normal subjects, in spite of the pathologically modified background of the EEG. During the action of neutral acoustic stimuli, reactions of the pure suppression type, as well as of the mixed type, were observed; in individual alpha frequencies the suppression varied 10–50% and involved three to six frequencies. In other frequency ranges, either a suppression of all frequencies or an intensification of the slow part of the spectrum (in a reaction of the mixed type) was recorded.

In all control patients, the presentation of meaningful sounds provoked a pure-suppression type of reaction, the magnitudes of which were within the normal bounds (ranging on the average from 29 to 53% per suppressed alpha frequency and involving four to six frequencies, Figure 5). These changes took place against the background of a decline in the energy of other frequencies in the spectrum (Figure 5B). During the repetition of the meaningful sounds, a predominant decrease of the magnitude of the suppression and the emergence of an enhancement of one or two alpha frequencies was observed. Analysis of the behavior of separate alpha frequencies did not reveal any selective character in the reactivity of higher and lower alpha range frequencies.

Particularly noteworthy are the results obtained from an investigation of three patients with lesions of the posterior parts of the brain. In these patients, the reactivity of the alpha range frequencies was obviously reduced: during the counting of the acoustic signals, the suppressive reaction was considerably weakened (ranging on the average from 17 to 19% per frequency and involving not more than three frequencies in the reaction). The effect of the presentation of the meaningful sounds was even weaker than that of the neutral sounds. In other frequency ranges, either a decline of the amplitude of the beta frequencies or a rise in the amplitude of all frequencies occurred. During the subsequent presentations of the meaningful sounds, a reaction of the retarded activation type predominated.

Just as in patients with frontal lobe lesions, the investigation of the control patients did not reveal any connection between the character of the pathological

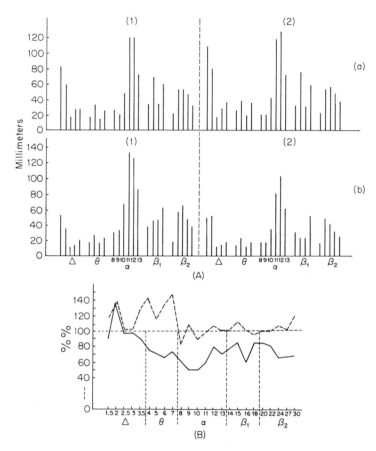

FIG. 5. Modifications in the EEG frequency spectrum in response to the presentation of indifferent and signaling acoustic stimuli in patient K (diagnosis: a tumorous process in the right parietal region). Designations as in Fig. 2. Explanation in the text.

process and the degree of increased intracranial pressure on the one hand, and the indications of the activation reaction, in response to neutral stimuli on the other. It should be noted that most control patients, in contrast to patients with frontal lobe lesions, exhibited a different reactivity of the alpha frequencies, which was close to that of normal subjects. Only three patients were exceptions; these suffered from lesions of the medial divisions of the temporal lobe, basal formations, and structures situated in the wall of the ventricles, i.e., structures that belong to the limbic system and so directly participate in the realization of the activation processes (Pribram, 1960; Lindsley, 1966).

IV. DISCUSSION

The analysis of the initial EEG indices of normal subjects showed in all cases a normal EEG background that did not exceed the bounds of normal EEG variations described by other researchers (Davis, 1941; Danilova, 1963; Zhirmunskaya, 1969). In all the subjects investigated, the alpha wave range clearly predominated over other frequencies. This alpha range predominance was observed against the background of a small and relatively regular amount of energy in other frequencies of the spectrum or was accompanied by a slight rise in that of the low and high frequencies. A similarity of the general configuration of the spectra of various parts of the cerebral hemispheres was observed, although the energy characteristics of the frequencies, chiefly in the alpha range, were at times different. The interhemispheral asymmetry in the spectra of nonfrontal or non-limbic subjects, as a rule, was within the bounds of normal physiological asymmetry.

In accordance with data published in the literature (Walter, 1959, 1966), we found that the frequency composition of the alpha rhythm was not uniform: in the overwhelming majority of subjects, the alpha rhythm contained all the frequency components, from 8 to 13 Hz. The amplitudes of the various components were different, indicating a considerable complexity of the main rhythm of the EEG. Stable characteristics of the alpha range for each subject were obtained; namely, the type of alpha rhythm (polyrhythmic or monorhythmic), its general configuration, which frequencies predominated over others (as a rule, the higher over the lower), and the maximum amplitude of individual alpha frequencies (the peak of the alpha rhythm). The amplitude characteristics of the alpha rhythm were quite constant over the course of a single experiment but changed from one experiment to another, a result found by other investigators (Danilova, 1963; Grindel, 1964; Ilyanok, 1965).

The main modification of the alpha rhythms in response to acoustic stimuli was a decline of the energy of the alpha frequencies (a pure suppression), which either developed against the background of a suppression of other (higher and lower) frequencies of the spectrum or was accompanied by a shift toward the higher frequencies. This type of modification of the alpha rhythm in normal subjects may be regarded as a shift of the functional state of the cerebral cortex in the direction of activation (Sokolov, 1958; Danilova, 1959). It should be noted, however, that in response to acoustic stimuli, some other modifications of the alpha rhythm were observed in individual cases; they were expressed as an enhancement of one or two alpha range frequencies and a suppression of all other alpha frequencies (reaction of the mixed type). This type of alpha rhythm modification apparently indicates a weak activating effect of the stimuli and a decline of the functional state of the brain caused by the extinction of the activation reaction (Dongier et al., 1957; Sokolov, 1960; Kogan, 1962).

Two types of activation reactions were noted in normal subjects. One of these was observed during passive perception of acoustic stimuli (during the presentation of neutral sounds); the other was observed under conditions of strained attention, during the accomplishment of a task (counting, i.e., during the presentation of meaningful stimuli). The difference between these two types of activation reactions was apparent in all the investigated parameters of the alpha frequency modifications, such as the depth, extent, duration, and composition of the suppressive reaction, as well as in the nature of the alpha frequency modifications and in correlations with modifications in other ranges of the frequency spectrum. Thus, in response to meaningful stimuli, the suppressive reaction was more intense, involved a greater number of frequencies, and lasted longer than the response to neutral stimuli. Moreover, higher alpha frequencies participated in the suppressive reaction than in the reaction to neutral stimuli. Modifications of the alpha frequencies during the perception of meaningful acoustic stimuli predominantly assumed the character of pure-suppression reactions, whereas during the presentation of neutral sounds, mixed-type reactions were often observed.

Investigations of patients with lesions of the frontal lobes (or of the limbic systems) showed that in these cases the background initial indexes of the EEG differed from those of normal subjects. The frequency composition of the EEG background in most patients was characterized by the absence of a dominant alpha rhythm and by a distinct intensification of the slow frequencies, most often in the zone of the focus or on the side of the lesion. (Mayorchik, 1956, 1960; Grindel, 1963). The alpha range proved variable in configuration and amplitude in various parts of the hemispheres. In the parietooccipital region of the normal hemisphere a considerable variability of the amplitude was observed not only from one experiment to another, but also in the course of a single experiment. Characteristic of the alpha range was a predominance of the lower alpha frequencies over the higher ones. All these indices reflected a *lower and more unstable functional state* of the brain of frontal patients in comparison with other patients and normal subjects.

The activation reaction in response to acoustic stimuli was also quite different in these patients. The stimuli did not call forth any distinct shift of the functional state in the direction of activation, as it was seen to do in normal subjects; this could be judged by the modifications of the alpha and other frequencies of the spectrum. Whereas in normal subjects reactions of the pure-suppression type predominated, in frontal patients, the reaction of the alpha frequencies often assumed the form of a weakly pronounced suppression of certain alpha frequencies and simultaneous enhancement of others, or the form of enhancement exclusively. These modifications, as a rule, took place against the background of a shift of the frequency spectrum toward the lower frequencies and suppression of the higher ones. According to some investigators, such modifications indicate a decline in the functional state of the brain and a weak activating effect of the stimuli (Sokolov, 1960; Kogan, 1966).

A comparison of both types of the activation reaction disclosed that in response to neutral and, particularly to meaningful sounds, the modifications of the EEG frequencies and, above all, of the alpha range *essentially differed from the normal indexes.* This was reflected in an insignificant magnitude of the suppressive reaction, the participation of a small number of frequencies in the reaction, a rapid extinction of the reaction, retarded activation phenomena, an increased number of reactions of the mixed- and pure-enhancement types, and the predominant participation of the lower alpha frequencies in the reaction.

The most appreciable difference between the results obtained from frontal patients and the indices shown by normal subjects consisted of a predominant disturbance of the second type of activation reaction in the frontal lobe lesioned patients, i.e., a disturbance of the special type of activation that accompanies the fulfillment of some task and emerges during the presentation of meaningful stimuli.

As stated earlier, the activation reaction was of a particularly pronounced character in cases where the medial cortex of both frontal lobes proved pathologically changed. It is characteristic that in these cases there was no direct correlation between the weakening of the activation reaction and the degree of expression of the syndrome accompanying increased intracranial pressure.

In some patients, in spite of the gravity of the lesion and the pathologically modified EEG background, no marked disturbances of the activation reaction were revealed during the action of either neutral or meaningful stimuli. This fact may be explained by the localization of the pathological process in the white matter of the frontal lobe or affected areas situated on the border of the frontal, parietal, and temporal parts of the brain.

The above data confirm the hypothesis that definite parts of the frontal lobes bear a relation to the activation reaction; apparently, these are the medial and basal divisions of the frontal cortex. This is proved by control investigations carried out on patients with lesions localized in other parts of the frontal lobes and outside the frontal divisions of the brain.

REFERENCES

Artemieva, E. Y., and Homskaya, E. D. (1966). Changes in asymmetry of EEG waves in different functional states in normal subjects and in patients with lesions of the frontal lobes. *In* "Frontal Lobes and Regulation of Psychological Processes" (A. R. Luria and E. D. Homskaya, eds.), pp. 294-313. Moscow Univ. Press, Moscow (in Russian).

Baranovskaya, O. P. (1966). Depression of the alpha-rhythm under the action of indifferent and signaling stimuli in patients with lesions of the frontal lobes. *Int. Congr. Psychol., 18th, 1966* Symp. No. 10.

Bekhtereva, N. P. (1959). Bioelectrical activity of the cerebral hemispheres in supratentorial tumours. Thesis, University of Leningrad (in Russian).

Danilova, N. N. (1959). The orienting reflex and the reaction of reconstruction of the bioelectrical activity of the brain to a rhythmical light stimulus. Collected articles. *In* "The Orienting Reflex and Problems of the Higher Nervous Activity." Moscow.

Danilova, N. N. (1963). Concerning individual peculiarities of the electrical activity of the human cerebral cortex. Collected articles. *In* "Typological Peculiarities of the Human Higher Nervous Activity," Vol. III.

Davis, P. (1941). Technique and evaluation of the electroencephalogram. *J. Neurophysiol.* **4**, 92.

Dietsch, G. (1932). Fourier-Analyse von Electroencephalogrammen des Menschen. *Pfluegers. Arch. Gesamte Physiol. Menschen Tiere* **230**, 106-113.

Dongier, S., Gastaut, H., and Dongier, M. (1957). Enregistrement polygraphique de l'effet de surprise. Relation entre les résultats obtenus et les données clinique. *In* "Conditionment et réactivité en EEG." Masson, Paris.

Drift, J. H. (1957). "The Significance of Electroencephalography for the Diagnosis and Localisation of Cerebral Tumours." Leiden.

Filippycheva, N. A. (1963). Significance of the frontal lobe in the process of elaborating and switching motor stereotypes. *Comm. 20th Conf. Probl. Cent. Nerv. Syst. 1963* p. 247.

Filippycheva, N. A. (1966). Neurophysiological mechanisms of disturbances of motor reactions in lesions of the frontal lobes. *In* "Frontal Lobes and Regulation of Psychological Process" (A. R. Luria and E. D. Homskaya, eds.), pp. 398-430. Moscow Univ. Press, Moscow (in Russian).

Gibbs, F. A. (1939). Cortical frequency spectra of schizophrenics, epileptics, and normal individuals. *Trans. Amer. Neurol. Assoc.* **60**.

Gibbs, F. A. (1942). Cortical frequency spectra of healthy adults. *J. Nerv. Ment. Dis.* **95**, 4.

Grass, A. M., and Gibbs, F. A. (1938). Fourier transform of the electroencephalogram. *J. Neurophysiol.* **1**, 25.

Grindel, O. M. (1963). Analysis of the EEG frequency spectrum in focal changes of the cerebral cortex. *J. Higher Nerv. Activity* **13**, 4.

Grindel, O. M. (1964). Some data on the analysis of a human EEG. *J. Physiol. USSR* **50**, No. 1.

Homskaya, E. D. (1960). Influence of a verbal instruction on the vascular and skin-galvanic components of the orienting reflex in different local lesions of the brain. *Proc. Acad. Pedagog. Sci. R.S.F.S.R.* Pap. No. 6 (in Russian).

Homskaya, E. D. (1961). Influence of a verbal instruction on the vegetative component of the orienting reflex in different local lesions of the brain. Communications II and III. *Proc. Acad. Pedagog. Sci. R.S.F.S.R.* Pap. Nos. 1 and 2 (in Russian).

Homskaya, E. D. (1966). Vegetative components of the orienting reflex to indifferent and significant stimuli in patients with lesions of the frontal lobes. *In* "Frontal Lobes and Regulation of Psychological Processes" (A. R. Luria and E. D. Homskaya, eds.), pp. 190-253. Moscow Univ. Press, Moscow (in Russian).

Ilyanok, V. A. (1960). Methods of studying high-frequency potentials of the EEG. *Biophysics (USSR)* **5**, No. 4.

Ilyanok, V. A. (1962). Frequency spectra of man's EEG and their modification under the action of a light stimulation. Thesis, University of Moscow (in Russian).

Ilyanok, V. A. (1965). Frequency spectra of the EEG of different areas in the human brain. *J. Higher Nerv. Activity* **15**, No. 5.

Kogan, A. B. (1962). Reflection of the processes of higher nervous activity in the electrical potentials of the brain cortex in conditions of the animal's free behaviour. *In* "Electroencephalographic Investigation of the Higher Nervous Activity." Moscow.

Kogan, A. B. (1966). The physiological nature of the electrical correlates of the reaction of attention. *Int. Congr. Psychol., 18th, 1966* Symp. No. 5.

Lindsley, D. B. (1966). The role of thalamo-cortical systems in attention and perception. *Int. Congr. Psychol., 18th, 1966* Symp. No. 5.

Livanov, M. N. (1938). Analysis of bioelectrical oscillations in the brain cortex of mammals. *Proc. Inst. Brain* Nos. III and IV.

Mayorchik, V. E. (1956). Reflection of the dynamics of the nervous processes in the EEG depending on the initial functional state of the brain cortex. *J. Higher Nerv. Activity* **6**, No. 4.

Mayorchik, V. E. (1957). Electrophysiological analysis of the functional properties of the cerebral cortex in the zone of the pathological focus. *J. Physiol. USSR* No. 3.

Mayorchik, V. E. (1960). General and local modifications of electrical activity of the cortex and subcortical structures during neurosurgical operations on different levels of the central nervous system. Thesis, University of Moscow (in Russian).

Mayorchik, V. E., and Rusinov, V. S. (1954). Some questions of the theory and practice of electroencephalography in focal lesions of the brain. *Probl. Neurosurg.* **1**, No. 1, 38-46.

Mundy-Castle, A. C. (1957). The electroencephalogram and mental activity. *Electroencephalogr. Clin. Neuropathol.* **9**, No. 4.

Pribram, K. H. (1960). A review of theory in physiological psychology. *Annu. Rev. Psychol.* **11**, No. 5.

Small, I., Bagchi, B., and Kooi, K. (1961). Electro-clinical profile of deep cerebral tumors. *Electroencephalogr. Clin. Neuropathol.* **13**, No. 2.

Sokolov, E. N. (1958). "Perception and the Conditioned Reflex." Moscow.

Sokolov, E. N. (1960). Concerning the reflection of the orienting reflex in the electroencephalogram in man. Collected articles. "Problems of Electrophysiology and Electroencephalography."

Sokolov, E. N., and Danilova, N. N. (1958). The EEG frequency spectrum as characteristics of the functional state of the human brain cortex. *Comm. Conf. Probl. Electrophysiol. Cent. Nerv. Syst., 1958* pp. 120-123.

Sokolova, A. A. (1957). Influence of afferent stimulations of the focus of pathological activity in the EEG of patients with tumors of the cerebral hemispheres. *Probl. Mod. Neurosurg.* **1**, p. 31.

Spilberg, P. I. (1941). Harmonic analysis of the EEG of man. *Physiol. J. USSR* **30**, No. 5.

Walter, W. G. (1943). Appendix on a new method of electro-encephalographic analysis. *J. Ment. Sci.* **89**, No. 375.

Walter, W. G. (1950). The function of electrical rhythms in the brain. *J. Ment. Sci.* **96**, 31.

Walter, W. G. (1957). "The Living Brain." Norton, London and New York.

Walter, W. G. (1959). Intrinsic rhythms of the brain. *In* "Handbook of Physiology" (Amer. Physiol. Soc., J. Field, ed.), Section I, Vol. I, pp. 279-298. William & Wilkins, Baltimore, Maryland

Walter, W. G. (1966). "The Living Brain." Moscow.

Walter, W. G., and Dovey, F. J. (1944). Electroencephalography in cases of sub-cortical tumor. *J. Neurol. Neurosurg. Psychiat.* **7**, pp. 57-65.

Zhirmunskaya, E. A. (1969). Variants of man's electroencephalograms and standardization of the methods of their determination. *In* "The IX Congress of the USSR Society of Physiology. Biochemistry and Pharmacology," Vol. I, pp. 198-202. Minsk.

Part Three

THE NATURE OF THE ELECTRICAL ACTIVITY OF THE FRONTAL CORTEX IN MAN

Chapter 5

CORRELATION OF BIOPOTENTIALS IN THE FRONTAL PARTS OF THE HUMAN BRAIN

M. N. LIVANOV, N. A. GAVRILOVA, and A. S. ASLANOV

*Institute of Higher Nervous Activity
Academy of Sciences of the U.S.S.R.
Moscow, U.S.S.R.*

At present, there is considerable clinical material, based upon case reports from neurological and psychiatric clinics, and experimental data that convincingly show that frontal lobe lesions lead to disturbances of conscious behavior. Although the formal intellect of such patients may remain relatively intact, they are unable to properly interact with their cultural environment. The delicate components of their mental activity are lost, their critical faculties are violated, and they become spontaneous. Their ability to work out programs of proper behavior, as it were, becomes lost.

Specific features of activity of the human frontal lobes may best be investigated under conditions where the patient must fulfill some special task. In particular, it is important to study the changes that develop in the frontal lobes of the brain during mental exercise.

Electrophysiological methods have been used, along with other methods, to determine the exact nature of this difficulty in mental activity. A majority of the experimenters who have investigated the bioelectric activity during mental work have taken the suppression of the alpha rhythm as a criterion of the EEG changes (Lindsley, 1944; Walter, 1957; Jasper, 1958; Peimer, 1960; Glass, 1964). These researchers concluded that the suppression of the alpha oscillations is connected with mental activity. Jasper even advanced the hypothesis that the suppression of the alpha rhythm during mental work is a manifestation "of higher psychic processes and acute emotional states." Grey Walter associated the dynamics of suppression of the alpha rhythm with the peculiarities of thinking in each individual and on this basis divided people into groups with either predominant verbal or visual modes of thought. A suppression of the alpha oscillations as the result of protracted mental strain (9–12 hr) and their replacement by slow oscillations was observed by Kiryakov (1964).

Some investigators do not associate the alpha rhythm suppression directly with mental activity. For example, Bujas *et al.* (1953) recorded a decline of the alpha index only at the very beginning of mental work, which can probably be regarded as a manifestation of the orienting reaction. A suppression of the alpha waves during the orienting reaction was also recorded by Sokolov (1959). Walsch (1953) and de Lange *et al.* (1962) did not observe any suppression of the alpha rhythm at all during mental activity.

In an attempt to resolve these discrepancies and to provide a more sophisticated analysis of the electrical activities of the cortex, data were obtained from multichannel recordings of biopotentials. The records of normal subjects and of patients with a delusional form of schizophrenia subjected to special tasks were compared.

I. METHODS

The investigations were carried out on eight normal adults and on seven schizophrenic patients; each subject was tested three to five times. A 50-channel electroencephaloscope was used to study the electrical activity of the cerebral cortex. Fifty carbon electrodes, each of which was a Plexiglas hollow cylinder with a carbon rod installed in it, were evenly placed on the surface of the head. An indifferent electrode was placed on the chin. The electrical activity of the subject's cerebral cortex was recorded in a state of relative rest, after a 5 min adaptation of the subject to a dark soundproof chamber. The biopotentials were recorded from the electroencephaloscope screen on film at the speed of 24 frames per second for 10 sec. The filming was done repeatedly within a period of 10–30 min. Then the subject was asked to solve an arithmetical problem of definite complexity; depending upon the subject's education, he was asked to multiply two- or three-figure numbers. The recording of bioelectrical activity during this mental arithmetic first lasted 25 sec, beginning 4–5 sec after the commencement of work; it was then repeated for another 25 sec, every other minute, until the subject was ready with his answer or admitted failure. The minimum time taken for analysis was 3 sec (separately for the background and for the mental activity); the maximum was 9 sec.

Not less than three or four arithmetical problems were offered to the subject in each experiment, and sections of the recordings of bioelectrical activity obtained at different periods during the mental arithmetic were analyzed.

If the subject had been given medicine (in particular, Aminasine), the biopotentials were recorded 30, 60, and 75 min after its administration.

The electroencephaloscope provides such a considerable amount of information that its processing requires the application of special methods. In view of this, we, like Farley et al. (1957) and Tunturi (1959), used electronic computers to analyze the correlations of biopotentials in various parts of the brain. This work was carried out by T. A. Korolkova and G. D. Kuznetsova, scientific workers in our laboratory, as well as by E. V. Glivenko, a scientific worker of the Institute of Electronic Control Machines.

We investigated paired correlations of bioelectrical activity in 50 leads. As a criterion for biopotential similarity we considered the coincidence of the biopotential changes from moment to moment for each pair of leads.

For this, we first determined the sign $(+, -, 0)$ of the first time derivative for each lead. Then, with the aid of an electronic computer, we compared for each pair of leads the signs of the derivatives from moment to moment for a fixed period of time and calculated the percentage of coincidence of these signs. This procedure corresponded to an approximate calculation of correlations for a pair of selected leads. We confined ourselves to measurements from the film taken of the biopotential oscillations recorded on the electroencephaloscope screen. Under conditions optimum for analysis, the frequency band ranged from 3 to 10 Hz.

As a result of the computing operations, we obtained a tabulogram consisting of 1225 numbers, each of which characterized the degree of similarity of a given pair of leads for a certain period of time.

The correlations between the bioelectric activity of each of the 50 cortical points investigated and the remaining 49 points were not necessarily equal. Proceeding from Gauss' theoretical curve, we selected a group of independently functioning points that correlated within 23–44% of the time analyzed. When the biopotentials showed unidirectional changes over the whole period analyzed (1.5 sec or 36 frames), we considered it 100% correlation. We also designated as positive a group of points with a low correlation, but where pairs of leads changed unidirectionally over 45–75% of the time analyzed as well as those of high correlation when unidirectional changes of the biopotential oscillations were manifest in 76–100% of the whole period analyzed. Finally, we designated as no correlation those biopotentials in which unidirectional changes occurred in less than 22% of the time under investigation.

II. RESULTS FROM NORMAL SUBJECTS

The study of biopotential correlations in various parts of the cerebral cortex yields a number of interesting results.

The data obtained by us graphically show how the relationships among various parts of the human brain (especially in the frontal lobes) change depending on their different states.

A. The State of Rest

In the state of rest in the cerebral cortex of normal humans, the biopotentials of approximately one half (50.5%) of the recorded areas function independently of one another, i.e., do not produce any positive correlations. A rather considerable number of cortical areas produce intercorrelations of biopotentials 47–75% of the time analyzed. Here, just as in independently functioning cortical areas, intercorrelations were observed between diffusely situated points. Intercorrelations could also occur between areas that are located side by side or at some distance from each other, within the same hemisphere, or in different hemispheres, as well as in the two opposite poles of the cortex, frontal and occipital. Approximately 45% of the cortical points had low correlations.

The human cerebral cortex has regions the biopotentials of which intercorrelate over 75% of the whole analyzed period. In the state of rest, the number of such regions was low in most of the sections investigated, the intercorrelations not exceeding 2–3% of the total 1225 possible correlations. As a rule, a high percentage of intercorrelations was observed between points situated immediately next to each other. Apart from the fact that these points are very rare and can be located in any region of the cortex, it should be emphasized that none of them ever shows a high percentage of correlation with distant points of the cortex; as a rule, they related to only one, and sometimes two or three other distant points.

We subjected five EEG sections to mathematical analysis in the course of experiments carried out on subjects NG and SE in the state of rest. Among these sections, only one period was observed when the number of positive correlations was considerable over a high percentage of the time. Subject NG was highly illustrative in this respect. In her, certain points situated at some distance from each other (e.g., in the right and left temporal zones) produced positive correlations. Moreover, separate points (e.g., point 24, in the region of the motor analyzer) were intercorrelated with six other points. However, the state of rest was of a relative character in this subject, as she was a worker in our laboratory and all the time mentally monitored the experiment. Therefore, this phenomenon may not be characteristic of the state of rest.

The above data show that in the state of rest the correlating points are diffusely situated in the human cerebral cortex and that there exists no zone that is predominantly involved. Furthermore, during rest a high percentage of correlations between biopotentials is nearly absent.

TABLE 1
Distribution of Biopotential Correlations of Various Cortical Points in Different Zones During Rest and Mental Arithmetic
(Correlations over a High Percentage of Time)[a]

Point number	Name	Prefrontal zones		Motor zones		Posterior-parietal zones		Occipital zones	
		Back-ground	Mental work	Back-ground	Mental work	Back-ground	Mental work	Back-ground	Mental work
1	NG	12	86	12	48	11	42	5	36
2	AS	5	74	2	32	6	21	2	9
3	BE	2	82	1	57	2	36	2	30
4	AA	0	11	0	2	0	4	0	2
5	VA	9	102	6	43	5	33	5	9
6	YCh	7	36	5	4	6	10	4	3
7	AM	4	4	6	23	15	13	18	14
8	SE	1	87	11	27	5	2	0	4

[a] The table contains data from films containing maximum correlations for a given subject.

B. During Mental Exercise

During mental work the correlations of the cortical electric activity undergo a sharp change. A marked segmentation of intercorrelations between different cortical points takes place. Points interrelated over longer periods of time begin to predominate (such changes were recorded in six out of eight patients). At the same time, the number of points with a low degree of synchrony increases, while the number of independently functioning points shows a considerable decline.

The greatest number of correlations arising during mental exercise fall in the frontal areas of the brain. Here the net of functional connections is particularly dense, whereas in the posterior regions of the cortex (posterior-parietal and, especially, occipital) the number of intercorrelations is much smaller. Table 1 shows the distribution of positive correlations in various regions of the cortex in the state of rest and during mental arithmetic.

It is evident from Table 1 that the greatest number of correlations fall in the prefrontal regions of the cortex. For example, in subject SE during mental arithmetic (multiplying 87 × 136), the anterior cortical regions produce 174 positive correlations, the motor zone produces 53, and the posterior regions produce 11 (the occipital region produces seven and the posteroparietal produces four).

Problem solving required considerable mental strain on the part of the subject. In his verbal account of his experiences, he stated that this had been a difficult mental operation. It lasted about 2 min.

Figure 1 shows the temporal correlations between biopotentials of certain points in the cerebral cortex in subject SE during the fifth second of problem solving. A concentration of intercorrelating points in the prefrontal areas of the cortex can be observed. In the posterior areas, correlating points are almost absent; only two correlations can be seen in the parietal zone and four in the occipital zone.

A similar law-governed phenomenon, a very high degree of involvement of the prefrontal cortical zones in the process of mental computation (according to the number of positive correlations), is also manifest in other subjects. In subject BA, for example, the anterior zones produce 102 correlations, against 33 correlations produced by the posterior-pariental zones and nine by the occipital zones. In subject AA the numbers of these correlations are 11, 2, and 2, respectively.

An increase of the number of positive correlations during mental work in comparison with the background can be explained by an increase in the total number of points showing a high percentage of correlation (as mentioned above) and by the fact that each cortical point intercorrelated with many others. Whereas in the state of rest each point correlated with one or two other points, during mental arithmetic each was connected with five to ten other points or more.

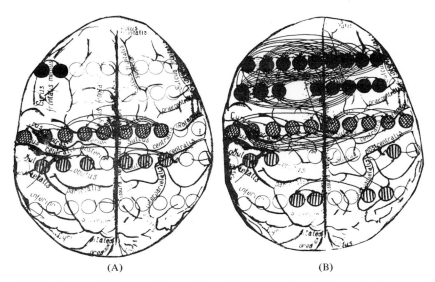

FIG. 1. Temporal biopotential correlations between points of the cerebral cortex in normal subjects SE. (A) In the state of rest. (B) During the fifth second of mental arithmetic. Circles on the map of the brain cortex indicate electrodes applied to the head surface. Electrodes situated in the prefrontal parts are marked in plain black; in the area of the motor analyzer, in checks; and in the posterior parts of the cortex, in stripes. In all other drawings the indications are the same. Arrows connect cortical areas that correlated a high percentage of the time.

Correlations easily occurred between cortical points located along the frontal lines. Here, not only points situated side by side become interconnected (for example, points 1 and 2, 2 and 3, in Figure 1B), but also points removed from each other (for example, points 9 and 11, 13 and 26, in Figure 1B).

Likewise, cortical points situated in opposite hemispheres invariably produce intercorrelations in all subjects, and intercorrelations of biopotentials between cortical points situated more caudally and rostrally were consistently observed. However, such connections develop predominantly between cortical points that are situated close to each other.

There were few correlations between biopotentials from points situated in the frontal and occipital zones, i.e., few correlations passing sagittally through almost the whole cortex.

The constellation formed by cortical points with a high percentage of correlations does not remain permanent; it changes its structure from moment to moment. Moreover, during mental work, the number of correlations between the cortical points fluctuates markedly. At given moments, this fluctuation may bring the number of correlations down to the background value.

Points situated in the sensory-motor area of the cortex play an essential part in correlations during mental work. They correlate both with one another and, to a considerable degree, with points situated in the prefrontal zones. This indicates a close functional connection between the anterior parts of the frontal lobes and the area of the motor analyzer proper. Constant participation of the anterior central gyrus in biopotential changes caused by mental work is indicative of the presence of ideomotor acts.

During mental arithmetic an increase of the number of correlations with a lower percentage of interaction also occurs. Such weak correlations were observed not only between those situated far from each other, for example, in the occipital zones.

Figure 2 shows that during mental work there takes place an increase of correlations between the fifth and seventh points, the biopotential oscillations of which are unidirectional during 45–75% of the time analyzed. Whereas in the state of rest (the top row of pictures) the fifth point correlated with 12 other points, during mental arithmetic it began to interact with 41 points. The seventh point correlated in the background with 15 points, whereas during mental work it correlated with 29 points. As may be seen from the analysis of these correlations, the points that produce them are situated diffusely through the entire cortex. These correlations probably reflect the emergence of a higher functional level (in comparison with the state of rest).

An increase of intercorrelations for low and high synchrony and a decrease of the number of points functioning independently were distinctly observed in six of the eight subjects investigated. Data obtained from investigations of subject AM, who was given the task of coding a number of figures for a computing machine, showed unusual constancy from experiment to experiment and showed no changes in comparison with the background. The task, which seemed difficult to the experimenter, was very simple for the subject, who solved it instantly and without any strain. This gave rise to the assumption that the correlations between various points of the cortex increase only when the mental work is comparatively complicated, whereas automatic mental processes do not produce such a result. The correlations of biopotentials of various cortical points show an increase only when the tasks given to the subjects are difficult for him.

The coincidence of the direction of the arising functional correlations with the topographomorphological peculiarities of the conducting commissural, short, and long associative pathways of the human brain (Dzugayeva, 1965) is of definite interest. Our data show that the functional correlations that occur in a high percentage of time generally correspond to the direction of commissural and short associative fibers. The correlations that arise only via low percentage of the time correspond not only to the direction of the commissural fibers but also, and predominantly, to the direction of the long associative fibers.

C. Action of Aminasine

We investigated biopotential correlations during mental work against the background of the action of Aminasine, a drug that affects chiefly the reticular

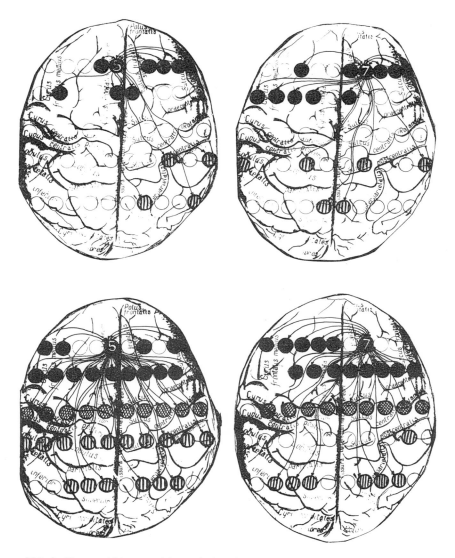

FIG. 2. Temporal biopotential correlations between single points of the cerebral cortex in normal subject S. Arrows connect cortical areas that correlated a low percentage of the time.

formation and the anterior parts of the brain. The investigations showed that Aminasine, when administrated in therapeutic doses of 25 or 50 mg, weakens the spatial synchronization of the biopotentials. The weakening appears as a decrease in the number of points that function with high and low coefficients of

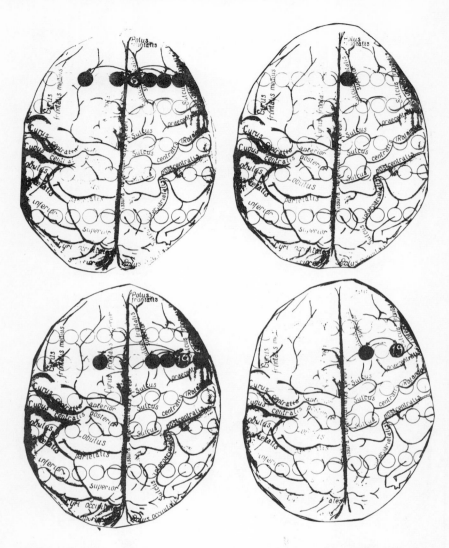

FIG. 3. Changes of temporal biopotential correlations between single points of the cerebral cortex in subject SE during mental arithmetic, before (1 and 2) and after (3 and 4) administration of Aminasine. Arrows connect cortical areas that correlate a high percentage of the time.

correlation and in a reduction in the number of intercorrelations between each of the recorded points and the others.

The effect of Aminasine on spatial synchronization shows up particularly well during tasks of mental arithmetic (Figure 3); Using individual points, for example, 6 and 19, the characteristic changes in correlation caused by Aminasine can be clearly seen. Point 6, which prior to the administration of Aminasine correlated with five other points (subject SE), did not show any correlation whatsoever (in a high percentage of the time) under the action of Aminasine. Point 19 produced only one positive correlation after Aminasine, whereas it had exhibited four correlations with other points before administration of the drug.

In subject BE, the total number of positive correlations during mental arithmetic was 24 prior to the administration of Aminasine. (This was the smallest number of interconnections recorded in an EEG section during mental computation; as is shown in Table 1, the maximum number of correlations in patient BE amounted to 205.) Under the action of Aminasine the number of positive correlations declined to one. Similar changes of a more or less pronounced character were observed in other subjects.

Thus, Aminasine, when applied in a therapeutic dose, changes the correlations of the bioelectrical processes going on in various points of the cerebral cortex of a normal subject; it weakens the positive correlations. However, the action of Aminasine is not confined exclusively to quantitative changes in the correlations; it also affects the topography of the positive correlations.

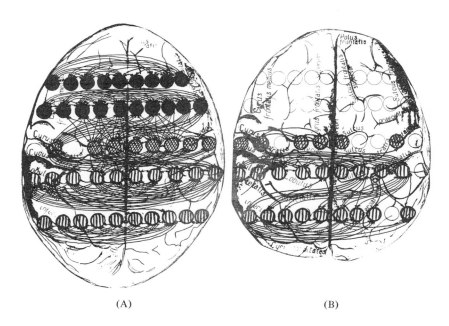

FIG. 4. Temporal biopotential correlations between points of the cerebral cortex after the administration of Aminasine. (A) During mental arithmetic before administration of Aminasine; (B) when doing an analogical sum 1 hr 15 min after administration of Aminasine.

Figure 4 presents correlations between separate cortical points in subject NG during mental arithmetic. It can be seen that the number of cortical point biopotentials that underwent unidirectional change during 75% and more of the time analyzed was quite high.

Cortical points situated in the anterior frontal gyrus, the anterior central gyrus, and the parietal region of the cortex are involved in the process of mental work. Figure 4B shows correlations between various cortical points in subject NG doing a sum 1 hr and 15 min after the administration of Aminasine. The figure shows that the number of pairs of points whose biopotentials manifested a high degree of intercorrelation appreciably decreased under the action of Aminasine. A particularly considerable reduction of these points took place in the anterior parts of the cortex. In subject NG, the frontal areas of the brain were fully excluded from this kind of synchronous activity, whereas in other areas (for example, in the parietal area) the synchronous activity was preserved, although in a weakened form.

Thus, the effect produced by Aminasine on the cortex manifests itself unevenly: different cortical areas are not equally sensitive to this preparation. The anterior parts of the cortex are most sensitive to the action of Aminasine. The sharp decrease of positive correlations in the frontal cortex fully accords with the widely current opinion that Aminasine blocks the tonic influence of the ascending activating system, which is involved mostly with the anterior parts of the brain (Dell *et al.*, 1954; Rothballer, 1956; Agafonov, 1956; Rinaldi and Himwich, 1956; Anokhin, 1958; Waldman *et al.*, 1961; Voronin *et al.*, 1961.

It should be pointed out that although mental work is accomplished by the subject taking Aminasine, such work requires more time than before its administration. Thus, the decline in the level of the functional state, in evidence in the background prior to the action of Aminasine, is reflected in the quality of the mental work. The functional activity of the cortex becomes reduced as a result of the pharmacological blocking of its frontal parts.

III. RESULTS FROM SCHIZOPHRENIC PATIENTS

Attempts to apply electrophysiological methods to the study of physiological mechanisms responsible for disturbances of brain functions in schizophrenia have been made since the 1930's (Berger, 1931; Davis, 1940; Kennard and Levy, 1952). However, the data obtained were contradictory and by no means promising. The pathological electroencephalographic changes recorded in schizophrenics, slow rhythms, acute waves and various kinds of modification of the alpha oscillations, were also found in a number of other kinds of mental diseases (Davis, 1940; Tokareva, 1953).

It has become quite evident that any search for characteristic disturbances in the waveforms is unjustified, and the functional approach has now replaced this search. Examination of EEG traces while the patient is performing various functional tasks has made it possible to detect disturbances in the basic neural processes occurring in the cerebral cortex of schizophrenics (Butorin, 1954; Shakin and Peimer, 1961; Vishnevskaya and Kaminskaya, 1962; Gamburg and Denisova, 1964; Borisova and Talavrinov, 1964).

We, too, have always advocated a functional approach to the analysis of cortical electrical activity. Having attained the possibility of spatially investigating the distribution of cerebral electrical activity by studying biopotential correlations, we deemed it advisable to apply our method to patients suffering from a delusional form of schizophrenia. We investigated seven patients, all characterized by pronounced systematized delusions, along with disturbances peculiar to any form of schizophrenia.

The biopotential correlations in these patients were of a highly distinctive character. In the cerebral cortex of these patients, the number of points producing a high degree of potential correlation (in the absence of external stimulation) was higher than in normal subjects.

It was found that the number of pairs of interconnected points is particularly great in patients with a strongly pronounced delusions. The number of pairs fluctuates at different periods of time, and their pattern is dynamic: sometimes biopotentials of a particular cortical point are synchronous with one point, sometimes with another.

The points that exhibit a high degree of correlation are more often encountered in the region of the anterior central gyrus and in the parietal region.

When the patients are given functional tasks connected with mental arithmetic, connections between the prefrontal cortical parts become scarce.

It was observed that the number of connections is generally small in comparison with those of normal subjects and that it is particularly small in the anterior frontal parts in comparison with all other cortical areas. Table 2 illustrates this phenomenon.

It may be seen that the number of pairs of positively correlating points in the anterior frontal parts (where the first and second arcs of the electrodes are situated) is smaller than in the posterior areas (Table 2).

As an example, patient KA (27 years old, suffering from schizophrenia for 7 years, with a delusion of physical coercion predominate in the clinical picture), in the course of multiplying 72 × 14, showed 51 positive correlations in the anterior frontal parts, 75 in the area of the motor analyzer, and 63 in the parietal areas.

In subject S (23 years old, suffering from schizophrenia for 5 years, with a delusion of physical coercion and persecution most pronounced in the clinical picture), there were observed 15 positive correlations in the anterior frontal

TABLE 2
The Number of Positively Correlating Points in Various Zones of the Cerebral Cortex of Schizophrenics at Rest and Doing Mental Arithmetic[a]

Patient	At rest					Mental arithmetic				
	Frontal region		Motor area	Parietal area	Occipital area	Frontal region		Motor area	Parietal area	Occipital area
	1st arc	2nd arc				1st arc	2nd arc			
KA	17	15	68	29	8	18	33	75	63	16
ZYa	9	17	10	43	20	Failed to solve the problem				
DA	9	6	10	6	10	5	4	10	22	1
VA	16	2	3	8	5	1	1	5	5	2
FA	0	6	15	12	13	Failed to solve the problem				
RV	0	1	11	4	0	6	8	14	19	4
GA	0	2	1	6	10	No problem was offered				

[a]The table presents one time period from each patient.

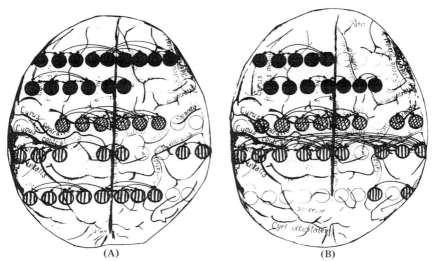

FIG. 5. Temporal biopotential correlations between points of the cerebral cortex in a patient with the delirious form of schizophrenia. Arrows connect cortical areas that correlated a high percentage of the time. (A) During rest. (B) During mental arithmetic.

parts, 10 in the motor area, and 22 in the parietal area (See Figure 5). The same distribution of positive correlations was evident in all other patients. It should be noted that the number of positive correlations in the occipital zones is small, but this is also characteristic of normal subjects. However, as stated in Section II,B, the number of positive correlations shows a sharp increase in the anterior frontal cortex of normal subjects during the accomplishment of a mental task and this was not found to be the case in schizophrenic patients.

Three patients (DA, VA, and RV in Table 2), in addition to showing fewer pairs of positive correlations in the frontal cortex than in other parts, showed few functional interconnections. Each of the points was interconnected with only one or two other points.

In the frontal area, positive correlations more often occurred between points situated close to each other. Moreover, during mental arithmetic, the total number of positive correlations in the cerebral cortex of schizophrenics may sometimes be even smaller than during rest, another contrast with normal subjects.

The above data show that the correlations arising in the cerebral cortex of patients suffering from schizophrenia and in normal persons during mental work greatly differ. This becomes particularly obvious when correlations of the frontal parts of the cortex are analyzed.

What, then, is responsible for the changes of biopotential correlations in the cerebral cortex of schizophrenics? In particular, how might the scarcity of positive correlations in the frontal zones be explained?

Because positive correlations are possible in the frontal parts of the cortex and because they arise a different moments between different cortical points, the structural units in the cortex and the ability of the cortex to function must be intact. In view of this, it can be assumed that no organic lesion exists here. At the same time, the extreme scarcity of correlations indicates a change in the functional state of the cortex.

The changes of correlations in patients suffering from schizophrenia to some degree resembe those occurring in normal subjects under the action of Aminasine: in both cases the greatest decline of correlations is observed in the prefrontal parts of the cortex.

It may therefore be supposed that one and the same mechanism is responsible for these changes. As is known, Aminasine blocks the reticular formation. Since the frontal parts of the cortex have the closest connections with the reticular substance, they suffer most. Owing to this, the fact that the frontal parts prove to be most affected in patients suffering from schizophrenia gives reason to assume a disturbance of the corticosubcortical relationships.

IV. CONCLUSION

The above data objectively demonstrate that the frontal lobes play a most important role in human mental activity. Mental strain is of particular importance here. When the frontal lobes become nonfunctional, mental activity markedly slows down. Moreover, our data, in full accord with present day concepts of neurophysiological research, show that the functional state of the cortical frontal lobes and their participation in mental activity are to a considerable degree determined by the corticosubcortical correlations and, apparently, above all by influences exerted by the reticular formation of the midbrain.

REFERENCES

Agafonov, V. G. (1956). Inhibiting influence of aminazine on the central afferentation of the pain. *J. Neuropathol. and Psychiatry S. S. Korsakoff* 56, Part 2, 94 (in ·Russian).

Anokhin, P. K. (1958). "Electroencephalographic Analysis of the Conditional Reflex." "Medizina," Moscow (in Russian).

Berger, H. (1931). Über der Electroencephalogramm des Menschen Dritte Mitteilung. *Arch. Psychiatr.* 91, 16.

Borissova, T. P., and Talavrinov, V. A. (1964). Spatial synchronization of the alpha waves in the cortex of schizophrenics. *J. Neuropathol. and Psychiatry S. S. Korsakoff* 64, Part 3, 420 (in Russian).

Bujas, L., Petz, B., and Krokovica (1953). Electricna aktivnost mozga u toky duzec intellectualna roda. *Arch. Zahik i Koda* 48, 419-421.

Butorin, V. I. (1954). "Problems of Psychoneurology. A Review of Studies of 1949–1951." Medgiz, Moscow (in Russian).

Davis, P. A. (1940). Evaluation of the EEG of schizophrenic patients. *Amer. J. Psychiatry* **96**, No. 4, 851.

de Lange, J. V. N., Storm van Lewen, V., and Werre, P. E. (1962). Correlations between the psychological and EEG phenomena. *In* "Electrophysiological Studies of the Higher Nervous Activity," p. 396. Acad. Sci. of the USSR, Moscow (in Russian).

Dell, P., Kubell, Z., and Bonvallet, M. (1954). Action de la Chlorpromazine au niveau du systeme nerveux central. *Semaine Hôpitaux Paris* Nos. 36–37, 2346.

Dzugayeva, S. B. (1965). The ontogenesis of the conduction paths of man. *In* "The Development of the Child's Brain" (S. A. Sarkissov, ed.). "Medizina," Moscow (in Russian).

Farley, B. Fishkopf, L., Clark, W., and Gilmore, J. (1957). *Tech. Rep. Res. Electron. Lab. MIT 337/165* **IV**, 101.

Gamburg, A. L., and Denissova, A. M. (1967). The influence of some drugs on the bioelectric activity of the brain in schizophrenics. *J. Neuropathol. and Psychiatry K. K. Korsakoff* **64**, Part 1, 116 (in Russian).

Glass, A. (1964). Mental arithmetic and blocking of the occipital alpha rhythm. *Electroencephalogr. Clin. Neurophysiol.* **16**, No. 6, 521.

Jasper, H. H. (1958). Recent advance in our understanding of ascending activities of the reticular system. *In* "Reticular Formation of the Brain" (H. H. Jasper *et al.*, eds.). Little, Brown, Boston, Massachusetts.

Kiryakov, K. (1964). Some electroencephalographic criteria of fatigue in mental work. *J. Higher Nervous Activity* **14**, No. 3, 412 (in Russian).

Lindsley, D. B. (1944). "Personality and Behavior Disorders." Ronald Press, New York.

Peimer, I. A. (1960). Electrophysiological studies of the higher nervous activity in man. Thesis, Leningrad University (in Russian).

Rinaldi, F., and Kimwich, H. (1956). *In* "Brain Mechanisms and Drug Action" (W. S. Fields, ed.). Thomas, Springfield, Illinois.

Rothballer, A. B. (1956). Studies on the adrenaline-sensitive component of the reticular activating system. *Electroencephalogr. Clin. Neurophysiol.* **8**, 603.

Shakin, M. I., and Peymer, J. D. (1961). EEG and the association experiment in paranoid syndromes. *Trans. Leningrad Soc. Neurol. and Psychiatry* **7**, 384 (in Russian).

Sokolov, E. N., ed. (1959). "Orienting Reflex and Problems of the Higher Nervous Activity." Acad. Pedagog. Sci., Moscow.

Tokareva, V. A. (1953). Changes of the bioelectric activity of the cortex in schizophrenics. Candidate Dissertation thesis, University of Moscow (in Russian).

Tunturi, A. (1959). Statistical properties of spontaneous electrical activity in the auditory cortex of the anesthetized dog. *Amer. J. Physiol.* **196**, No. 6, 1175.

Vishevskaya, A. A., and Kamenskaya, V. M. (1962). Comparative study of EEG in acute and protracted schizophrenia. *J. Neuropathol. and Psychiatry S. S. Korsakoff* **62**, Part 10, 156 (in Russian).

Voronin, L. G., *et al.* (1961). Electrophysiological study of the mechanisms of the aminazine influence. *J. Neuropathol. and Psychiatry K. K. Korsakoff* **61**, Part 2, 208 (in Russian).

Waldman, A. V., *et al.* (1961). On the influence of aminazine on the ascending reticular formation. *Physiol. J. USSR* **47**, Part 7, 852.

Walsch, E. (1953). Visual attention and the alpha rhythm. *J. Physiol. (London)* **120**, 115-159.

Walter, W. G. (1957). "The Living Brain." Norton, London and New York.

Chapter 6

HUMAN FRONTAL LOBE FUNCTION IN SENSORY–MOTOR ASSOCIATION

W. GREY WALTER

Burden Neurological Institute
Bristol, England

Since the experiments of Fritsch and Hitzig in 1870, the human frontal cortex anterior to the motor strip has been regarded as a silent area. This means simply that electrical stimulation of these extensive regions in a conscious person produces neither motor reaction nor subjective sensation nor autonomic response. These observations were made on patients undergoing acute brain operations but during the last few years we have been able to extend much investigations with multiple electrodes implanted in the brain for the treatment of mental disorder, epilepsy, and dyskinesia. As many as 80–100 electrodes can be inserted in the brain and left in place for months or even years, and during this time the patient is free to move about and even to go home and carry on his normal work between periods of treatment.

Although the purpose of these implantations is therapeutic, the preliminary investigation of the patients, before treatment can be started, involves a variety of physiological procedures necessary to identify the precise location of each electrode and to analyze the functional state of the brain in its neighborhood. During these investigations, we have been able to collect a large amount of information about the functions of the frontal lobes in man, and we have used this information to help in the interpretation of electrical records taken with electrodes on the scalp (as in the conventional electroencephalograph, EEG) in normal subjects.

The methods and scope of investigation have been described elsewhere (Walter and Crow, 1964) and the clinical results have been surveyed by Crow *et al.* (1961, 1963). Briefly, the multiple electrode sheaves provide contact over a few millimeters of brain and, being made of gold wire, permit polarographic recording of local oxygen availability ($O_2 a$) and of the bioelectric activity of the brain tissue. This is mentioned here because, with such a system, variations in oxygen tension around a gold electrode modify its effective impedance so that if there is any small potential difference between this electrode and the reference electrode (as there is certain to be), slow voltage fluctuations will appear. These may be confused with slow bioelectric potentials, and indeed this effect probably accounts for some of the slow potential changes described, for example, by Aladjalova (1962), in particular, her Figure 25. Failure to appreciate the origin and significance of these slow potential changes can introduce misleading ideas because the local oxygen tension does sometimes vary with the functional state of the cortex and also often shows slow spontaneous rhythmic fluctuations, at a rate of about six per minute, which can easily be mistaken for changes in potential of the brain tissue itself. These effects have been described in detail (Walter, 1962; Cooper, 1963) and are mentioned here only to indicate a subtle but common source of error in the interpretation of intracerebral records.

One of the most important preliminary procedures in intracerebral therapy is the judicious electric stimulation of each electrode. The purpose of this is primarily to identify by exclusion those electrodes that are in white matter because these are generally the target for therapeutic polarization. The criterion is that an electrode that shows no after-discharge when stimulated by optimal pulses at a level of 8–10 V is not in cortex. Conversely, an electrode from which an after-discharge can be recorded at a voltage of 4–6 V is certainly in gray matter and should not be polarized in the later stages of treatment.

The systematic application of this technique has allowed us to observe the effect of intracerebral stimulation at about 2000 electrodes in conscious people, and we can assert quite confidently that, with very few exceptions, the evocation of an after-discharge by local stimulation has no subjective or objective effects whatever. This confirms the early reports but adds to their interest because the use of this technique allows observations of the electrical effects produced by the stimulation and at the same time provides a simple explanation for the surprising ineffectiveness of a powerful physiological procedure. The first important factor in this explanation is that the effect of local stimulation remains local. Even when an after-discharge at 1 mV is produced at an electrode and lasts several minutes, the actual volume of brain involved is very small—only a few cubic millimeters. Electrodes 4 or 5 mm away from the first electrode show no trace of the imposing paroxysmal discharges but continue with their own intrinsic activity. The waveforms of these local after-discharges reproduce quite precisely all the appearances associated in the conventional scalp EEG with epileptic seizures, but the patient shows no sign of any such disturbance; he continues to converse with the attendants, reads a book, or even dozes peacefully, quite unaffected by the intense but strictly confined electrical tornado.

These observations illustrate an essential character of frontal lobe function; even quite abnormal disturbances have no general importance provided that they affect only a very limited zone of cortex. This fact should be borne in mind in considering observations on the sensory functions of the frontal lobes, which involve very large areas; it is the global participation of the frontal structures that gives them their unique character. The same conclusion is reached, of course, from the study of the effect of lesions, whether accidental or deliberate; small discrete cortical wounds, however serious, interfere little with frontal function, but a small degree of damage over a wide area has a disastrous effect on mentality and all higher functions.

We may now proceed to consideration of the new facts about frontal function that the most recent techniques have revealed. These techniques include various forms of record analysis, which, again, have been described elsewhere (Walter, 1961). The simplest and most useful of these is the extraction of significant information from records of evoked potentials by automatic direct (on-line) averaging of a series of traces. In this way, the responses to a variety of stimuli can be measured quite accurately, even when they are submerged in a background activity of considerably greater amplitude. The disadvantage of this method is that it implies a statistical assumption about the distribution and variation of the responses and, of course, does not clarify the appearance of any particular response. In many laboratories, the desire for a clean record has induced experimenters to work with averages of very large numbers of signals—over 1000 in some cases. Unfortunately, such large samples cannot be used for studying the frontal lobe functions because the responses in these regions never remain constant over so many repetitions and the extra technical clarification is cancelled by physiological variation. In our experiments, the number of responses included in the average presentation is adjusted between 8 and 32 and for the majority of situations eight has been found a suitable number. This provides a minimum signal—noise gain of $\sqrt{8}$, that is, about 2.8.

In spite of its other limitations, the method of automatic averaging lends itself particularly well to the general procedure used for the experiments to be described here, because the various stimuli are presented in sets at irregular intervals and each set can be treated as a unit from the standpoint of semantic analysis.

Before considering the results in detail, it will be helpful to discuss the nature of the experiments and their relation to other approaches to these problems.

Our early observations suggested, not surprisingly, that the human frontal lobes are intimately concerned with the selection and storage of sensory information leading to relevant action. The procedure we developed for the detailed study of these processes derives directly from the principles established by I. P. Pavlov and analyzed in terms of a statistical theory of conditional learning (Walter, 1953, 1960). Physiological stimuli in any modality, particularly vision, hearing, and touch, are presented by an automatic programer linked with the two-channel average computer. The intervals between presentations are irregular, from 3 to 10 sec. First, isolated stimuli in the chosen modalities are given, in sets of 12 for averaging. This is for following the development and decline of the Orienting or Novelty responses. After 30 or 40 such presentations (that is, three or four averages of the responses) or when an appreciable degree of habituation or extinction has occurred, stimuli in two modalities are presented in pairs, again at random intervals. The responses to the associated stimuli are also averaged over a few dozen presentations. At this stage, the subject may be asked or allowed to perform some action in response to the second of the paired stimuli. Alternatively, if the second stimulus is such as to evoke an unconditional reflex (for example, an eye blink by corneal stimulation from a controlled puff of air) a conditioned response may have appeared already. In this way the involvement of both primary and secondary signaling systems can be studied and compared.

The designation of the stimuli and the responses to them is of great importance and unfortunately is still not standardized. The terms *first* and *second* are too ambiguous and the traditional Pavlovian *Conditional* and *Unconditional* are scarcely adequate. During the course of several years' experimentation with human subjects, we have come to adopt what one might call a grammatical or syntactic terminology; we are really communicating with the subject in a physiological language, of which the stimuli are the substantives and their associations the conjunctions. Thus, in the first set of presentations, when a flash of light or a click is given irregularly in isolation 30 or 40 times, we are saying in effect: "The light flashes," or "There is a click." We would call this an *Indicative* stimulus. Later, when for example flashes of light are followed 1 sec later by clicks, the grammatical equivalent is: "If there is a flash, there will be clicks." The flashes would quite properly be called *Conditional* as in the Pavlovian sense, and the clicks are still merely Indicative. When we tell the subject to act on the clicks, however, for example by pressing a button, we assign a new context to the auditory stimuli and they become *Imperative*. Our statement is now: "If there are flashes, there will be clicks, so press the button!" In these experiments we arrange the button circuit so that it will interrupt the Imperative stimuli provided it is not pressed before they start. The subject can therefore tell how quickly and accurately he has responded on each occasion. The attitude of the subject and the details of the response to the Imperative stimuli are very important to the development of the frontal responses, so to make the position quite clear a simple analogy may be given. We may compare our laboratory procedure with learning to drive a motor car. Once the controls have been identified, the learner must acquire experience of real traffic and of the various formal signals that warn him of impending problems. Thus, the other cars on the road, the curb, and so forth are at first merely indicative, they just exist; but quickly they become imperative—an obstacle in front of the car means "Stop!" But there are also Conditional signals, for example, the lights at an intersection. These mean: "*If* the light is red, there will be a crossing, so start to slow down so as to stop at the crossing." The driver does not stop as soon as he sees the Conditional light at a distance, but at another arbitrary Imperative stimulus, the crossing itself, and his action avoids the necessity to stop suddenly at a truly imperative stimulus, another car ahead. The sequence of Conditional and arbitrarily Imperative signals insures an economical, accurate action, and we shall see later that the development of this capacity can be followed quite accurately in the interaction of sensory responses in the frontal lobes. We may add here that, as we should expect from everyday experience and from the elegant laboratory experiments of Luria and his colleagues, the secondary signaling systems have an enormous influence on the rate of development of these cerebral processes at all stages. Language, in the sense of verbal utterance, is not the only component of the secondary signaling system; there are gestures, expression, posture, and general atmosphere, so we prefer to refer to these factors as *Social Influences* as compared with *Direct Experience*. They include instruction, reassurance, praise, admonition, reproof, and competition. The most exciting feature of this work is that the effects of such Social Influences can now be compared objectively and quantitatively with those of direct physiological experience.

The last feature of these experiments that should be explained before describing the results is the statistical or probabilistic approach. These terms are unattractive in their associations, but our working hypothesis is simply that the brain acts continually as a computer of contingency. It first selects and then stores information from the senses on the basis of the improbability of isolated events and the probability of the association between particular sets of events being greater than chance expectation. The brain must then in effect decide what action is worth taking on the basis of the computed probability. This is a purely materialistic or physicalistic hypothesis; the only assumptions made are that behavior as a whole must be economical of time and energy and that some criterion of significance can be established from experience. Some people find this concept a little difficult and the mathematical expressions are still confusing, but the everyday application is perfectly straightforward. In ordinary life, any event may or may not happen and the probability of two events happening together frequently is less than that of their happening separately. Reverting to the illustration of the traffic problem, an intelligent savage without experience or instruction, could learn to drive a motor car the hard way by Direct Experience, by discovering, for example, that some red lights meant stop and others only indicated the position of obstacles. He might acquire in his brain a set of probabilistic rules that would approximate more and more closely to the official regulations. In the modern jungle of the streets, however, he would stand a small chance of survival without Social Influences and his behavior, like that of a child, would appear delinquent and irresponsible until he had accumulated enough information to permit establishment of a set of statistical criteria for decisive action. This would be possible, though hazardous, because traffic situations are frequently repeated, with few exceptions or anomalies. If the experience of our savage chauffeur were to include moving frequently without warning from Britain with a left-hand rule, to Russia with a right-hand rule, he would probably develop an experimental neurosis. Again, as we know, Social Influences can mitigate such effects by simple warnings and explanations to such an extent that a complete reversal of rule can be effected in a few minutes without confusion or distress.

Having now described our experimental procedure and the simple concepts underlying it, we may proceed to the actual discoveries relating to frontal lobe function.

1. Dispersive Convergence

All physiological stimuli evoke responses over a wide area of the frontal cortex. In any particular area, the proportion of nervous tissue involved in these responses is only about 1% of the total excitable tissue in that area, but the responses are so widely dispersed and synchronized that they can be seen clearly even on the scalp in some subjects. This illustrates a very important general point about the relation between the electrocorticogram (ECG) and the scalp EEG; the amplitude of potential differences on the scalp depends directly on the dispersion and synchrony of events in the cortex. Another way of putting this is that the tissues between a scalp electrode and the cortex act as a spatial averaging circuit.

The essential conclusion is that the human frontal lobes are an integral part of the sensory system; the classical division between anterior-motor and posterior-sensory cortical localization can no longer be maintained.

2. Modality Signature

Although essentially similar, the responses to stimuli in the various modalities have characteristic signs; that is, it is possible to tell from the latency and waveform of a response whether it arose from a visual, an auditory, or a tactile stimulus.

An important feature of frontal lobe responses is their short latency. In orbital and lateral cortex, the latency for the first or positive responses to auditory stimuli is rarely greater than 30 msec, and for visual and tactile responses about 40 msec. The latency to the peak of the larger negative component is usually about 100 msec. These figures are not very different from those for responses in specific sensory areas, indicating a rapid relay of signals from the thalamic nuclei through the ascending reticular system to the frontal cortex.

3. Idiodromic Projection

Idiodromic projection simply means that signals in the various sensory modalities reach the frontal cortex by private lines; there is no intermodality occlusion. The effect of this, in fact its conclusive proof, is that when, say, a visual stimulus is followed by an auditory one at a critical interval of about 60 msec the positive and negative components of the later response appear to cancel with the negative and positive components of the earlier one. If the responses were propagated along the same pathway, this would mean that the first event was affected by the later one, which is absurd. The appearance of cancellation is due to the arrival of similar signals along separate channels with time relations such that their components are out of phase, leading to a purely instrumental cancellation.

The physiological inference is that the relays from the specific thalamic nuclei retain their specificity and identity right up to the upper layers of the frontal cortex. From the operational standpoint, it is obviously essential that the system should work in this way, because otherwise signals in one modality would mask or jam those in another; this would make sensory association less efficient for simultaneous signals than for successive ones, which again would be absurd.

4. Habituation

This term has become familiar in Western neurophysiology, but it is not a particularly good one (except in French where the meaning is "to become accustomed to something"). The connotation is the same as the Russian "Extinction of the Orienting Response" but this too is confusing because the habituated responses are by no means extinct; they reappear at once when a monotonously repeated stimulus is changed slightly in character or context. The essence of the effect is that the frontal responses are a part of the Orienting or Novelty response; if a stimulus is repeated regularly, frequently, and monotonously without association or relevant action, the frontal responses diminish slowly over about 50 presentations and sometimes disappear finally below the intrinsic noise level. The rate and degree of this progressive attenuation varies over a wide range between individuals and even from time to time. The extent of attenuation is usually greatest for tactile stimuli, moderate for visual ones, and least for auditory stimuli.

Experiments with this effect have demonstrated that the frontal responses are practically independent of the intensity of the stimuli. In fact the responses to very weak stimuli are sometimes larger and more persistent than those to intense ones. Furthermore, when a certain degree of attenuation has been established to intense stimuli, the response can often be restored by reducing the intensity of the stimuli.

A fact of great importance should be mentioned here; we have found no sign of habituation in records from primary specific receiving areas in human subjects. The original claims that specific pathways were under corticofugal influence in animals seem to have been based on inadequately controlled experiments, and in scalp records from human subjects the responses are always contaminated by components from association areas of cortex that do show habituation. When records are made from electrodes in human primary visual cortex, responses to visual stimuli continue unchanged during regular stimulation at a high rate over several hours, by which time the responses in nonspecific parietal and frontal cortex have entirely disappeared.

It would seem that between the specific thalamic relays and the frontal cortex there are special mechanisms that respond to the improbability rather than the intensity of signals. A signal that is repeated by itself regularly and frequently becomes increasingly probable and evokes a diminishing response. Any change in the quality or context of the stimulus is improbable and the response accordingly increases. The so-called extinction of the Orienting or Novelty response, that is, the phenomenon of habituation, may thus be described as associated with increase in expectancy with monotonous repetition.

The concept of the brain, and particularly the frontal lobes in man, as a computer of probability, or more properly contingency, is not a new one, but these experiments are probably the first to show that even the massive electric responses from the scalp indicate the operation of such a mechanism within the brain.

The decline of responses during monotonous repetition of stimuli is the first step toward the computation of contingency, because isolated unassociated stimuli are as meaningless as a series of random numbers; they become a component of the background noise level. However, two important factors intervene: the regularity or rhythm of presentation and the distribution of the stimuli in time. In general, habituation occurs most rapidly when the stimuli are rhythmic and rapidly repeated, but if the repetition rate is too high, the quality of the stimulus inevitably changes. A series of clicks at 100 per sec or more acquires a musical pitch, flashes at more than 4 or 5 per sec are flicker, and so forth.

In the experiments described here, habituation is very slow when stimuli are presented at intervals longer than 10 sec but appears rapidly when the interval between presentations is constant at about 1 sec.

5. Contingent Interaction

This term is used to include all the variations in the frontal responses observed when two or more stimuli are associated and involve an operant or conditioned response on the part of the subject.

When two stimuli in different modalities are presented frequently in pairs with, say, an interval of 1 sec between them, the responses to the first stimuli are usually augmented and those to the second are attenuated. This is described as Contingent Amplification and Contingent Attenuation. This effect is really another example of the probabilistic relation between stimuli and frontal lobe responses, because when associated in pairs, the first stimulus becomes less probable because of its association, whereas the second stimulus becomes more probable or expected, as it is always preceded by the first stimulus.

If paired neutral stimuli are presented in this way without comment or instruction, the responses to the associated signals habituate as with isolated stimuli. The effect is quite different, of course, if the second stimulus is such as to evoke an unconditional reflex response. In human subjects, the range of such stimuli is limited unless the experimenter is prepared to starve or alarm his subjects. The easiest reflex to study in this way is the corneal blink reflex, which can be evoked by a brief stimulus, such as a puff of air. If puffs of air to the eye are regularly preceded by another stimulus, such as a click or flash of light, the frontal response to the conditional stimulus shows marked contingent amplification that persists indefinitely and, moreover, other secondary effects appear, which will be described in detail later.

The most interesting and significant changes appear when the subject is instructed to respond in some way to the second stimulus, which thereby becomes what we may call *Imperative*. The instruction to respond obviously assigns to the imperative stimulus a special significance as a signal for action, but the first conditional stimulus also acquires a unique importance as a warning for action. This importance is reflected in the appearance of a new electro-cortical effect, a wave of negative change that begins immediately after the conditional response and extends up to the moment of the response to the imperative stimulus. This secondary negative wave can be seen clearly only with special recording techniques because its spatial extent and long duration require wide electrode placements and very long coupling time constants or better direct coupling. It is almost invisible with the conventional bipolar montage and 0.3 sec time constant used for ordinary EEG's, which is why it has not been noticed before. This secondary wave reflects very precisely the subjective probability of the triple associations: conditional stimulus–imperative stimulus–operant response. We have called it the *Contingent Negative Variation* or CNV. A more graphic term is *expectancy wave* or *E wave*. A still better designation may be found later when all the details of the functional relations of this phenomenon have been worked out; at present, the CNV is related most closely to the expectancy of stimulus association by the subject and to his intention to respond promptly to the Imperative stimulus.

Similar prolonged changes in cortical potential have been reported in animals during conditioning and arousal, notably by Shvets (1958), Caspers (1959), Rusinov (1960), and Rowland and Goldstone (1963), and the origin of the human expectancy wave is probably the same as that suggested from the animal studies—the feltwork of apical dendrites in the plexiform layer of cortex.

We have been able to record this effect also in patients with chronic intracerebral electrodes, and in such experiments expectancy waves arise in many regions of the frontal cortex. A technical difficulty under such conditions is that the intrinsic time constant of the electrodes limits the extent to which slow potential changes can be recorded. Metals with a low (ohmic) resistance (notably silver) are toxic to the brain and the noble metals (especially gold) that we use for chronic implants have an intrinsic time constant of about 0.1 sec. Nevertheless, the contingent expectancy waves can be recorded, particularly with subdural electrodes over the frontal cortex, and in several cases we have observed that the waves tend to sweep from the frontal pole back toward the premotor zone during the 0.5 sec or so between the conditional and imperative responses.

The correlations between the CNV or expectancy wave and the psychological situation have been studied in some detail in adults. They may be summarized as follows.

(1) The CNV grows slowly over the first few dozen presentations of associated stimuli after instructions to respond to the imperative stimuli.

(2) Once established, the CNV persists indefinitely as long as the subject retains an interest and concern with his response. In some subjects, identical records of the CNV have been obtained day after day over a period of several months.

(3) When the imperative stimuli are withdrawn (extinction trials), the CNV subsides slowly to zero over the first 20—50 trials. When the imperative stimuli are restored, the CNV reappears after about 12 presentations. This process of alternating extinction and restoration can be repeated indefinitely in strong, well-balanced, normal adult subjects.

(4) When the subject is told beforehand that the imperative stimulus is going to be withdrawn or that he need not make the operant response, the CNV subsides at once. One can say that a single accurate and trustworthy social instruction is equivalent in probability to 20—50 direct experiences of association or extinction. This "figure of merit" naturally depends on the social relation between the experimenter and the subject, because instructions from a stranger or anyone likely to trick or tease the subject may have an inverse effect on the CNV. This is particularly evident in children between the ages of about 5 and 15 years, in whom the development of the CNV is exquisitely sensitive to the social influences that pervade the experimental situation. Even in young adults, the social factors may be more powerful determinants of the development of the CNV, particularly when they are studied in a group situation. The actual words or gestures and the tone of voice used in the instructions are often as effective as the physical association between the stimuli. In the extreme case, subjects in deep hypnotic trance may be persuaded to respond entirely illogically, e.g., to ignore highly significant objective associations in favor of entirely imaginary subjective ones, and the CNV (unlike the intrinsic brain rhythms) follows the suggestion rather than the objective reality.

(5) The probabilistic relation between stimulus association and the CNV in the brain is best demonstrated by diluting the probability of association with a

known proportion of unreinforced conditional stimuli. Thus, if sets of 12 clicks followed by 12 flicker stimuli have been given repeatedly, and the subject is instructed to press a button when he seen the flashes, the probability of the association between the clicks and flashes rises rapidly to a figure approaching certainty as the total number of experiences increases. The CNV is usually fully developed after about 20 trials, although the growth rate is a highly individual factor. If clicks alone are now interspersed between the associated pairs, the degree of probability dilution may be indicated by a fraction in which the numerator is the number of associated pairs and the denominator the total number of conditional stimuli. For example, the probability of association after the first 20 clicks followed by 20 flashes may be given as 20/20, that is, 1.0; then, if 20 more clicks are given but only 10 are followed by flashes, the probability is reduced to a total of

$$\frac{20 + 10}{20 + 20} = \frac{3}{4} = 0.75.$$

This process of partial reinforcement may be described as *progressive equivocation*; if it is continued for a further series of 20 clicks, of which again only half are reinforced, the probability of association would be

$$\frac{20 + 10 + 10}{20 + 20 + 20} = \frac{2}{3} = 0.66$$

and so on, with the probability converging to 0.5 when the total of equivocal experiences greatly outnumbers that of strict associations. At this level, of course, the Conditional stimuli are no longer conditional at all, they have lost their significance in relation to the Imperative stimuli and are simply a part of the background noise.

As already implied, the amplitude of the contingent expectancy wave reflects very accurately such variations in objective probability. Again, there are considerable individual variations, even within the normal range, and remarkable correspondences with anomalies of mental state in clinical conditions; but in normal stable young adults, the CNV follows experimental dilutions of significance with great fidelity. In most such subjects, dilutions to 0.8 have little effect; two exceptions in ten are accepted, even if this ratio is maintained for a long time. If the dilution is greater, however, the CNV declines rapidly and disappears at the even-chance level of 0.5. This is in accord with common sense, but there are many factors that must be considered along with the numerical ratio of reinforcement. One of these is the sequence and interval of presentation in real time. In our experiments, the presentations are generally made at irregular intervals of between 3 and 10 sec and the pause between the Conditional and Imperative signals is either 1.0, 1.5, or 2 sec without variation during a set of trials. If the intervals between presentation or the gaps between stimuli are much longer, the development of the CNV is very much slower. We have not yet explored the limits of duration of the CNV during association, but in extinction trials, when there is no imperative stimulus to terminate the negative wave, it may persist for 10 sec or more, and the level of steady potential, presumably indicating partial cortical depolarization, can last for several minutes in certain subjects. It may be a coincidence, but the subjects who have shown such prolonged persistence of the CNV during extinction have all been very suggestible people trained for hypnotic experiments. The results of hypnotic suggestion in such subjects have been reported elsewhere (Black and Walter, 1965).

(6) The abrupt termination of the CNV has already been alluded to. Even in averages of 12 or more presentations the termination is often extremely sharp and coincides with the moment at which the negative components of the Imperative response would be at their peak. This indicates that the CNV may act in effect as a primer for the discharge of the neuronic elements involved in the performance of the operant response. In accord with this, the reaction time of the subject for the operant response to the imperative stimulus is usually very much shorter when the CNV has developed. For example, the reaction time for pressing a button in response to the visual stimuli is rarely less then 200 msec. However, when the subjective expectancy of association between an auditory conditional stimulus and the visual signals, as indicated by the CNV, has fully developed, the reaction time is often halved and in some subjects has dropped to 50–80 msec. This was at first assumed to be a time reflex based on the constant interval between the conditional and imperative stimuli, but two facts contradict this assumption. First, the distribution of latencies is not normally distributed around the mean. The operant circuit is arranged so that the subject cannot terminate the Imperative stimulus by pressing the button before the stimulus but must wait for the first flash or click. If the abbreviated response latency were due to a precise judgement of time following the conditional stimulus, the errors should be distributed normally around the target point. Second, when the Imperative stimulus is withdrawn, as during extinction, or when the conditional stimuli are present occasionally without the imperative ones, as in equivocation, the subjects do not in general make a false response at the instant when the imperative stimulus would be expected. Records are always taken of the electromyogram (EMG) of the operant muscle groups and these rarely show any sign of a conditional response when there is no imperative stimulus.

The inference is that the depolarization of the cortex, seen on the surface as the CNV, accelarates the voluntary action. The muscular activity involved in the response is also abbreviated and diminished when the CNV has developed, suggesting improved cerebral economy in operation as well as an acceleration of response time.

(7) As already mentioned, Social Influences are extremely potent in the control of the CNV and this can be used to study the effect of purely verbal or

symbolic signals as compared with physiological stimuli. The development of the CNV may be promoted by a single word from the experimenter, provided that the subject intends to perform some action in reply, even to make a suitable verbal answer, such as: "Ready"–1 sec–"Now!" Recently, experiments have been made with visual stimuli in which there is no change in energy level at all, and these have been found as effective in the experimental situation as when the stimuli are flashes of light or clicks, provided that they are significant to the subject, either because of their unexpectedness or their implication.

The importance of social and symbolic reinforcement is seen most clearly in children between the ages of 3 and 10. Without some explanation or encouragement, very few children in this age group show any sign of a CNV, but after suitable instruction or demonstration the conditional interactions often become almost adult, at least for a while. The spatial distribution of the contingent components is different in young children also; in general the frontal nonspecific areas are less involved and the specific association regions play a larger part, particularly in the case of visual conditioning. Whenever social reinforcement is effective, it is the frontal components that are augmented and consolidated, suggesting that one of the essential functions of these regions is the transposition and integration of social and direct information to form coherent patterns of relevant action.

An important question raised by these observations is: What is the nature of the physiological processes reflected in the CNV? A simple description is that they are essentially inhibitory and this is the explanation suggested, for example, by Roitbak (1963) in the Sechenov Centenary Conference for his observation of the prolonged surface-negative waves evoked by electrical stimulation of the cortex. He assumed that these "reflect depolarization of apical dendrites resulting from the activation of glia around them." He suggests that the presynaptic inhibition of pyramidal neurons involved in this process may be the basis for certain kinds of cortical inhibition. This is a reasonable conjecture, but whether the features that we have seen in the human frontal lobes may be usefully described as inhibition is another matter. We have demonstrated that the CNV is intimately related to the timing of voluntary responses and the establishment of involuntary conditioned reflexes. In this sense, the CNV is like an alarm clock, set by the conditional stimulus and triggered by the imperative or unconditional one. Indeed, an alarm clock could be considered as an inhibitory device because it prevents the sleeper from awakening too soon! It seems doubtful whether the term inhibition is useful in this paradoxical sense, however; simple concepts of excitation and inhibition are probably too ingenuous for the analysis of complex functions such as those we are considering here. All coordinated actions must involve both increases and decreases of activity, and above all, the synchronization of action must require precise deferment of activity until critical moments, determined, as we have seen, by the statistical relation of afferent signals and the nature of the relevant action.

In our experiments, the operant action is ineffective unless it is deferred until after the presentation of the imperative stimulus, but the conditional stimulus can act as a warning signal by which the precise timing of the motor response may be adjusted. The mechanisms seen as the CNV seem to mediate this process. It is relevant here to mention that the CNV is equally prominent when the subject is instructed to perform an inhibitory act, such as to stop pressing a button.

The foregoing account provides only an outline of recent advances in the study of frontal lobe function. Perhaps the most exciting feature is the way in which the conjectures and predictions of I. P. Pavlov are being confirmed and extended. This is particularly gratifying to me personally as a pupil of Pavlov's, but the degree of concordance is far greater than I ever expected, even with the bias of a convinced disciple. The role of the frontal cortex in man cannot be adequately described in a phrase, but there can be no doubt now that, with its immensely rich connections to all sources of sensory information, its capacity for economical storage and cross-correlation of information, and its exquisite sensitivity to subtle shades of social and semantic implication, the frontal cortex provides the essential link between the elementary stages of structural evolution and the present phase of psychosocial development of the human organism.

REFERENCES

Aladzalova, N. A. (1962). "Slow Electrical Processes in the Brain." Academy of Science, Moscow.
Black, S., and Walter, W. G. (1965). *J. Psychosomatic Res.* 9, 48.
Caspers, H. (1959)."The Nature of Sleep." Churchill, London.
Cooper, R. (1963). Local changes of intra-cerebral blood-flow and oxygen in humans. *Med. Electron. Bio. Eng.* 1, 529-537.
Crow, H. J., Cooper, R., and Phillips, D. G. (1961). Controlled multi-focal frontal luecotomy for psychiatric illness. *J. Neurol., Neurosurg. Psychiat.* 24, 353.
Crow, H. J., Cooper, R., and Phillips, D. G. (1963). Progressive leucotomy. "Current Psychiatric Therapies." Grune & Stratton, New York.
Roitbak, A. I. (1963). Sechenov Centenary Conference, Moscow (Progress in Brain Research, Vol. 22, Hoeber, New York, 1967).
Rowland, V., and Goldstone, M. (1963). *Electroencephalogr. Clin. Neurophysiol.* 15, 474.
Rusinov, V. S. (1960). *Electroencephalogr. Clin. Neurophysiol* 13, Suppl., 309.
Shvets, T. B. (1958). "Electrophysiology of Higher Nervous Activity," p. 138. Moscow.
Walter, W. G. (1953). "The Living Brain." Duckworth, London.
Walter, W. G. (1960). A statistical approach to the theory of conditioning, Moscow colloquium. *Electroencephalogr. Clin. Neurophysiol.* 13, 377.
Walter, W. G. (1961). *In* "Computer Techniques in EEG Analysis" (M. A. B. Brazier, ed.), p. 14. Elsevier, Amsterdam.
Walter, W. G. (1962). *In* "Neural Physiopathology" (R. G. Grenell, ed.), (Progress in Neurobiology), Vol. 5. Hoeber, New York.
Walter, W. G., and Crow, H. J. (1964). Depth recording from the human brain. *Electroencephalogr. Clin. Neurophysiol.* 16, 68-72.

Part Four

THE NATURE OF ELECTRICAL ACTIVITY IN THE FRONTAL CORTEX OF NONHUMAN PRIMATES

Chapter 7

WHILE A MONKEY WAITS

E. DONCHIN*

Neurobiology Branch
NASA Ames Research Center
Moffett Field, California

D. A. OTTO,†
L. K. GERBRANDT, and
K. H. PRIBRAM

Department of Psychology
Stanford University
Stanford, California

As a working hypothesis, the slow potential variations described here may be considered as an outward sign of the electro-chemical processes underlying the capacity of the non-specific frontal cortex to integrate information from sensory sources with internal stores. In this sense they may reflect the operation of short-term memory as an essential component in the formation of decisions and the planning of relevant action (Walter, 1967).

Ever since Jacobsen (1936) showed that lesions in the frontal granular cortex of monkeys produce severe deficits in the performance of delayed response tasks, the role of the frontal lobes in short-term memory has been under active investigation. Pribram and Tubbs (1967) added an intriguing dimension to the problem by showing that frontally lesioned monkeys could learn to perform a delayed alternation task if the temporal sequence of events was appropriately parsed. The results suggested that the frontal lobes assist in coding and programing rather than in simply maintaining a memory trace and that short-term memory processes were dependent on such coding operations. Results of another study (Spinelli and Pribram, 1967), in which electrical stimulation of the frontal cortex speeded visual recovery cycles in afferent channels, also implicated the frontal lobes in the active processing of neural signals rather than in any simple maintenance of a memory trace.

Steady cortical potentials recorded from the scalp of humans or directly from the cortex of animals give promise of providing yet another method of studying

---

*Present address: Department of Psychology, University of Illinois, Urbana, Illinois.
†Present address: Environmental Protection Agency, Clinical Environmental Research Laboratories, University of North Carolina, Chapel Hill, North Carolina.

frontal lobe function. Walter *et al.* (1964) reported that a sustained negative-going potential shift can be recorded from the human cortex during the anticipatory interval preceding a motor response or decision. Walter designated this waveform *contingent negative variation* (CNV) or the *expectancy-wave*. Low *et al.* (1966b) and numerous other experimenters have confirmed the existence of CNV in humans. Walter *et al.* (1964) concluded that CNV originated in the frontal lobes and swept back across the cortical mantle, although the evidence was inferential. Attempts were therefore made to link the CNV with a wide variety of psychological and physiological processes, among them, short-term memory (Walter, 1967).

Low *et al.* (1966a) and Borda (1970) reported that CNV-like potential changes could be recorded directly from the cortex of Rhesus monkey, paving the way for a careful investigation of the phenomenon in nonhuman primates in whom short-term memory processing deficits have been studied.

The present study was undertaken to test the frontal origin of the contingent negative variation. We intended subsequently to use the CNV as an index of frontal lobe function in monkeys during various classical delayed response tasks that have or have not produced performance deficits in frontally lesioned animals. The relationship between electronegative patterns and task parameters might then shed light on both the functions of the frontal cortex and the mechanism of short-term memory. In order to determine the anatomical source of CNV, saggital arrays of transcortical electrodes were implanted in Rhesus monkeys trained to perform a variety of foreperiod reaction-time tasks, each of which incorporated a fixed delay interval.

I. METHODS

The methods used have been reported in greater detail elsewhere (Donchin *et al.*, 1971). Seven *Macaca mulatta* monkeys weighing between 5 and 7 lbs were chronically implanted with platinized platinum transcortical electrodes. A typical array of five electrodes was placed equidistantly from the anterior frontal granular cortex back to the occipital striate cortex. Histologically verified electrode placements for all monkeys are shown in Figure 1. Each bipolar electrode consisted of a coiled platinum screen positioned directly on the dura and was referenced to a platinum wire extending through the coil into deep cortical layers on subjacent white matter. Nichrome wires were implanted in the neck and above the supraorbital ridge to monitor head and eye movements. Physiological signals were channeled into Brush AC amplifiers set at a band pass of 0.1–30 Hz and recorded with appropriate event markers on an Ampex FR 1300 magnetic tape recorder. The data were subsequently analyzed with an IBM 1800 computer.

During testing sessions, monkeys were placed in restraining chairs in the training chamber shown in Figure 2. Visual stimuli were generated by a xenon

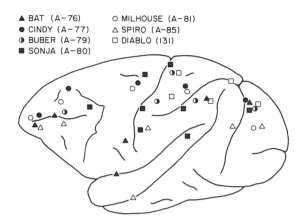

FIG. 1. Electrode placements in all monkey; see key in figure. (From Donchin *et al.*, 1971.)

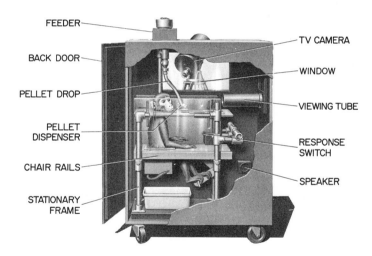

FIG. 2. Testing chamber. Laboratory noises were masked by a white-noise generator. (From Donchin *et al.*, 1971.)

arc lamp and presented on a circular ground glass screen. Brightness of the 2.5° light patch at the viewing screen was 2.18 log foot lamberts. A 1000-Hz tone of moderate intensity was presented via a small speaker mounted inside the chamber. A response key was available within access of the animal's right hand. To depress the key, the monkey had to fully extend his hand, grasp the handle, and depress the key over a distance of 2 in. Small banana-flavored reinforcing pellets were delivered through a chute to a plastic tray directly in front of the animal's mouth. The behavior of animals in the chamber was continuously monitored via closed-circuit television.

An Iconix 136 preprogramed logic system was used to control stimulus scheduling. Amplitude-coded event markers (Donchin and Pappas, 1970) corresponding to each stimulus and response were recorded on one tape channel to facilitate signal averaging of data. Figure 3 provides a diagram of the testing, recording, and data processing apparatus used in the experiment. The sequence of stimulus–response events in each training task is depicted in Figure 4. Methods and results of each task will be described separately.

II. EXPERIMENT I: A SIGNALED DOUBLE RESPONSE TASK

The typical situation used to elicit CNV in humans incorporates a waiting period, delineated by two stimuli, followed by a decision or response. The first stimulus is called a *warning* stimulus, and the second, an *imperative* stimulus. To guarantee that the monkeys were actively waiting during the interstimulus interval (ISI), we designed the initial task (A and B in Figure 4) so that animals

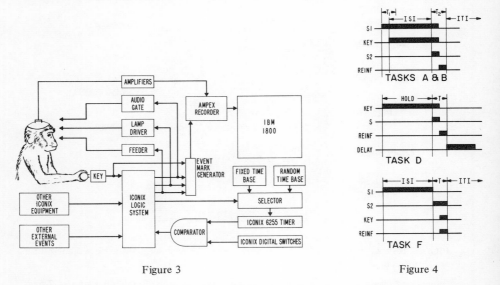

Figure 3

Figure 4

FIG. 3. Flow chart of testing, recording, and analysis procedures. Analog data was digitized and averaged offline with an IBM 1800 computer.

FIG. 4. Diagram of basic training tasks. S_1 and S_2, first and second stimuli; T_1 and T_2, time intervals during which key press and release were required in order to receive food pellet; ISI, interstimulus interval; ITI, intertrial interval. Tasks A and B are designated *signaled double response* tasks; and F is called a *single response* task in the text. (From Donchin *et al.*, 1971.)

were required to press a response key within T_1 msec after the warning stimulus S_1, hold the key down until the imperative stimulus S_2, occurred, and then release the key within T_2 msec in order to receive a small banana pellet. In A, S_1 consisted of a tone and S_2, a flash. These stimuli were reversed in B. A fixed 10 sec intertrial interval (ITI) separated successive trials. If a monkey failed to respond within T_1 or T_2 msec or responded during the ITI, the trial immediately aborted without reinforcement and a new ITI commenced. Monkeys were trained 6 days per week, about 100 trials daily, until key press and key release reaction times were reduced to 400 msec. Most monkeys reached this criterion within 4 weeks, at which stage monkeys generally aborted less than 5% of the trials.

Results of Experiment I

The most dramatic waveform during the waiting interval (ISI) in these tasks was observed in the postcentral region (Figure 5). This pattern consisted of a series of positive and negative shifts with a prominent negative component peaking about 1000 msec (which we will designate N1000) after stimulus onset, followed by a slow positive shift incrementing throughout the remainder of the ISI. A surface negative peak at 200 msec (N200) and positive peak at 400 msec (P400) after S_1 onset were also observed. A small notch between the P400 and N1000 components also appeared regularly. The amplitude of the postcentral pattern ranged from 50 to 75 μV. The same configuration, but at lower amplitude, also appeared in the precentral motor cortex. We have designated the N200–P400–N1000 sequence observed in the pre- and postcentral cortex the *transcortical negative variation* (TNV). No consistent patterns were observed in the frontal or the occipital cortex of highly trained animals during the signaled double response task, although frontal negativity was present during early stages of training in the one animal examined during this period and briefly following major changes in stimulus configuration in all animals. Frontal negativity habituated as performance improved and the stimulus pattern became familiar.

The TNV was observed with remarkable intrasubject consistency (Figure 6) in six monkeys trained to perform this task. Three naive monkeys were implanted prior to training to study the evolution of the TNV during conditioning. Figure 7 depicts the emergence of the TNV in monkey A-81. Note that the amplitude and slope of the P400 and N1000 components increased dramatically as training progressed, whereas the latency of the N1000 component slowly decreased. The effect of varying the interstimulus interval also modified the slope of the positive going ramp following the N1000 component, as shown in Figure 8.

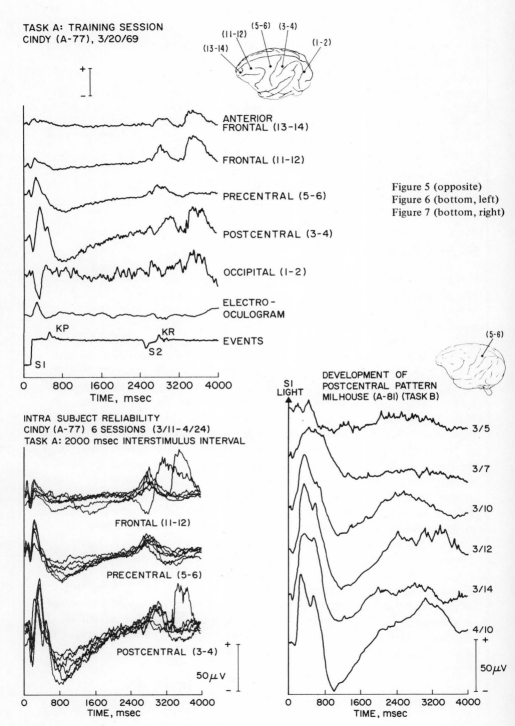

Figure 5 (opposite)
Figure 6 (bottom, left)
Figure 7 (bottom, right)

III. EXPERIMENT II: AN UNSIGNALED DOUBLE RESPONSE TASK

The results of Experiment I were surprising in that the topography of the TNV showed a maximum in the postcentral cortex and no appreciable negativity in the frontal cortex once the task had been mastered. In order to determine whether this topography was dependent on the presentation of an initial stimulus, we designed an additional task (D), in which the animal was required to voluntarily depress the key in the absence of any initial external stimulus. In this unsignaled response task, the monkey was required to hold the key down for an interval equivalent to the ISI used in the original tasks and then to release it within T msec following the presentation of an imperative stimulus (S), as shown in Figure 4. Successive trials were separated by a 2–4 sec forced delay during which the monkeys were not permitted to respond. Release of the key prior to the warning stimulus also aborted the trial and initiated a new intertrial delay.

Results of Experiment II

Electrocortical patterns observed at frontal, precentral, and postcentral sites during the unsignaled task are displayed in Figure 9. Data averaged forward and backward from the key press have been placed in continuity in this figure. The largest amplitude waveform again appeared in the postcentral cortex, although similar patterns of lower amplitude were now seen in both precentral and frontal cortex. These patterns closely resembled the N1000 component and slow positive-going ramp of the TNV. The N200 and P400 components were not distinguishable in task D averages. Clearly the N1000 components of the TNV is not stimulus determined.

IV. EXPERIMENT III: A SINGLE RESPONSE TASK

The TNV observed in Experiments I and II showed a maximum response in the parietal cortex and was not determined by the initial signal. Could this

---

FIG. 5. Transcortical averages obtained during the signaled double response task. Stimulus onset (S_1 and S_2), key press (KP), and key release (KR) are depicted in the bottom line. Surface positivity is upward in all figures. Averages of 25 trials. (From Donchin et al., 1971.)

FIG. 6. Superimposed averages at three recording sites from six training sessions spanning 1.5 months indicate the remarkable consistency of the transcortical patterns observed. Each trace is an average of 25–50 trials. (From Donchin et al., 1971.)

FIG. 7. Emergence of postcentral transcortical pattern during training. Training consisted of about 100 trials daily, 6 days per week. Averages of 25–50 trials. (From Donchin et al., 1971.)

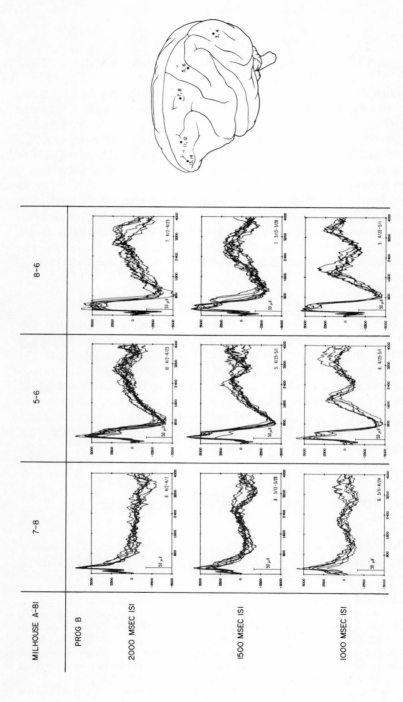

FIG. 8. Transcortical (electrodes 7 and 8 and 5 and 6) and surface–surface (electrodes 8 and 6) patterns observed at varying interstimulus intervals during the signaled double response task. Figures in the lower right corner of each box indicate the number of tracings superimposed in each record and the inclusive dates of the recording periods. Each trace is an average of 25–50 trials. (From Donchin et al., 1971.)

component be caused by the demand for a motor response during the waiting period? In order to test this possibility, a single response task (F), more similar to that ordinarily employed in human experiments, was given to the monkeys. The sequence of events in a single response task is outlined in Figure 4. The design is very similar to the original signaled double response task with the exception that monkeys were required to wait until the imperative stimulus (S_2) occurred, and then respond within T msec. Reinforcement was contingent only upon a correct key press at the time of S_2 in this case, whereas reinforcement in previous tasks required pressing, holding, and releasing the lever within specified time intervals. Successive trials were separated by 10 sec intervals. Three monkeys were tested on this task. Two were highly trained on previous tasks, whereas the third had failed earlier to learn the signaled double response task.

Results of Experiment III

Transcortical patterns generated at frontal, precentral, and postcentral sites during the single response task are shown in Figure 10. Note that the postcentral negative waveform that appeared prominently during the foreperiod of tasks A, B, and D was clearly absent during the interstimulus interval but reappeared in conjunction with the response following the imperative stimulus. TNV's were evident, however, in both frontal and precentral cortices during the foreperiod. An average of the electrooculogram is included in Figure 10 to demonstrate that transcortical patterns were not an artifact of eye movement.

V. DISCUSSION

This study was undertaken to determine the neuroanatomical source of the CNV first described by Walter *et al.* (1964). In order to assess the topographical distribution of electronegative patterns, saggital arrays of transcortical electrodes were implanted in Rhesus monkeys trained to perform a variety of foreperiod reaction time tasks, each of which included a waiting interval and delayed response. The results demonstrated an exquisite degree of specificity in the shape and distribution of electrocortical patterns contingent on the precise stimulus and response parameters employed. Transcortical negative variations (TNV's) were observed in the postcentral region during the waiting period when a motor response was required at the beginning of the interval (tasks A, B, and D). but not when response was withheld until the end of the interval (task F). However, TNV's were seen in the frontal region when the waiting interval was

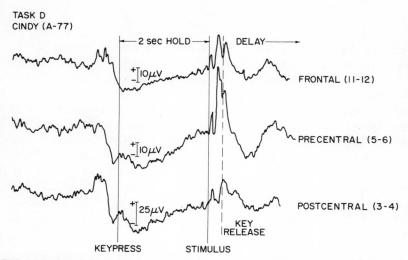

FIG. 9. Transcortical patterns obtained during the unsignaled double response task in which the monkey voluntarily initiated a 2 sec waiting interval. Averages of 50 trials. (From Donchin *et al.*, 1971.)

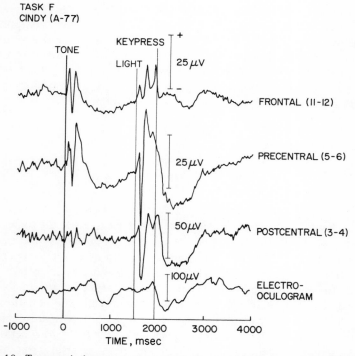

FIG. 10. Transcortical patterns obtained during the single response task in which lever pressing was withheld until the second stimulus. Note that sustained ISI negativity occurred at the frontal and precentral, but not the postcentral, sites. Averages of 40 trials. (From Donchin *et al.*, 1971.)

initiated by either a signaling stimulus or a response alone (tasks F and D, respectively) but the TNV's appeared to habituate when a stimulus—response pair occurred at the start of the interval (tasks A and B). In the precentral motor-arm area, TNV's were recorded during all waiting tasks.

The largest and most reliable TNV pattern appeared in the postcentral region when a sustained key press was required during the interstimulus interval. The presence of the prominent postcentral pattern during the waiting period was unexpected, but even more surprising was the disappearance of the frontal TNV as performance in the signaled double response task improved. The frontal TNV habituated when the waiting period entailed a prolonged motor response but not when waiting was unaccompanied by such overt behavior (single response task). Identical patterns of habituation of frontal negativity have been observed in human scalp recordings (Otto, 1972).

Frontal negativity during the holding interval of the signaled double response task thus appears to index an orienting reaction. The importance of the frontal lobes in the regulation of orienting has been noted elsewhere (Luria et al. 1964; Kimble *et al.* 1965). In these studies, frontal lesions severely depressed the orienting reaction. Again, as in the double response task under discussion here, orienting was investigated in situations (e.g., galvanic skin response to repetitive beeps) where no overt response was required.

A postcentral pattern was observed during task D, in which each trial commenced with a voluntary key press in the absence of an external stimulus. This result indicates that the TNV is not an elaboration of late components of an average evoked response to a stimulus and suggests that it is probably associated with the motor response of key pressing. This hypothesis was confirmed in the single response task where the motor response was withheld until the end of the interstimulus interval. In the latter task, no postcentral TNV was observed during the waiting interval. As might be predicted, a postcentral waveform similar to the TNV in all respects appeared immediately after the second stimulus in conjunction with the delayed motor response. It is probable, therefore, that the TNV seen in the postcentral gyrus reflects proprioceptive or kinesthetic feedback from (or, under some conditions, a feedforward to) muscle or joint receptors associated with the initiation of movement.

The TNV's observed in the precentral and postcentral regions appear to be closely related to the potentials preceding voluntary movement described by Kornhuber and Deecke (1965) and Vaughan, Costa, and Ritter (1968). While Donchin, Gerbrandt, Leifer, and Tucker (1972) and others have shown that a motor response is not a necessary condition for the CNV, the relation of the TNV to motor potentials must be considered. Vaughan *el al.* (1970) also recorded motor potentials in monkeys during unsignaled lever pressing. These investigators found that the negative component of the motor potential was greatly prolonged when monkeys were trained to make a sustained contraction rather than a brief flexion. Electronegative patterns observed in our monkeys during sustained foreperiod contraction in the unsignaled task bear a striking resemblance to recordings by Vaughan *et al.* (1970). Unfortunately, the latter investigators recorded from the precentral motor cortex only. Our results indicate that maximal negativity during sustained contraction occurs postcentrally. Similarly, Gerbrandt *et al.* (1970) found a strong correspondence in the topographical distribution of the abrupt negative component (the motor potential of Deecke *et al.*, 1969) and early components of the somatosensory evoked potential. In view of the fact that somatosensory afferents reach both the precentral motor cortex and the postcentral cortex via the ventrolateral thalamus (Malis *et al.*, 1953) this result is not altogether surprising.

The single response task closely resembles the standard delayed response task employed in most human CNV studies. Response or decision in human studies is always delayed until the end of the interstimulus interval, a behavioral feature incorporated in the single response task here. It is significant in this regard that a frontal TNV very similar to the human CNV appeared in monkeys during the foreperiod of this task. As performance stabilized, a precentral TNV, often larger in amplitude than the frontal pattern, was also seen during this task. The latter observation is consistent with human CNV studies in which maximal negativity is usually found at the vertex (see the review by Cohen, 1969). An electronegative pattern in the frontal and precentral regions of the macaque brain similar to the CNV in humans can thus be observed if the appropriate stimulus and response contingencies are used. However, other sources of cortical negativity are evident under other conditions. The topography and distribution of the TNV's are therefore task specific.

Further evidence for the stimulus and response specificity of cortical negativity has been obtained by Stamm and Rosen (1969). Steady negative cortical shifts were recorded in monkeys during the performance of a classical delayed response task. Negative shifts were found in the occipital cortex during the presentation of cue and reinforcement lights, in the motor cortex during instrumental response, and in the anterior frontal cortex during the delay interval. Stamm and Rosen did not observe any negativity in the postcentral cortex during the delay period, which was consistent with the results of the single response task in the present study.

Stamm and Rosen (1969) concluded that surface negative steady potential shifts index the participation of underlying cortex at functionally relevant points during the delayed response task. Results of the present study are consistent with this view. Transcortical negative variations were observed in the frontal granular cortex when response was delayed but habituated when response was sustained during a waiting interval. TNV's were also found in the precentral and the postcentral cortex, depending on the behavioral contingencies of the task.

Results of these experiments suggest that waveforms originating in multiple regions of the cortex summate in the vertex-negative CNV recorded on the human scalp. Borda (1970) and Jarvilehto and Frustorfer (1970) also reached this conclusion in related studies. Contingent negative variation, therefore, is not an exclusive index of frontal lobe function. Psychological processes appear to be coded not in the shape or amplitude of negative shifts in any single region of the brain but in the configuration of participating neural structures. The neural architecture of psychological processes may thus be sketched by carefully mapping the topographical distribution of negative variations in precisely defined experimental settings.

It appears that differential habituation patterns occur in the frontal cortex relative to task response parameters (Otto, 1972). Thus temporal, as well as spatial, characteristics of the CNV and related potentials should be assessed. Difficulties inherent in the interpretation of volume-conducted scalp potentials underscore the critical need for chronic intracerebral studies of CNV-related phenomena in laboratory animals to further elucidate the functional significance of this intriguing wave form.

ACKNOWLEDGMENTS

This chapter is based on work reported in detail by Donchin et al. (1971) conducted in the laboratories of the Neurobiology Branch at Ames Research Center. The authors are indebted to Dr. Eric Ogden and Dr. William Mehler for their support. E. Donchin held a National Research Council Associateship and L. Gerbrandt a NIMH postdoctoral fellowship. The study was supported in part by NIMH Grant MH 12970 and NIMH Career Award MH 15214 to Dr. K. H. Pribram. The authors thank Elsevier Publishing Company, Amsterdam for permission to reproduce Figures 1, 2, and 4-10.

REFERENCES

Borda, R. P. (1970). The effects of altered drive states on the contingent negative variation in Rhesus monkeys. *Electroenceph. Clin. Neurophysiol.* **29,** 173-180.

Cohen, J. (1969). Very slow brain potentials relating to expectancy: The CNV. *NASA Spec. Publ.* **NASA SP-191,** 143-198.

Deecke, L., Scheid, P., and Kornhuber, H. H. (1969). Distribution of readiness potential of the human cerebral cortex preceding voluntary finger movements. *Exp. Brain Res.* **7,** 158-160.

Donchin, E., and Pappas, N. (1970). An event-coder for evoked potential studies. *Behav. Res. Methods Instrum.* **2,** 142-144.

Donchin, E., Otto, D. A., Gerbrandt, L. K., and Pribram, K. H. (1971). While a monkey waits: Electrocortical events recorded during the foreperiod of a reaction time study. *Electroenceph. Clin. Neurophysiol.* **31,** 115-127.

Donchin, E., Gerbrandt, L. K., Leifer, L., and Tucker, L. (1972). Is the contingent negative variation contingent on a motor response? *J. Psychophysiol.* **9,** 178-188.

Gerbrandt, L. K., Goff, W. R., and Smith, D. B. (1970). Topography of the averaged movement potential. Paper presented to the American EEG Society.

Jacobsen, C. F. (1936). The functions of the frontal association areas in monkeys. *Comp. Psychol. Monog.*, 13, 3-60.

Jarvilehto, T., and Frustorfer, H. (1970). Differentiation between slow cortical potentials associated with motor and mental acts in man. *Exp. Brain Res.* 11, 309-317.

Kimble, D. P., Bagshaw, M. H., and Pribram, K. H. (1965). The GSR of monkeys during orienting and habituation after selective partial ablations of the cingulate and frontal cortex. *Neuropsychologia* 3, 121-128.

Kornhuber, H. H., and Deecke, L. (1965). Hirnpotentialänderungen bei Wilkurbewegungen und passiven Bewegungen des Menschen: Bereitschaftspotential und reafferente Potentials. *Pflügers Arch. Ges. Physiol.* 284, 1-17.

Low, M. D., Borda, R. P., and Kellaway, P. (1966a). Contingent negative variation in rhesus monkeys: An EEG sign of a specific mental process. *Percept. Motor Skills* 22, 443-446.

Low, M. D., Frost, J. D., Borda, R. P., and Kellaway, P. (1966b). Surface negative slow potential shift associated with conditioning in man. *Neurology* 16, 771-782.

Luria, A. R., Pribram, K. H., and Homskaya, E. D. (1964). An experimental analysis of the behavioral disturbance produced by a left frontal arachnoidal endothelioma (meningioma). *Neuropsychologia* 2, 257-280.

Malis, L. I., Pribram, K. H., and Kruger, L. (1953). Action potentials in "motor" cortex evoked by peripheral nerve stimulation. *J. Neurophysiol.* 16, 161-167.

Otto, D. A. (1972). Slow potential changes in the brain of man and monkey during the reaction-time foreperiod. Ph.D. Dissertation, Stanford University, Stanford, California.

Pribram, K. H., and Tubbs, W. E. (1967). Short-term memory, parsing and the primate frontal cortex. *Science* 156, 1765-1767.

Spinelli, D. N., and Pribram, K. H. (1967). Changes in visual recovery function and unit activity produced by frontal cortex stimulation. *Electroenceph. Clin. Neurophysiol.* 22, 143-149.

Stamm, J. A., and Rosen, S. C. (1969). Electrical stimulation and steady potential shifts in prefrontal cortex during delayed response performance by monkeys. *Acta Biol. Exp. (Warsaw)* 29, 385-399.

Vaughan, H. G., Jr., Costa, L. D., and Ritter, W. (1968). Topography of the human motor potential. *Electroenceph. Clin. Neurophysiol.* 25, 1-10.

Vaughan, H. G., Jr., Gross, C. G., and Bossom, J. (1970). Cortical motor potential in monkeys before and after upper limb deafferentiation. *Exp. Neurol.* 26, 253-262.

Walter, W. G. (1967). Slow potential changes in the human brain associated with expectancy, decision and intention. *Electroenceph. Clin. Neurophysiol. Suppl.* 26, 123-130. 123-130.

Walter, W. G., Cooper, R., Aldridge, V.-J., McCallum, W. C., and Winter, A. L. (1964). Contingent negative variation: An electric sign of sensorimotor association and expectancy in the human brain. *Nature (London)* 203, 380-384.

Chapter 8

THE LOCUS AND CRUCIAL TIME OF IMPLICATION OF PREFRONTAL CORTEX IN THE DELAYED RESPONSE TASK

JOHN S. STAMM and *STEVEN C. ROSEN*

Department of Psychology
State University of New York
Stony Brook, New York

The discovery by Jacobsen (1935) that removal of the prefrontal cortex in primates results in severe and long-lasting impairments on delayed response (DR) tasks has been confirmed by many investigators and has led to recognition of this cortical area as an essential structure for correct performance on DR and the related task of delayed alternation (Brutkowski, 1965). The search for a clear formulation of the specific role of the prefrontal cortex in behavior, however, has met with only limited success, and the difficulty appears related to the complexity of the DR problem. This task typically involves exhibition of a cue or reward in one of two locations, a delay period during which the subject is prevented from viewing the cue, the animal's instrumental response, delivery and consumption of the reward, and an intertrial interval. These events in the DR trial may be related to a sequence of psychological processes that are required for solution of the task, namely: perception of the spatial location of the relevant cue; establishment, storage, and retrieval of the transient memory; elicitation of the appropriate instrumental response; recognition of the reinforcement; and, finally, extinction of the short-term memory.

The implication of the prefrontal cortex in any one or in several of these component processes might be determined by utilization of the technique of electrocortical stimulation. With this technique, reversible impairments have been demonstrated on delayed alternation tasks (Stamm, 1961; Weiskrantz *et al.*, 1962), when the monkeys' performance became impaired during application of the electrical stimulus and returned to criterion level after its termination. In the present experiments, we utilized this experimental technique by stimulus applications to prefrontal cortex for relatively brief periods during the different portions of the DR trial. The significance of the findings of impaired performance with stimulus applications during only some of the components of the DR trial could then be further evaluated by systematically varying the parameters of the task components; i.e., the durations of cue presentation and of the delay period. With this experimental approach it was possible to specify more clearly than has thus far been possible those psychological processes required for the DR task that are mediated by the prefrontal cortex.

A further advantage of the stimulation technique is that with multiple electrode implantations in the same animal it may be possible to further delineate the crucial locus, or possibly several loci, for mediation of the DR task. Evidence has been presented during recent years in support of more specific functional localization between portions of the dorsolateral prefrontal cortex and different tasks. Thus, the cortical segments surrounding, but not within, the principal sulcus have been found essential for solving tasks in which the monkey must approach a food cup that is located in the direction opposite to that of the cue (Stepien and Stamm, 1970a), whereas the cortex in the depth and banks of the principal sulcus seems crucial for mediating tasks that involve a temporal separation between cue presentation and the instrumental response (Mishkin, 1957; Stepien and Stamm, 1970b; Weiskrantz et al., 1962). In the present experiments, stimulus applications through different electrode pairs located on the prefrontal cortex made it possible to delineate even smaller segments of prefrontal cortex as the essential structures for correct DR performance.

The behavioral impairments that have been described for brain-damaged animals have generally been obtained only after bilateral resections of cortical tissue and consequently, have led to the conclusions of functional equivalence of the two hemispheres. The validity of this concept may now be investigated more directly with symmetrical implantation of stimulating electrodes in both hemispheres and with unilateral stimulation of either prefrontal cortex. We applied this experimental approach to problems of interhemispheric differences and possible cerebral dominance in monkeys.

I. METHOD

A. Subjects and Electrodes

This report presents the results obtained with seven immature monkeys (*Macaca speciosa*) of 6 to 9 lb body weight.

The stimulating electrodes consisted of arrays of four to eight stainless steel spheres, approximately 0.5 mm in diameter, that were mounted on a thin polyethylene sheet. The electrode points were arranged in two rows, 10 mm apart. The wires from the electrode points were brought together in a cable and soldered to an Amphenol connector.

During surgery the skull was opened with rongeurs, the dura was cut, and the electrode assemblies were placed on the appropriate cortical surfaces. Four monkeys, tested in Experiment I (Stamm, 1969) had electrode assemblies placed on the prefrontal cortex contralateral to the preferred hand, so that the two rows of electrode points straddled the principal sulcus. The assemblies for three of these subjects contained three electrodes in each row with 10 mm between adjacent points (Figure 1, S153), whereas that for S158 contained four electrodes in each row with 8 mm between adjacent points. The third electrode pair was placed in different monkeys from 4 mm anterior to 2 mm posterior of the branches of the arcuate sulcus. In these monkeys, the prefrontal cortex ipsilateral to the preferred hand was ablated by subpial suction from the frontal pole to the depth of the arcuate sulcus, medially to the midline, and laterally to the orbital ridge, including the banks and depth of the principal sulcus. The three monkeys in Experiment II had electrode assemblies implanted bilaterally on the prefrontal cortex. In two monkeys, six electrode points were arranged in each assembly, as illustrated in Figure 1 for S153. One of these subjects (S199) had additional electrode points that were placed on the cortex between the midline and the electrodes located superior to the principal sulcus. In S124, the prefrontal assemblies contained four electrodes each, with 10 mm between adjacent points, and assemblies consisting of four electrodes each, arranged in a square with 8 mm between adjacent points, were also implanted on each inferotemporal cortex. Each array was placed on the inferior temporal convexity, ventral to the superior temporal sulcus and anterior to the occipitotemporal sulcus. After placement of each electrode assembly, the dura was sutured over the polyethylene sheet and the skull opening covered with stainless steel screening. The Amphenol plugs were tied to screws over the occipital bone and cemented to the skull. Muscles, fascia, and skin were sutured in layers so that only the top of the electrode plugs protruded from the skin.

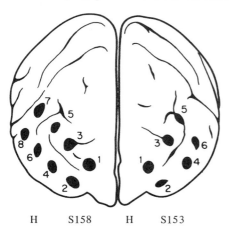

FIG. 1. Illustration of the locations of electrode points on the frontal cortex; left side for S158 and right side for S153. The markers have diameters approximately 10 times those of the electrode points. (Adapted from J. S. Stamm, Electrical stimulation of monkey's prefrontal cortex during delayed-reaction response. *Journal of Comparative Physiology and Psychology,* 1969, 67, 535-546, by permission of the American Psychological Association.)

B. Testing Apparatus

During testing sessions the monkey was placed in a bucket seat that was hinged to the frame of a portable restraining chair (Figure 2). A cuff, which was attached by a chain to the frame of the chair, was clamped to the wrist of the subject's nonpreferred arm. Molded plastic shields were fitted around the monkey's face so that it could not reach the Amphenol plug or the connecting cables. The chair was placed securely in front of a vertical panel in which two circular display windows, 3.5 cm in diameter, were mounted at the subject's eye level, with 6.5 cm between centers. In front of each window was a transparent plastic disk, which, when pressed lightly, activated a microswitch. Two transparent food cups, 16 cm apart, were mounted beneath the display windows.

On the DR task the cue was presented as a bright white field either in the left or the right window for 3 sec in Experiment I and for 2 sec in Experiment II.

FIG. 2. Monkey in a restraining chair. Note the chain attached to the monkey's right wrist and the plastic head restrainers. (Reprinted from J. S. Stamm, Electrical stimulation of monkey's prefrontal cortex during delayed reaction response. *Journal of Comparative Physiology and Psychology,* 1969, 67, 535-546, by permission of American Psychological Association.)

This was followed by the delay period, when both windows were not illuminated, after which blue fields were projected in both windows. When the subject then pressed on either window the illumination was extinguished for an 8 sec intertrial (time-out) interval. The correct response, a press on the window on which the cue had been projected, resulted in delivery of a 45 mg dextrose pellet to the food cup beneath the correct window and a 2 sec illumination of that cup from the rear. On successive trials the cue was presented in the left or right window according to a random schedule, modified so that in every block of ten trials each side was correct five times. Dim overhead illumination was provided throughout each testing session.

C. Stimulation Apparatus

The implanted electrodes were connected to a relay panel that could be switched to either an electroencephalograph or a stimulator. Bipolar stimulation across two electrode points was obtained with a Grass S4 square-wave stimulator, a stimulus isolation unit, and a constant-current regulator. Stimulation consisted of 2 sec trains of 1 msec pulses, at the rates of 20 pulses/sec in Experiment I, and 50 pulses/sec in Experiment II. The onset of the stimulus train was triggered by a timer that was activated by the subject's instrumental response. Electrocorticographs (ECGs) were recorded with a 12-channel Grass EEG apparatus.

II. PROCEDURE

A. Preliminary Training

Each monkey was first adapted to sitting in the restraining chair and its hand preference was determined by presenting it with peanuts while both of its hands were free. Training on the DR task then commenced with the subject's nonpreferred hand restrained. Each session consisted of 100 trials. At first the delay was zero sec and then it was gradually increased. After a monkey responded on 8 sec DR at 80% correct during three successive sessions, surgery was performed. Training resumed on 8 sec DR until the subject responded at 85% correct.

B. Prefrontal Stimulation

Electrical stimulation was applied during DR testing through electrodes located across the principal sulcus of the cortex contralateral to the responding hand. The 2 sec stimulus train started 1 sec before the start of the delay period of each trial, and the current strength was gradually increased after every block of ten trials until major motor convulsions were elicited or until the maximum available current was applied. If the monkey continued to respond correctly until the convulsive threshold was reached, stimulation was applied during the following session through a different electrode pair. When stimulation resulted in markedly impaired DR performance, stimulating currents at this magnitude were applied during the subsequent testing sessions. During each 120 trial session, the onset of stimulation was shifted after each block of ten trials, except for the first and last blocks, which were given without stimulation. When responses dropped to below 80% correct for a ten trial block, the stimulus onset was shifted to a portion of the DR trial where correct responses were above this level. This procedure was continued until adequate samples were obtained for stimulation of each portion of the DR trial. During some of the sessions the EGGs were recorded between periods of stimulus applications. For evaluations of normal performance scores occasional testing sessions were given without any stimulation.

The above procedure was subsequently replicated with stimulation across other pairs of electrode points. Additional stimulation procedures were conducted with some of the subjects. These included presentation of cues for 4 and 6 sec, intermanual transfer to the nonpreferred hand, training to 20 sec DR, and prefrontal stimulation during training on visual discriminations. These procedures are described in the subsequent sections of this chapter.

III. RESULTS

A. Stimulation during the Delay Period

Stimulation through electrodes straddling the midsegment of the principal sulcus resulted in impaired DR performance in every monkey when the stimulus train occurred during the onset of the delay period. As shown by Figures 3–6, stimulus applications at that time disrupted correct performance more effectively than did stimulation during any other portion of the DR trial, including the later portion of the delay. The response scores for the four monkeys in Experiment I were evaluated statistically. The t tests for differences between means indicate that the group mean of 58% correct with stimulus applications during the first 4 sec of the delay was significantly ($p < .01$) below the mean score obtained with stimulation during any of the other portions of the DR trial. Although stimulus applications after the initial portion of the delay period did not severely impair correct responses in any monkey, there were individual differences with regard to both the highest correct DR scores and the portion of the delay period when these scores were obtained. Thus, one monkey (Figure 3)

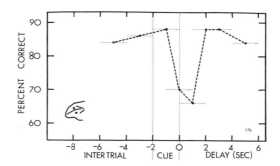

FIG. 3. Performance on an 8 sec DR task by S176, with stimulation consisting of 2 sec trains of 1 msec pulses at 50 pulses/sec and 4.0 mA applied across prefrontal electrodes (locations shown in insert). The time scale is with reference to the start of the delay period. The horizontal lines indicate periods of stimulation. Each point on the graph represents a mean score for 50–110 trials.

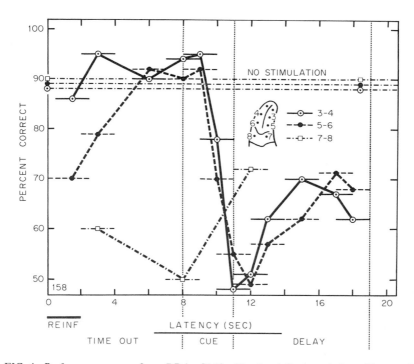

FIG. 4. Performance on an 8 sec DR by S158 with stimulation consisting of 2 sec trains of 1 msec pulses at 20 pulses/sec through electrodes 3 and 4 at 4.6 mA, 5 and 6 at 4.0 mA, and 7 and 8 at 1.8 mA. The time scale is with reference to the instrumental response. The horizontal lines indicate periods of stimulation. Each point on the graph for prefrontal stimulation represents the mean score for 50–90 trials. See key and insert on figure.

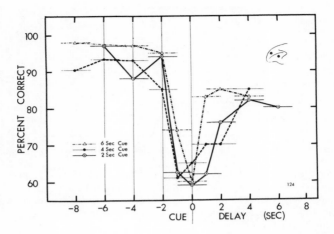

FIG. 5. Performance on 8 sec DR by S124 with cue presentations of 2, 4, and 6 sec. Stimulation was with 2 sec trains of 1 msec pulses at 50 pulses/sec and 5.5 mA applied across the left prefrontal electrodes (locations shown in insert). The time scale is with reference to the start of the delay period. The horizontal lines indicate periods of stimulation. Each point represents a mean score for 50–130 trials. See key on figure.

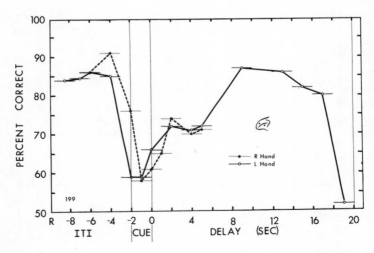

FIG. 6. Performance on 20 sec DR by S199 with right and then with left hand. Stimulation across left prefrontal electrodes (locations shown in insert) was with 2 sec trains of 1 msec pulses at 50 pulses/sec and 2.0 mA. The time scale is with reference to the start of delay. The horizontal lines indicate periods of stimulation. Each point represents a mean score for 70–300 trials. See key on figure.

responded at near criterion level for stimulation that started only 1 sec after onset of the delay, whereas the majority of subjects, as shown by Figure 4 (electrodes 3 and 4) and Figure 5 (2 sec cue), responded at 70 to 85% correct with stimulation beginning 3 sec after the start of the delay. Of further interest are the results obtained with the monkey that was trained on a 20 sec DR. As seen by Figure 6 (monkey's left hand), stimulus applications during 1 to 6 sec after the start of the delay resulted in performance scores below 75% correct, whereas correct responses remained unaffected with stimulation during the subsequent portion of the delay period. When the end of stimulation coincided with the end of the delay period (Figures 4 and 6), correct responses again dropped markedly to near chance level. The slightly deficient response scores obtained with stimulation that terminated a few seconds before the end of the delay period may be related to cortical after-discharges, which were seen in the poststimulus ECGs of several monkeys, especially S199 (Figure 6).

B. Stimulation during Cue Presentation

Prefrontal stimulation during presentation of cues of 3 sec (Experiment I) or 2 sec (Experiment II) duration did not affect correct performance for any of the subjects,* provided the stimulus train terminated at least 1 sec before the start of delay. In order to ascertain whether the effect of stimulation was time-locked to the onset of the cue or was related to the start of the delay period, monkeys 124 and 199 (right hand) were systematically tested with cues of differing duration. When cues were presented in different sessions for 2, 4, or 6 sec, it was found (Figure 5) that stimulation did not disrupt correct DR performance, except when it occurred during the final second of cue presentation. Only when the stimulus train coincided with the final 2 sec of cue presentation were response scores affected by the duration of the cues. For this experimental condition, the mean response scores for the two subjects were 58%, 63%, and 73% correct, respectively, for cues of 2, 4, and 6 sec duration. This finding indicates that presentation of the cue for several seconds before the onset of the stimulus provides the monkey with sufficient information for the subsequent stimulation to be less effective in disrupting its correct responses.

C. Stimulation of Inferotemporal Cortex

When unilateral stimulation was applied across the posterior pair of the inferotemporal electrodes, DR performance became markedly impaired for

---

*The one exception was S199, which, when tested with its left hand, obtained low response scores with stimulation during the first second of cue presentation (Figure 6). The error responses are correlated with 1 to 2 sec after-discharges, which were seen in the poststimulus ECG traces. During testing with its right hand, the incidence of ECG after-discharges was less regular and the subject's response scores were higher.

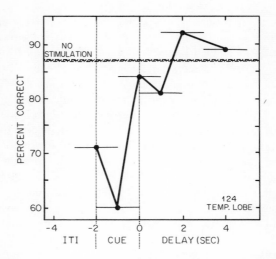

FIG. 7. Performance on an 8 sec DR by S124 with unilateral stimulation of the posterior segment of the inferotemporal cortex. Stimulation was with 2 sec trains of 1 msec pulses at 50 pulses/sec and 6.0 mA. The time scale is with reference to the start of the delay period. The horizontal lines indicate periods of stimulation. Each point represents a mean score for 50–150 trials.

stimulus applications during cue presentation (Figure 7), but not for stimulation during any portion of the delay period. The lower response scores for stimulation during the first second of the cue are related to high-voltage after-discharges, which were generally seen for several seconds in the poststimulus ECG traces from both inferotemporal cortices. The high response scores with stimulation starting 1 sec after onset of the cue would indicate that the first second of cue presentation provides sufficient information to the monkey for correct responding. These results may be contrasted with the findings obtained for prefrontal stimulation in the same monkey (Figure 5).

D. Functional Localization within the Prefrontal Cortex

The effects of stimulation across the principal sulcus depended upon the electrode location along its anterior–posterior dimension. In the perfused brains, the distance between the frontal pole and the bend of the arcuate sulcus was measured as approximately 30 mm. Reconstructions of the electrode locations indicate that the results presented thus far were obtained with stimulus applications to the middle third of the prefrontal cortex. Stimulation across the anterior third of the principal sulcus (Figure 1, electrodes 1 and 2) during any portion of the DR trial did not appreciably affect correct DR performance, even with the maximum available currents of at least 9 mA. Under these conditions the poststimulus ECG tracings revealed persistent high-voltage waves (2–4 per sec) from the stimulating electrodes. Propagation of these ECG patterns to the more posteriorly located electrodes, which was occasionally observed in one monkey, was associated with slightly impaired performance scores.

Stimulation of the posterior third of the principal sulcus resulted in the behavioral effects illustrated by Figure 4, electrodes 5 and 6. Correct performance became markedly disrupted with stimulus applications during the early portion of the delay and also shortly after the response, i.e., during the 2 sec illumination of the food cup, when the reward was delivered. The latter effect could not be attributed to any motor impairment, because the monkeys were observed to pick up and eat the sugar pellets without difficulty. Chewing artifacts in the ECG traces indicated that the pellet was generally swallowed within 3 sec after the response. These marked differences in behavioral scores between stimulation of the posterior and stimulation of the middle segments of the principal sulcus for stimulus applications during the reinforcement period were obtained with the two subjects in Experiment I and S199 in Experiment II, who were systematically tested under these two stimulating conditions.

Stimulation of the cortex along the branches of the arcuate sulcus (3 mm anterior and posterior) and of the premotor cortex generally did not interfere with correct DR performance until motor convulsions were elicited. In one monkey stimulation of premotor cortex (Figure 4, electrodes 7 and 8) with currents just below convulsion thresholds disrupted correct performance, regardless of the period of stimulus application during the DR trial.

Different behavioral effects were also obtained between stimulation of cortical segments located superior and inferior to the principal sulcus. In two monkeys, bipolar stimulation between the anterior and middle electrodes located along the superior bank of the principal sulcus did not affect DR scores until convulsions were elicited at approximately 3.0 mA. The same results were obtained in the monkey (S199) in which a third row of electrodes was placed between the midline of the frontal lobes and the electrodes above the superior bank of the principal sulcus. By contrast, stimulation along the lateral banks of the principal sulcus resulted in behavioral effects similar to those obtained with stimulation across the sulcus, except that the effective current strengths were somewhat higher (3.5 to 6.0 mA).

E. Hemispheric Differences

In the monkeys with intact frontal lobes (Experiment II), the effects of stimulation were obtained most clearly and consistently by unilateral stimulus application across the principal sulcus through electrodes corresponding to points 3 and 4 in Figure 1 of the cortex contralateral to the preferred hand. During subsequent testing with the monkeys' nonpreferred (left) hand, stimulation through these electrodes yielded essentially the same results that had been obtained with the preferred hand, at approximately the same current settings (see Figure 6). However, stimulus applications through symmetrically located electrode points in the prefrontal cortex ipsilateral to the preferred hand did not affect correct responding, even with currents that elicited motor convulsions. This finding was obtained for testing with either the left or right hand. The convulsion thresholds were approximately the same for symmetrical electrode locations in the left and right hemispheres. These results would indicate interhemispheric differences, because DR performance could be affected only by stimulation of the cortex contralateral to the hand with which the monkey had received extensive training.

IV. DISCUSSION

The present findings permit greater specification of the involvement of the prefrontal cortex in delayed response performance than has been possible with the method of cortical ablation. We found that electrical stimulation across the principal sulcus disrupted correct DR performance most severely when it was applied at the onset of the intratrial delay period and, to a lesser degree, after the instrumental response. Stimulation with identical parameters during other portions of the DR trial, including most of the delay period, had only minimal or no effects on correct DR performance. Furthermore, we found that DR performance was affected only when stimulation was applied across the posterior two-thirds of the principal sulcus, contralateral to the monkey's trained hand, even during transfer testing with the ipsilateral hand. The possibility of even more restricted functional localization is suggested by the finding that stimulation during the reinforcement period was effective only when it was applied to the posterior segment of the principal sulcus.

The specific implication of prefrontal cortex in DR performance during these two relatively brief portions of the trial is confirmed by the findings from electrophysiological recording experiments with monkeys that had chronically implanted electrodes. Fuster and Alexander (1971), using microelectrode techniques, found a substantial population of single neurons in principalis cortex which discharged during the cue and delay periods of the DR trial. Many of these cells exhibited accelerated rates of firing, particularly during a brief period at the onset of the delay. In other investigations with chronically implanted transcortical nonpolarizable electrodes, Stamm and Rosen (1969, 1972) found surface negative steady potential (SP) shifts from the principalis cortex that started during cue presentation of the DR trial, reached maximum amplitude at the onset of the delay, and returned to the pretrial base line during the subsequent portion of the delay period. A second surface negative shift started after the instrumental response and returned to base line during the intertrial interval. The latter shift appeared closely related to the reinforcement, since omission of the sugar pellet after correct responses resulted in marked attenuation of this SP shift. The magnitude of the first SP shift was not affected by changes in either the duration of cue presentation (from .06 to 8 sec) or the delay period (4 to 20 sec). However, correlation coefficients between prefrontal SP shift magnitudes and scores of correct responses were highly significant, with a range of .74 to .90 for different monkeys, but were insignificant and variable for SP shifts from other cortical electrode locations. In view of the agreement between the time course of prefrontal unit discharges and surface negative SP shifts, we interpret these shifts as expressions of increased cortical excitation, and hence of specific prefrontal cortical involvement in the DR task.

The findings from recording experiments, furthermore, provide evidence for interhemispheric differences (Stamm and Rosen, 1969). During extensive training with the preferred hand the SP shifts were found to be of substantially greater magnitude in the contralateral than the ipsilateral prefrontal area and subsequent intermanual transfer testing resulted in increased surface negativity from both prefrontal areas. Whether these hemispheric differences are the consequence of the extensive training that the monkeys had received with the preferred hand or are indeed related to endogenous hemispheric dominance, can only be answered by further experimentation.

The present findings substantiate the important involvement of principalis cortex in the short-term mnemonic process required by the DR task. Since the monkey must distinguish between the locations of the left and right cues and respond in the appropriate direction in order to obtain a reward, the short-term memory appears to be specific for spatial differentiations. The question of whether prefrontal cortex is uniquely implicated in spatial short-term mnemonic processes, or is more generally involved in the performance of tasks that require short-term memories was investigated in further experiments with electrocortical stimulation. In one experiment (Cohen, 1972) monkeys were trained on a delayed visual discrimination task, in which a 3 sec display of two identical patterns was followed by an 8 sec intratrial delay, and then by illumination of both windows with identical colors. Responses during the latter portion of the trial were rewarded, if they were to the left window when one pair of patterns had been displayed, and to the right window for another pattern pair. It was found that correct performance could be disrupted by 3 sec stimulation, either of inferotemporal cortex during the pattern display or of principalis cortex during the early portion of the delay. This task requires visual discrimination during the pattern display, but short-term spatial memory during the delay. In the second experiment (Kovner and Stamm, 1972) monkeys were trained on a visual delayed matching-to-sample task. On this problem the monkey must remember the sample pattern that is first displayed and, after the intratrial delay (matching period), discriminate between the sample and a differing pattern. Here, a short-term visual memory is required for correct responding. Inferotemporal stimulation resulted in markedly impaired performance when it was applied during the delay or matching portions of the trial, whereas stimulation across the principal sulcus during any portion of the delay did not affect correct responding, and mild performance decrements were obtained only for stimulus applications during the matching period. These findings provide evidence for dissociations of cortical functions in the mediation of short-term memories, with specific implications of inferotemporal cortex in visual, and of principalis cortex in spatial short-term memories.

Taken together, the corroborative findings of stimulation and recording experiments suggest that the prefrontal cortex does not perform a unitary role in mediation of the DR task, but that at least two functional components may be identified, namely the processes at the onset of the delay, and those following the response. Interpretations of these findings in terms of the psychological processes that are required for solution of the DR task indicate that the prefrontal cortex is crucially implicated in the establishment of a short-term spatial memory, but is not essential to either the perceptual aspects of the task, or to the functions of mnemonic storage and retrieval. The second functional component would implicate the principalis cortex in the processes of evaluation of the consequences of the response, i.e., in registering the response strategy that results in the highest rate of reward. This interpretation is consonant with Pribram's (1961) evaluations of the nature of the frontal lobe impairments. His explanation (Pribram et al., 1964) of the impairments in terms of a "flexible noticing order" accounts for the prefrontal functions in relation to stimulus–response–reward contingencies. The present findings may point to clearer specifications of the nature of such contingencies.

ACKNOWLEDGMENTS

This research was supported by National Science Foundation Research Grants GB-5256 and GB-6911. The authors express their appreciation to T. Aranow, H. Gould, and T. Lidsky for their assistance in the experimentation.

REFERENCES

Brutkowski, S. (1965). Functions of prefrontal cortex in animals. *Physiol. Rev.* **45**, 721-746.

Cohen, S. M. (1972). Electrical stimulation of cortical-caudate pairs during delayed successive visual discrimination in monkeys. *Acta Neurobiol. Exp.* **32**, 211-233.

Fuster, J. M., and Alexander, G. E. (1971). Neuron activity related to short-term memory. *Science* **173**, 652-654.

Jacobsen, C. F. (1935). Functions of the frontal association area in primates. *Arch. Neurol. Psychiat.* **33**, 558-569.

Kovner, R., and Stamm, J. S. (1972). Disruption of short-term visual memory by electrical stimulation of inferotemporal cortex. *J. Comp. Physiol. Psychol.* **81**, 163-172.

Mishkin, M. (1957). Effects of small frontal lesions on delayed alternation in monkeys. *J. Neurophysiol.* **20**, 615-622.

Pribram, K. H. (1961). A further experimental analysis of the behavioral deficit that follows injury to the primate frontal cortex. *Exp. Neurol.* **3**, 432-466.

Pribram, K. H., Ahumada, A., Hartog, J., and Roos, L. (1964). A progress report on the neurological processes disturbed by frontal lesions in primates. In "The Frontal Granular Cortex and Behavior" (J. M. Warren and K. Akert, eds.) pp. 28-55. McGraw-Hill, New York.

Stamm, J. S. (1961). Electrical stimulation of frontal cortex in monkeys during learning of an alternation task. *J. Neurophysiol.* **24**, 414-426.

Stamm, J. S. (1969). Electrical stimulation of monkeys' prefrontal cortex during delayed-response performance. *J. Comp. Physiol. Psychol.* **67**, 535-546.

Stamm, J. S., and Rosen, S. C. (1969). Electrical stimulation and steady potential shifts in prefrontal cortex during delayed response performance by monkeys. *Acta Biol. Exp. (Warsaw)* **29**, 385-399.

Stamm, J. S., and Rosen, S. C. (1972). Cortical steady potential shifts and anodal polarization during delayed response performance. *Acta Neurobiol. Exp.* **32**, 193-209.

Stepien, I., and Stamm, J. S. (1970a). Impairments on locomotor tasks involving spatial opposition between cue and reward in frontally ablated monkeys. *Acta Biol. Exp. (Warsaw)* **30**, 1-12.

Stepien, I., and Stamm, J. S. (1970b). Locomotor delayed response in frontally ablated monkeys. *Acta Biol. Exp. (Warsaw)* **30**, 13-18.

Weiskrantz, L., Mihailovic, L. J., and Gross, C. G. (1962). Effects of stimulation of frontal cortex and hippocampus on behavior in the monkey. *Brain* **85**, 487-504.

Part Five

**THE RELATIONSHIP BETWEEN FRONTAL CORTEX
AND SUBCORTICAL BRAIN FUNCTION**

Chapter 9

TRANSIENT MEMORY AND NEURONAL ACTIVITY IN THE THALAMUS

JOAQUIN M. FUSTER

*Brain Research Institute and Department of Psychiatry
School of Medicine
University of California at Los Angeles
Los Angeles, California*

One of the best substantiated experimental results in primate neurobiology is the deficit produced by ablation of the prefrontal cortex in delayed response performance, a time-honored procedure for testing transient memory in animals. There is, however, considerable controversy over the identity of the aspect or component of this kind of performance that is critically impaired by prefrontal damage. It is not clear whether the functional integrity of the prefrontal cortex is essential for the acquisition, the storage, or the retrieval of information by the organism. Furthermore, the role of the anatomical links between the prefrontal cortex and other structures of the brain is largely unknown. Of particular interest is the obscure functional significance of the well-demonstrated connections between the cortex of the convexity of the frontal lobe and the nucleus medialis dorsalis of the thalamus (Akert, 1964; Walker, 1966). Because these connections are profuse and bidirectional, it has been postulated that lesions of this nucleus can also produce deficits of delayed response performance. The evidence for this is not as plentiful as that gathered in determining the importance of the prefrontal cortex for the same form of behavior. There is, however, a thorough study by Schulman (1964), showing that complete bilateral lesions of the nucleus medialis dorsalis produce, indeed, severe and enduring impairments of delayed response performance in monkeys. This result supports the concept that the nucleus medialis dorsalis and the prefrontal cortex form a functional unit that is the basis of the integration mechanisms for delayed responses.

In view of the long history of research on the effects of brain lesions on memory, one is struck by the scarcity of previous work on the bioelectrical phenomena associated with behavior during transient memory tests or associated with the behavioral alterations produced by lesion. It seems that much could be learned from such work about the role of cerebral structures in memory processes. In this respect, the electrical activity in the thalamofrontal sector is a subject of special importance. Thus, it was thought that a study of the activity of cells in the nucleus medialis dorsalis during performance of delayed responses would be a reasonable inquiry. This chapter deals with the initial results from this study.

Rhesus monkeys are used in our experiments because much work on the effects of frontal lesion in delayed response behavior has been conducted in such monkeys and also because the nucleus medialis dorsalis is particularly well developed in this species. A special method is utilized for recording single-cell activity with a movable microelectrode from chronically prepared animals (Fuster, 1961). Such activity is recorded extracellularly in the thalamus and, in addition, the electroencephalogram (EEG) from the cortex of the frontal lobe is monitored with implanted gross electrodes. The animals, restrained in chairs to allow convenient recording, have been conditioned in delayed response (DR) performance using a modified version of the Wisconsin General Test Apparatus. The monkey sits in a chair facing a glass window that separates the animal compartment from the manipulanda. These are two identical white blocks placed over corresponding food wells, one to the right and the other to the left, on a flat black surface. Two small trapdoors under the window allow the monkey to reach one object with each hand. A trial begins with the cue-presentation phase, in the course of which a piece of fruit is placed in one of the food wells and these are covered by the test objects. These two actions are executed in full view of the animal. A delay of up to 60 sec is then imposed, during which the trapdoors remain mechanically locked. At the end of this delay, the doors are unlocked and the animal is allowed to select one of the objects, retrieving the fruit piece if the baited object has been chosen. A trial leading to incorrect choice is terminated without reward. The position of the bait is changed in random order from trial to trial. Data presented here were obtained from two animals that had reached a high degree of proficiency in the DR task, making only occasional errors on 1 min delay trials.

The spontaneous activity of thalamic neurons in a resting animal habituated to the environment and not exposed to external stimuli soliciting its attention is one of the most distinctive modes of cell discharge that can be observed by microelectrode recording in the brain. The activity of a given unit in these conditions is characterized by the periodic occurrence of clusters or groups of spikes of a frequency loosely varying between two and five per second. Within these groups the spikes are separated by brief intervals of less than 10 msec. When arousal supervenes or the attention of the animal is attracted by an environmental event, grouped firing subsides forthwith and an irregular sequence of spikes ensues, which is correlated with a low-voltage fast activity pattern in the cortical EEG.

The cells of the nucleus medialis dorsalis (MD) are no exception to these general principles of thalamic unit behavior. In periods of rest, such as between trials in the DR test, these cells ordinarily exhibit grouped discharge associated with the presence of synchronous EEG activity in the alpha frequency range (7–12 Hz). Trains of spike clusters are most commonly seen during cortical EEG spindles, although there generally is a lack of correspondence between the frequency of the electrocortical rhythm within a spindle and the frequency of the concomitant spike clusters in the MD.

In the course of DR testing, the observation by the animal of the placing of food under one of the two objects is accompanied by clear changes in the electrical activity of the thalamus and the cortex. In the frontal cortex an EEG arousal reaction typically appears at the commencement of this operation and low-voltage fast activity persists for the duration of the cue-presentation phase. When this phase is terminated, namely, when the food is covered and the delay phase begins, the EEG usually reverts to synchronized rhythmic activity.

On initiation of the cue presentation, the pattern of cell discharge shows a diminution of grouped firing, if present, and a tendency to randomness of interspike intervals (Figures 1 and 2). This pattern change is observable in the MD as well as in other thalamic nuclei that we have been able to explore with microelectrodes. However, changes of firing frequency in the course of the DR testing are generally more conspicuous and characteristic in MD cells than in cells of surrounding nuclei. An augmentation of firing frequency during the cue-presentation phase is the most common feature of MD cells. It may appear at the start of this phase or a few seconds later, coinciding with the covering of the bait. A firing rate higher than that during control intertrial periods usually continues beyond the baiting operation into the delay period, at times persisting until the response of the animal, although firing during the delay rarely remains as high as it is during the last part of cue presentation. More characteristic, during delay, is a gradual return to low spike frequency and to grouped firing, interrupted at irregular intervals by brief and rapid trains of nongrouped spikes. Inhibition of firing is exceptional in any phase of DR performance, although a transient slowing may be seen in the activity of some units at the start of cue presentation. These observations are based on the study of records from 28 single units in addition to records of spikes obtained simultaneously from two or more units. All these units were within the confines of the MD, the great majority of them in the parvocellularis region of the nucleus.

There is considerable variability in the behavior of different MD units but, on attempting to find common characteristics, it becomes evident that any given unit is most likely to show a higher rate of firing during the last seconds of cue presentation and during the beginning of the ensuing delay. The records shown

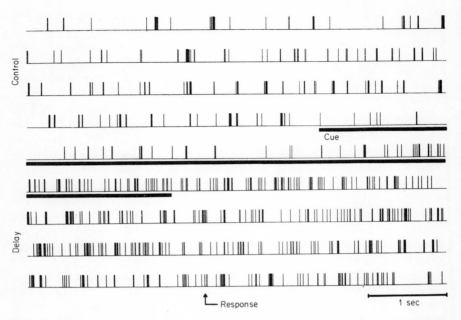

FIG. 1. Spike activity of a cell in the nucleus medialis dorsalis (MD) during a delayed response (DR) trial. From left to right and top to bottom, continuous record of spikes converted into standard marks by computer processing. The first part of the record represents pretrial activity as control. The cue-presentation period, marked by the thick horizontal line, is followed by a 15 sec delay.

in Figures 1 and 3 illustrate two different forms of unit behavior. In one, increased firing appears at the end of the cue-presentation interval, whereas in the other it appears already at the very beginning of this interval.

Some units show a transitory acceleration of discharge at the time of the motor reaction of the animal. The same units may not show any change in firing frequency during other movements, whether spontaneous or provoked. However, some of the most dramatic activations can be observed during the cue or delay periods, while the animal is quiescent. The activation on cue presentation shown, for example, by the unit presented in Figure 2 could not be reproduced by sudden noises or flashes of light, which obviously elicited startle reactions with gross movements of the eyes, head, body, and limbs. Activation invariably appeared, however, at the initiation of the baiting procedure and continued during the first part of the delay. The most subtle movements undoubtedly escape detection by the experimenter observing the animal through a one-way vision screen, but the coordinated examination of unit activity and electrical artifacts of movement generally shows a lack of correlation between movements and temporal gradients of firing except for some units during the delayed motor response.

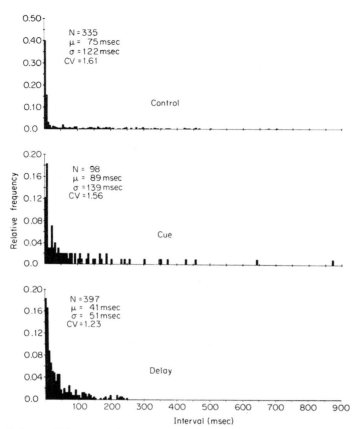

FIG. 2. Interval histograms from the spike record shown in Fig. 1. Grouped firing in the control period is reflected by a relatively high incidence of very short intervals. (For the control histogram, a longer portion of record was used than that of Fig. 1.) N, number of intervals; μ, mean interval length; σ, standard deviation; CV, coefficient of variation.

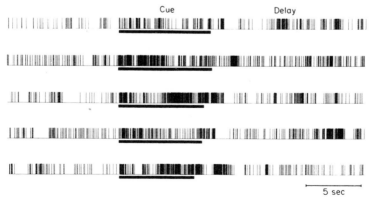

FIG. 3. Firing of a cell in the MD during five DR trials. The record of every trial is preceded by 10 sec of control; only the first half of a 30 sec delay is shown.

The change of firing pattern that thalamic neurons undergo during orienting reactions may have something to do with the prompt establishment of adequate conditions for the processing of sensory information. It may represent a preparatory transition that makes thalamic units more susceptible to fine modulation than they are while their spontaneous activity is grouped and rhythmic (Viernstein and Grossman, 1961). However, the activation that is exhibited, in addition to change of pattern, by MD cells during DR performance appears to be a more specific phenomenon, probably associated with the acquisition of information for short-term storage. The time and course of this activation in relation to the behavioral task support the validity of this hypothesis. Of singular importance is the fact that the period during which we observe the highest incidence of increased firing among MD cells coincides with the period of reception of critical information that will permit the animal to maximize the probability of reward. It is also during this period that extraneous influences can be most disruptive for DR performance. Stamm (1969) has demonstrated that electrical stimuli to the prefrontal cortex interfere markedly with such performance if applied during the last part of the cue-presentation phase and the beginning of the delay phase. Interestingly, this usually seems to be the time of highest cell-firing rates in the thalamic nucleus that is most closely connected with that part of the cortex.

Electrical stimulation of the surface of the cortex around the sulcus principalis induces modifications in the impulse activity of MD units. These modifications are not uniform for all units and vary in accordance with the frequency of the stimulus. Single shocks produce the most uniform effect, namely transitory increases of discharge with a long latency (100–200 msec). These increases may be repetitive, a single shock giving rise to two or more successive firing peaks at regular intervals (Figure 4).

The long latency and the variability of MD responses to stimulation of the surface of the dorsolateral frontal cortex indicate that the connection between the tissue underlying the stimulating electrodes and the MD units is not a direct one and possibly involves several cortical and thalamic synapses. There are perhaps also intervening synapses in the caudate nucleus, which some studies have suggested is a part of the same neural system that includes the MD and the granular frontal cortex (Rosvold, 1968).

The suggestion is found in our data that MD cells participate not only in the acquisition of sensory information but also, to some degree, in the retention of this information in short-term storage. This is the possible significance of rapid trains of action potentials recurring during the early part of the delay in DR performance. Such activity may represent a manifestation, at the cellular level, of the continuance of attention beyond the presentation of the sensory cues, a manifestation of the form of attention that has been characterized as mnemonic or ideational in man and that has been shown to be fundamentally impaired by frontal lobe injury (Robinson, 1946). It is conceivable that attention to mne-

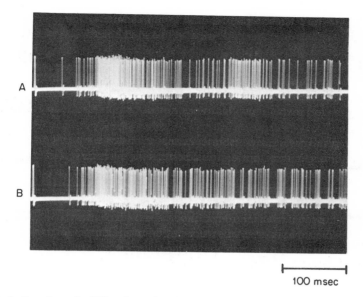

FIG. 4. Reaction of a MD unit to electrical stimuli applied to the surface of the frontal cortex. Oscillographic record of standard marks generated by the spikes of the unit. Stimuli are 3 V pulses of 0.5 msec duration, coinciding with the beginning, at left, of each oscillographic trace. (A) Superimposed traces following ten stimuli delivered at irregular intervals of 2 to 5 sec. (B) Superimposed traces after 10 stimuli at the frequency of 1 per sec.

monic contents for a short term is assured by phasic or tonic influences from the frontal cortex upon cells of the thalamus. Our observations and the anatomical studies point to the nucleus medialis dorsalis as likely recipient of such influences.

In summary, the single-cell phenomena observed in the course of delayed response performance are consistent with the postulate of a role of the nucleus medialis dorsalis in the acquisition and storage of information for short-term use by the organism. More evidence is expected from correlative studies in experimental conditions in which performance is reversibly impaired within the time of recording from one or several cellular units (Fuster and Alexander, 1970).

I. ADDENDUM

After this article was submitted to the editors, the essential findings of this study were confirmed with more data from thalamic units. In addition, the

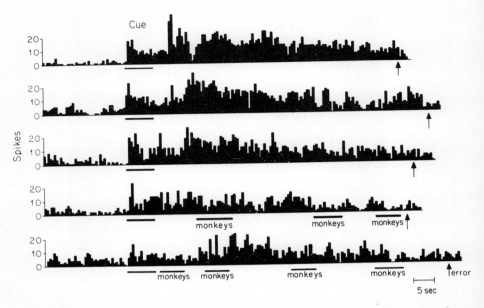

FIG. 5. Frequency histograms of the firing of a prefrontal cortical unit in the course of five trials with delays of approximately 1 min. (Arrow marks end of delay.) During the delay of the last two trials distracting stimuli were repeatedly used (monkeys); they consisted of monkey cries stereophonically recorded in the animal quarters and played back to the experimental animal through overhead loudspeakers. Note the lower degree of activation of the unit during the delay of these two trials, the last one leading to an incorrect response.

activity of units in the prefrontal cortex during performance of the delayed response task has also been investigated (Fuster and Alexander, 1971). The spontaneous activity of such units shows less tendency to grouped discharge than that of thalamic units under the same experimental conditions. Otherwise, during the short-term memory test, the reactions of cortical units show definite similarities to those of MD units, although in both cortex and MD some complex reaction patterns have been encountered that deviate somewhat from those described above. As in the MD, a substantial proportion of prefrontal cortical units exhibit elevations of firing frequency outlasting the cue period and in some instances persisting throughout the delay. One interesting characteristic of such reactions is that they appear related to the accuracy of performance by the animal. In some units normally activated during the delay, the magnitude of this activation has been observed to be lower in trials terminated by incorrect response. Errors of performance can be induced by distracting stimuli presented to the animal during the delay. Such stimuli have been seen to antagonize or attenuate normal unit reactions (Figure 5). These findings lend support to the view of a role by frontothalamic circuits in the attentive process involved in short-term memory.

ACKNOWLEDGMENTS

This work is supported by grant GB-24482X of the National Science Foundation. The author is the recipient of a Research Scientist Award (K5-25,082) from the National Institute of Mental Health. Computing assistance was obtained from the Data Processing Laboratory of the Brain Research Institute and the Health Sciences Computing Facility, UCLA, sponsored respectively by NIH grants NB-2501 and FR-3.

REFERENCES

Akert, K. (1964) Comparative anatomy of frontal cortex and thalamo-frontal connections *In* "The Frontal Granular Cortex and Behavior" (J. M. Warren and K. Akert, eds.), pp. 372-396. McGraw-Hill, New York.

Fuster, J. M. (1961). Excitation and inhibition of neuronal firing in visual cortex by reticular stimulation. *Science* 133, 2011-2012.

Fuster, J. M., and Alexander, G. E. (1970). Delayed response deficit by cryogenic depression of frontal cortex. *Brain Res.* 20, 85-90.

Fuster, J. M., and Alexander, G. E. (1971). Neuron activity related to short-term memory. *Science* 173, 652-654.

Robinson, M. F. (1946). What price lobotomy? *J. Abnorm. Soc. Psychol.* 41, 421-436.

Rosvold, H. E. (1968) The prefrontal cortex and caudate nucleus: A system effecting correction in response mechanisms. *In* "Mind as a Tissue" (C. Rupp, ed.), pp. 2-38. Harper, New York.

Schulman, S. (1964). Impaired delayed response from thalamic lesions. *Arch. Neurol. (Chicago)* 11, 477-499.

Stamm, J. S. (1969). Electrical stimulation of monkeys' prefrontal cortex during delayed-response performance. *J. Comp. Physiol. Psychol.* 67, 535-546.

Viernstein, L. J., and Grossman, R. G. (1961). Neural discharge patterns in the transmission of sensory information. *In* "Information Theory," pp. 252-269. The Universities Press, Belfast.

Walker, A. E. (1966). "The Primate Thalamus," pp. 116-127. Univ. of Chicago Press, Chicago, Illinois.

Chapter 10

THE ROLE OF THE BRAIN STEM IN ORBITAL CORTEX INDUCED INHIBITION OF SOMATIC REFLEXES

EBERHARDT K. SAUERLAND* and CARMINE D. CLEMENTE

Department of Anatomy and the Brain Research Institute
School of Medicine
University of California at Los Angeles
Los Angeles, California

I. INTRODUCTION

The orbital surface of the frontal lobe appears to be unique in its capacity for inhibiting both somatomotor and visceromotor activities. It has been demonstrated that electrical stimulation of this same region can induce states of behavioral inhibition (Brutkowski, 1965) and the onset of sleep (Kaada, 1951; Penaloza-Rojas *et al.*, 1964). Furthermore, electrical stimulation of the orbital-frontal cortex has been shown to inhibit a brain stem monosynaptic reflex (Clemente *et al.*, 1966; Sauerland *et al.*, 1966) as well as spinal reflexes (Kaada, 1951; Sauerland *et al.*, 1967a). Figure 1 summarizes results that in part, have been described elsewhere (Sauerland *et al.*, 1967a). Following a short train of pulses applied to the orbital gyrus or its immediate vicinity, test reflexes were electrically elicited, and as a result, a marked and consistent reflex inhibition of a variety of monosynaptic and polysynaptic reflexes was found at different levels of the neuraxis. This inhibition was diffuse and nonreciprocal when the concurrently recorded EEG was synchronized. Additionally, synchronization of the EEG and diffuse inhibition of spontaneous muscle tone were obtained by electrical stimulation of this same cortical area (Sauerland *et al.*, 1967a, b).

Magoun and Rhines (1946) were able to inhibit phasic and tonic reflex activity by stimulation of the ventromedial bulbar reticular formation. Our

---

*Present address: Department of Anatomy, The University of Texas, Medical Branch, Galveston, Texas.

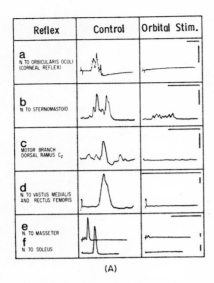
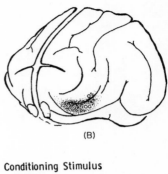
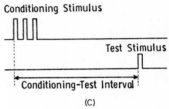

FIG. 1. Cortically induced inhibition of monosynaptic and polysynaptic reflexes at various levels of the neuraxis. (A) Reflexes were recorded either from the proximal portion of the severed motor nerve (a, b, c, e) or from the proximal end of the severed ventral root L_5 (d) or L_7 (f). N stands for nerve. Site of electrical stimulation for reflex elicitation: (a), cornea; (b) and (c), proximal end of cut dorsal root C_3 and C_2, respectively; (d) and (f), proximal end of severed motor nerve; (e), mesencephalic nucleus of the trigeminal nerve. The small first deflection in (e) is an antidromic response recorded along proprioceptive fibers in the masseteric nerve. In all cases the time lines correspond to 10 msec, and the amplitude calibrations to 200 μV. (B) Cortical inhibitory area in the cat, located on the orbital gyrus (OG) and in the vicinity of the orbital sulcus (OS). Reflex inhibition could be induced by cortical stimulation within the limits of the stippled region. (C) Electrical stimulation pattern. Simple control responses were elicited by test stimuli only. For orbital-cortical stimulation, a conditioning stimulus was used that preceded the test stimulus by the conditioning-test interval (approximately 40 msec). [Reprinted from Sauerland, E. K., Nakamura, Y., and Clemente, C. D. (1967). *Brain Res.* 6, 164-180.]

previous experiments, as summarized above, showed that strikingly similar inhibitory phenomena can be initiated from the orbital gyrus. It seemed, therefore, a logical step to investigate whether or not Magoun and Rhines' medullary inhibitory area plays a role in orbital-cortically induced reflex inhibition. The monosynaptic masseteric reflex, originally described by Hugelin and Bonvallet (1957a), offered itself as a particularly useful test reflex. Since its anatomical substrates are located rostral to the lower brain stem inhibitory area, it permits testing of orbital-cortically induced reflex alterations with or without an intact bulbar inhibitory region.

Furthermore, the nature of the orbital-cortically induced inhibitory mechanism is of great interest. Data will be presented here showing that both pre- and postsynaptic inhibitory processes play a role in cortically produced reflex inhibition. Parts of these data were presented elsewhere (Sauerland *et al.*, 1967a, b; Sauerland and Mizuno, 1969).

II. METHODS

Twenty-seven cats were used for the transection studies. Surgical procedures were performed under general anesthesia produced by intravenous injection of short-acting sodium methohexital (Brevital; initial dose 5 mg/kg, maintenance dose 2.5–3 mg/kg whenever necessary). During the entire experiment, wound edges and pressure points were repeatedly infiltrated by a solution of 2% procaine hydrochloride. Following all surgery, the cats were immobilized with gallamine triethiodide and maintained under artificial respiration. Volume and rate of respiration were carefully controlled as described elsewhere (Sauerland *et al.*, 1970). Isolated peripheral nerves and exposed surfaces of the brain and spinal cord were covered with warm (37°C) mineral oil. In most experiments, the cerebellum was removed by suction. Experiments involving lower brain stem transections were performed in *encéphale isolé* preparations with spinal cord transections at the level of C_2 under general sodium methohexital anesthesia.

Two different test reflexes were used, one at the upper and one at the lower portion of the neuraxis, respectively, the brain stem monosynaptic masseteric reflex and the monosynaptic soleus reflex. The masseteric reflex (Hugelin and Bonvallet, 1957a) was elicited by a single shock to the mesencephalic nucleus of the trigeminal nerve through a stereotaxically placed, bipolar strut electrode and was recorded from the proximal portion of the severed nerve to the masseter muscle. The soleus reflex was elicited by stimulation of proprioceptive fibers in the proximal portion of the cut nerve to the soleus muscle and was recorded from the proximal end of the severed ventral root L_7. Two small silver hooks were used as recording electrodes for the reflex discharges. For recording and stimulating purposes in the pontine and bulbar reticular formation, stereotaxically inserted bipolar strut electrodes were used. Their actual location was verified by histological examination of the electrode paths.

The method for reflex elicitation and cortical conditioning stimulation has been described in detail previously (Sauerland *et al.*, 1967a, b), and is illustrated in Figure 1C. Basically, control reflexes were elicited by test stimuli (0.1 msec, 0.4–2 V). For orbital-cortical stimulation, a conditioning stimulus (usually 3 pulses, 100/sec or 500/sec) was used, which preceded the test stimulus by the conditioning-test interval (approximately 35–45 msec). For the purpose of evoked potential studies, a single shock of short duration was delivered to the orbital gyrus.

After every brain stem transection, 50–150 reflex discharges (controls, and those under the influence of orbital-cortical stimulation) were recorded. Reflex amplitudes were measured and averaged. Actual potentials, whose amplitudes conformed with the computed average size, were selected for illustration.

Ten additional cats were utilized to establish the time courses of orbital-cortically evoked changes in amplitude of the masseteric reflex. The time courses were plotted within relatively short periods of time by utilizing a semiautomatic recording technique (Sauerland and Mizuno, 1969). Tektronix RM 122 preamplifiers were used in conjunction with a Tektronix RM 565 dual-beam oscilloscope.

Nomenclature of subcortical structures, brain coordinates, and histology were based on the stereotaxic atlas by Snider and Niemer (1961). The gyri and sulci of the brain surface were named according to the atlas of Winkler and Potter (1914).

III. RESULTS

A. Transection Experiments

The effectiveness of orbital-cortically induced inhibition of the masseteric reflex was found to be dependent on the functional integrity of the medulla

FIG. 2. The effect of lower brain stem transections on cortically induced inhibition of the monosynaptic masseteric reflex. Reflexes were recorded from the proximal portion of the severed right masseteric nerve. Total reflex inhibition was induced by ipsilateral or contralateral orbital-gyral stimulation. Transections of the medulla oblongata 2 mm caudal to the obex (P17) and lower (A, P25) did not diminish the efficiency of cortically induced reflex inhibiton. Higher transections led progressively to ineffectiveness and abolition of the cortically induced inhibitory influence (B–D, P15–P10). Transection at an even more rostral level (E, P8) resulted in facilitation of the masseteric reflex by stimulation of the same orbital-gyral sites. The inhibitory influence was more effective contralaterally than ipsilaterally. Transection of the medulla also resulted in a slight increase of the amplitude of the control reflex. PF = pontine facilitatory area; MI = medullary inhibitory area. Calibrations: Time bar, 10 msec; amplitude bar, 500 μV. [Reprinted from Sauerland, E. K., Nakamura, Y., and Clemente, C. D. (1967). *Brain Res.* **6**, 164-180].

FIG. 3. Effect of lower brain stem transections on cortically induced inhibition of the masseteric reflex; based on seven separate experiments. The level of transection (in millimeters) is plotted versus the computed average reflex amplitude under the conditions of ipsilateral or contralateral orbital-gyral stimulation. The curves show that with successively higher transections through the medullary inhibitory area, the effectiveness of the cortically induced inhibitory influence is progressively lost. At a given level of transection, contralateral orbital-gyral stimulation induces stronger inhibition than ipsilateral stimulation. Transections at the level of P5 result in reflex facilitation by ipsilateral or contralateral orbital-gyral stimulation. The facilitatory effect is considerably stronger on the ipsilateral side. [Reprinted from Sauerland, E. K., Nakamura, Y., and Clemente, C. D. (1967). *Brain Res.* **6**, 164-180.]

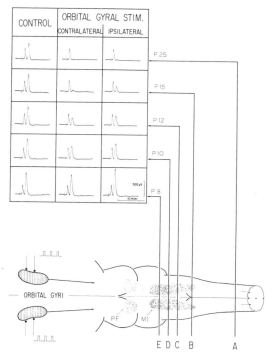

Figure 2

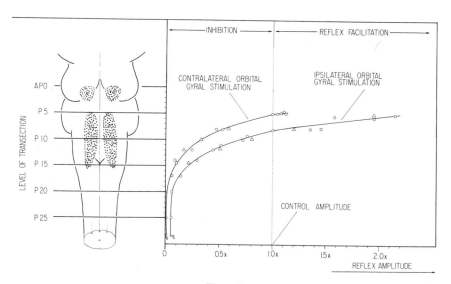

Figure 3

oblongata. If its function was impaired, the cortically induced inhibitory influence was reduced, as observed with remarkable consistency in 14 experiments (Figure 2). Transections 2 mm caudal to the obex and lower (Figure 2A, P25) did not diminish the efficiency of cortically induced reflex inhibition. Higher transections led progressively to ineffectiveness and abolition of the cortically induced inhibitory influence (Figure 2B, C, and D, P15–P10). Transections at an even more rostral level (Figure 2E, P8) resulted in facilitation of the masseteric reflex by stimulation of precisely the same orbital-gyral sites. Similar results were obtained in two experiments in which just the medial part of the medulla (2.5 mm bilateral from the midline) was severed.

A quantitative analysis of these findings, based on seven separate experiments, is shown in Figure 3. The level of transection is plotted versus the computed average reflex amplitude under the conditions of ipsilateral or contralateral orbital-gyral stimulation. The graph demonstrates that, at a given level of transection, contralateral stimulation induces stronger inhibition than ipsilateral stimulation. Transections at the level of P5 result in reflex facilitation by ipsilateral or contralateral orbital-gyral stimulation, with the facilitatory effect considerably greater on the ipsilateral side. Transections of the medulla oblongata also resulted in a slight increase of the amplitude of the control reflex (Figure 2, controls P8 and P10).

FIG. 4. Effects of electrical stimulation of the right and left medullary inhibitory area (MI) and pontine facilitatory area (PF) on the simultaneously elicited left masseteric reflex (upper trace) and left soleus reflex (lower trace). In this particular experiment, bipolar stimulating electrodes in the pontine area were positioned at P1.0, L2.1, and H −4.5 (nucleus reticularis pontis oralis). In the medullary area, the stimulating electrodes were located at P8.5, L1.5, and H −8 (nucleus reticularis gigantocellularis). The reflex potentials shown at A, C, and D were obtained with the same gain; those shown at B were recorded via repositioned electrodes and with slightly changed gain. Ipsilateral or contralateral stimulation of the MI leads to equally effective inhibition of both reflexes, whereas stimulation of the PF results in reflex facilitation. By comparison, however, ipsilateral (left) stimulation of the PF has a stronger facilitatory effect than contralateral (right) stimulation. Calibrations: time, 5 msec; amplitudes, 500 μV. [Reprinted from Sauerland, E. K., Nakamura, Y., and Clemente, C. D. (1967). *Brain Res.* 6, 164-180.]

FIG. 5. Potentials evoked by stimulation (single pulse, 0.03 msec) of the right or left orbital gyrus and recorded with bipolar electrodes from the pontine facilitatory area, PF (A and B), and the medullary inhibitory area, MI (C and D). Same preparation and same electrode positions as in Fig. 4. Numerous responses were superimposed at a rate of 100/sec. Short latency (0.4–0.5 msec) direct responses were obtained throughout the PF and the MI. Ipsilateral and contralateral responses in the MI were of approximately equal size, whereas in the PF, contralateral responses were consistently smaller (40–65% less) than ipsilaterally evoked potentials. The small arrows indicate the beginning of the square-wave stimulus to the orbital gyrus. [Reprinted from Sauerland, E. K., Nakamura, Y., and Clemente, C. D. (1967). *Brain Res.* 6, 164-180.]

ORBITAL CORTEX INDUCED INHIBITION

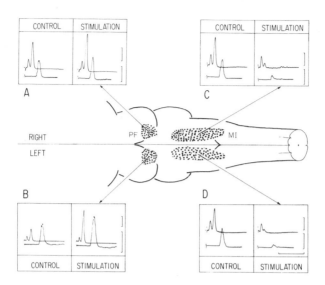

Figure 4

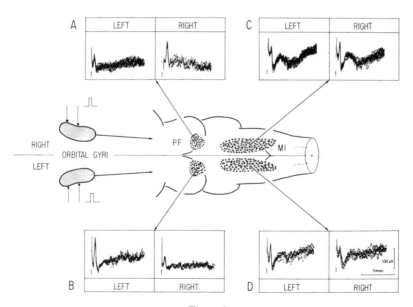

Figure 5

B. Stimulation of Brain Stem Reticular Formation and Evoked Potential Studies

In order to investigate the possible correlation between orbital-cortically induced reflex inhibition on the one hand, and the reflex inhibition induced by stimulating the ventromedial part of the bulbar reticular formation (Magoun and Rhines, 1946) on the other, the following experimental procedures were carried out:

Step I, bipolar electrodes were inserted into Magoun and Rhines' (1946) ventromedial reticular inhibitory region and surrounding areas. Points whose stimulation inhibited simultaneously the masseteric reflex at the upper and the soleus reflex at the lower portion of the neuraxis were mapped. Their total distribution is collectively referred to here as the medullary inhibitory area (MI), which extends from approximately P5 to P16 and includes the following structures: the nucleus reticularis gigantocellularis, the rostral part of nucleus reticularis ventralis, and the medial portion of nucleus reticularis lateralis. Strongest inhibition was obtained by stimulating the nucleus reticularis gigantocellularis and its immediate vicinity. Figure 4 (C and D) demonstrates simultaneous inhibition of the left masseteric reflex (upper trace) and the left soleus reflex (lower trace) by ipsilateral (left) or contralateral (right) stimulation of the nucleus reticularis gigantocellularis.

Step II, the single stimulus was applied to the rostral portion of the orbital gyrus (Figure 1B) and evoked activity was recorded from precisely the same points in the MI that previously (Step I) were found to be effective in reflex inhibition. Short latency (about 0.5 msec) responses were obtained throughout the MI; the largest responses were recorded from the nucleus reticularis gigantocellularis and its immediate vicinity. Figure 5 (C and D) shows typical medullary responses superimposed at a rate of 100/sec. The short latency and the responsiveness to high-frequency stimulation indicated that this evoked potential was directly mediated without synaptic relay. The conduction speed of the responsible fibers was calculated as 60–70 m/sec. Ipsilateral and contralateral responses were of approximately equal size. In addition to the direct response, an evoked potential of longer latency (about 4 msec), longer duration (25–30 msec), and higher amplitude (five to eight times the direct response) could be recorded from the MI. This response could be easily depressed by repetitive stimulation (10/sec), and was considerably diminished at a frequency of 30/sec.

In a similar experimental series, orbital-cortically induced reflex facilitation, which was observed after higher medullary transections (Figures 2 and 3), was investigated at the brain stem level. In Step I, bipolar electrodes were inserted into pontine reticular areas known to be involved in reflex facilitatory mechanisms (Rhines and Magoun, 1946). More rostral areas of the facilitatory reticular formation were not investigated. Points where stimulation facilitated simultaneously the masseteric and the soleus reflex were mapped. Their distribution, called collectively here the pontine facilitatory area (PF), extends from approximately P0.5 to P3.5 and includes the nucleus reticularis pontis oralis and the rostral portion of the nucleus reticularis pontis caudalis. Figure 4 (A and B) demonstrates simultaneous reflex facilitation of the left masseteric reflex (upper trace) and the left soleus reflex (lower trace) by ipsilateral (left) or contralateral (right) stimulation of the nucleus reticularis pontis oralis, with the facilitatory effect being stronger on the ipsilateral side. After discontinuation of pontine reticular stimulation, we occasionally observed for several seconds a depression of reflex potentials below the control level. In Step II, the orbital gyrus was stimulated by a single pulse, and potentials were recorded from precisely the same points in the PF that previously (Step 1) were found to be effective in reflex facilitation. Short latency (about 0.4 msec) direct responses were obtained throughout the PF, with maximal responses in the nucleus reticularis pontis oralis. A slow-wave potential could also be recorded from the PF; its amplitude, however, was very small. Figure 5 (A and B) shows typical direct pontine responses that were superimposed at a rate of 100/sec. Evoked responses elicited by contralateral orbital-gyral stimulation were consistently smaller (40–65% less) than ipsilateral evoked potentials.

C. Cortically Induced Changes of the Masseteric Reflex Amplitude

Figure 6A shows the time course of amplitude changes of the masseteric reflex induced by conditioning stimulation in the vicinity of the orbital sulcus. There are two phases of reflex suppression. The first has its peak at a conditioning-test interval of 10 msec. At this point of the relatively brief first phase, the masseteric reflex inhibition is complete. Following a brief facilitatory phase, there is a second suppressive phase with a peak at a conditioning-test interval of approximately 40 msec. This second suppressive phase is substantially weaker (i.e., there is less reflex inhibition) than the first one, but it is very prolonged. The two phases of reflex depression could be differentiated with the aid of pharmacological agents. Figure 6B illustrates the effect of picrotoxin on cortically induced amplitude changes of the masseteric reflex. Three minutes after slow intravenous injection of this drug, the second inhibitory phase disappeared completely. After 7 min, the facilitatory phase was remarkably prolonged and lasted for several hundred milliseconds. On the other hand, picrotoxin had no obvious effect on the first phase of the masseteric reflex suppression. In three experiments, the effect of strychnine (0.1 mg/kg) on cortically induced masseteric reflex inhibition was investigated. Strychnine had no appreciable effect on the second phase of reflex inhibition, whereas the first phase was clearly reduced or completely abolished by the pharmacological agent.

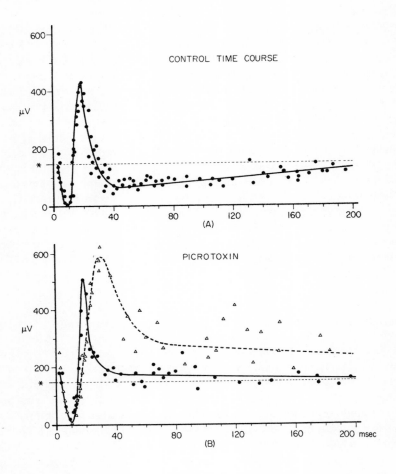

FIG. 6. Time courses of suppression and facilitation of the masseteric reflex induced by a cortical conditioning stimulus (3 pulses, 500/sec, 0.3 msec, 7 V) to the vicinity of the contralateral orbital sulcus. The asterisk (*) indicates the mean level of control amplitudes based on 33 control responses. Abscissae: conditioning-test interval in milliseconds. Ordinates: amplitudes of the cortically conditioned masseteric reflex. (A) Typical time course obtained immediately prior to drug application. Two phases of reflex suppression can be observed: the first one, with a peak about 10 msec, is complete and relatively brief; the second phase is less pronounced but very prolonged and shows its peak at about 40 msec. (B) Time courses 3 min (●) and 7 min (△) after slow intravenous injection of picrotoxin, 0.44 mg/kg. There is no change of the first phase of reflex suppression, whereas the second prolonged inhibitory phase has been completely eliminated. [Reprinted from Sauerland, E. K., and Mizuno, N. (1969). *Brain Res.* **13**, 556-558.]

IV. DISCUSSION

The results of our transection experiments and of our studies involving stimulation and recordings in the brain stem implicate medullary and pontine reticular regions in the neurophysiological mechanisms of orbital-cortically induced reflex inhibition and facilitation. Figure 7 illustrates schematically the conclusions drawn from our findings. These conclusions are presented and discussed in the following.

(1) Short latency, direct potentials, which could not be depressed at high stimulation frequencies, were recorded from the pontine facilitatory area (PF) and the medullary inhibitory area (MI) in response to electrical single shock stimulation to the orbital gyrus. Therefore, neurons located in this cortical area send axons directly and without synaptic relay to the PF and the MI.

(2) Direct evoked potentials of contralateral and ipsilateral orbital-gyral origin were recorded in the MI, and their amplitudes were found to be approximately equal in size. Therefore, it may be assumed that direct orbitomedullary fibers are distributed in equal quantities to the ipsilateral and contralateral inhibitory area, MI.

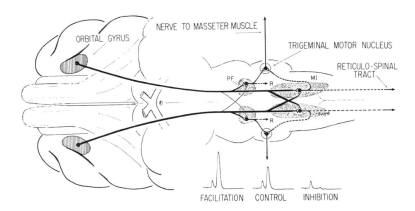

FIG. 7. Schematic diagram showing proposed direct orbitopontine and orbitomedullary connections to neurons of the reticulospinal tract. MI = medullary inhibitory area; PF = pontine facilitatory area; R = reticulospinal tract. The MI receives equally strong bilateral connections from the orbital gyri. The PF receives more connections from the ipsilateral than from the contralateral orbital gyrus. Collaterals of the reticulospinal tract, or axons of other reticular cells within the PF and MI, mediate excitatory and inhibitory influences to the trigeminal motor nuclei. For reasons of simplicity, the known bilateral distribution of the reticulospinal tract from either medullary area (Torvik and Brodal, 1957) has been omitted. [Reprinted from Sauerland, E. K., Nakamura, Y., and Clemente, C. D. (1967). *Brain Res.* 6, 164-180.]

(*3*) Direct evoked potentials of contralateral and ipsilateral orbital-gyral origin were recorded in the pontine facilitatory area, PF, but the size of the contralaterally evoked responses was 40–65% less than that of the ipsilaterally evoked potentials. Therefore, it may be assumed that direct orbitopontine fibers are bilaterally distributed, but that the majority of those fibers descend to the ipsilateral PF. This assumption is further supported by the fact that after impairment of the MI (transections), orbital-cortically induced facilitation affects significantly more the ipsilateral than the contralateral masseteric reflex (Figures 2 and 3).

Following our electrophysiological studies, experiments were carried out in an attempt to identify the anatomical substrates subserving orbital-gyral influences on somatic reflexes. Mizuno *et al.* (1968) investigated direct projections from the orbital gyrus and its vicinity to the brain stem by means of the Nauta and Marchi methods. The direct orbitopontine and orbitomedullary pathways proposed on the basis of our electrophysiological studies (Figure 7) could be firmly substantiated by these neuroanatomical findings (Mizuno *et al.,* 1968). Substantial preterminal fiber degeneration was found bilaterally in various regions of the brain stem, including the oral pontine (PF) and gigantocellular medullary (MI) reticular nuclei. There appeared to be a somewhat higher degree of degeneration on the side of the lesion in the pontine reticular nuclei and on the contralateral side in the gigantocellular nuclei. Our conclusions (Figure 7; points *1–3*) are also supported by the anatomical work of Rossi and Brodal (1956), who found that corticoreticular fibers to the nucleus pontis oralis and nucleus gigantocellularis were distributed in approximately equal numbers, and by the work of Kuypers (1958), who demonstrated diffuse bilateral projections to the central tegmental area of the medulla oblongata and also to the pontine tegmental region. The latter area seemed to receive more fibers from the ipsilateral frontal cortex. In addition, neurophysiological findings by Newman and Wolstencroft (1959) point to direct connections between the orbital gyrus (including its caudal portion) and the medulla oblongata, with maximal evoked responses in the medial part of the nucleus reticularis pontis caudalis and the medial and rostral portion of the nucleus reticularis gigantocellularis.

The fourth and fifth conclusions drawn from our findings (Figure 7) are as follows:

(*4*) Orbital-cortically induced direct potentials were recorded in the MI from precisely the same points that were effective in reflex inhibition. These points were concentrated in an area known to be the site of origin for the medullary portion of the reticulospinal tract (Torvik and Brodal, 1957). Therefore, it is assumed that orbitomedullary fibers connect directly or indirectly with neurons of the descending reticulospinal tract, and that axon collaterals of these neurons or axons of other reticular inhibitory cells within the MI ascend and mediate inhibitory influences to the trigeminal motor nuclei.

(5) Orbital-cortically induced direct potentials were recorded in the PF from precisely the same points that were effective in reflex facilitation. These points were concentrated in an area known to be the site of origin of the pontine portion of the reticulospinal tract (Torvik and Brodal, 1957). Therefore, it is assumed that orbitopontine fibers connect directly or indirectly with neurons of the reticulospinal tract, and that axon collaterals of these neurons or axons of other excitatory cells within the PF mediate facilitatory influences to the trigeminal motor nuclei.

Our assumptions (points 4 and 5) are based on the work by Torvik and Brodal (1957). On the basis of retrograde degeneration studies, these authors have distinguished two spatially separated sites of origin for the reticulospinal tract in the lower brain stem: the pontine portion (caudal part of nucleus pontis oralis, rostral part of nucleus pontis caudalis) and the medullary portion (entire nucleus gigantocellularis, medial part of nucleus reticularis lateralis, rostral part of nucleus reticularis ventralis). Brodal (1957) and Torvik and Brodal (1957) emphasized that there is a good correspondence between the bulbar inhibitory area described by Magoun and Rhines (1946) and the site of origin of reticulospinal fibers in the medulla. However, they pointed out the approximate congruity of the pontine portion of the facilitatory region of Rhines and Magoun (1946) with the site of origin of the pontine portion of the reticulospinal tract. The direct pathway from the orbital gyrus to the lower brain stem, as described above, projects precisely to the same regions; it is, therefore, reasonable to assume that descending orbital-cortical influences reach reticulospinal neurons. It is quite likely, however, that other reticular cell elements are also affected. Torvik and Brodal's (1957) degeneration experiments show that the fibers originating from the pontine portion of the reticulospinal tract descend ipsilaterally to the spinal cord, whereas medullary reticulospinal fibers are bilaterally represented. Our own studies (Figure 4) show that reflex facilitation could be obtained from both pontine areas (PF), although the facilitatory influence was stronger on the ipsilateral side. Earlier, Niemer and Magoun (1947) described that unilateral brain stem stimulation was effective in facilitating bilateral patellar reflexes. Rossi and Zanchetti (1957) suggested that this discrepancy between anatomical and physiological findings can be easily reconciled if one assumes the existence of short axon collaterals impinging upon contralateral reticulospinal cells.

Inhibitory influences from the MI (Figure 7) were thought to be mediated to the trigeminal motor nuclei, either by ascending axon collaterals of the reticulospinal tract or by axons of other inhibitory cells within the MI. That large cells of the medullary reticular formation send axons both rostrally and caudally has been demonstrated anatomically by the Scheibels (1958). These authors also observed collaterals of reticular axons to all cranial nerve nuclei; and on occasion a single axon supplied an appreciable number of these. Similarly, facilitatory influences from the PF are thought to be mediated to the trigeminal and other motor nuclei via axon collaterals of reticulospinal cells or by axons of other excitatory cells located within the pontine facilitatory area. That excitation of this pontine region, and of the nucleus reticularis pontis oralis in particular, has a profound effect on motor behavior has been demonstrated by Sterman and Fairchild (1966).

The sixth and last conclusion drawn from our experimental findings is the following:

(6) Our transection experiments (Figures 2 and 3) demonstrate that large portions of the medullary inhibitory area could be removed before orbital-cortically induced reflex inhibition was abolished. From this finding it could be inferred that descending orbital-gyral influences to the medullary inhibitory regions prevail over those to the pontine facilatatory areas.

It is important to realize that the orbital gyrus is predominantly but not strictly an inhibitory area. Under certain circumstances it possesses the capacity to enhance facilitatory mechanisms, or perhaps, to inhibit inhibitory processes. Sauerland et al. (1967a) described how under the conditions of a desynchronized EEG, orbital-gyral stimulation induced inhibition of the soleus reflex but facilitated the reflex to the antagonistic tibialis anterior. However, with synchronized EEG, diffuse inhibition of both reflexes was obtained. Thus, the problem of reciprocal inhibition, first raised by Sprague and Chambers (1954) and later by King and her associates (1955), seems to be extended into a predominantly inhibitory system, the rostral end of which is formed by the orbital gyrus and its immediate vicinity. This cortical area is a unique region that is marked by two important characteristics: First, it is here that inputs from bilateral somatic, auditory, and visual sources converge (Korn et al., 1966). Second, as stressed before, the orbital gyrus appears to be the rostral end of a powerful descending inhibitory and synchronizing system (Clemente et al., 1966; Clemente and Sterman, 1967; Sauerland et al., 1967a). We pointed out earlier (Sauerland et al., 1967a) that this apparent duality can be reconciled if one thinks in terms of recurrent loops serving predominantly inhibitory mechanisms. It seems reasonable that afferent information from all parts of the body would be a prerequisite for coordinating functions of a cortical area that exerts its complex regulatory influences on many levels of the neuraxis. Now, in this communication, the functional link between this neocortical region and the phylogenetically older lower brain stem inhibitory areas has been demonstrated. Hugelin and Bonvallet (1957b) have also postulated a neocortical inhibitory influence on the masseteric reflex, but they denied a participation of Magoun and Rhines' bulbar inhibitory region in this mechanism.

A more detailed analysis of orbital-cortically induced inhibition of somatic reflexes should focus on the nature of the effective inhibitory process. Nakamura et al. (1967) showed that postsynaptic inhibition plays a major role in cortically induced suppression of the masseteric reflex. These investigators (Nakamura et al., 1967) demonstrated that electrical stimulation of the orbital gyrus produced a hyperpolarizing potential in masseteric motoneurons, and that the time course of this IPSP resembled closely the time course of cortically induced masseteric reflex inhibition. Subsequently, Sauerland and Mizuno (1969) examined the possible existence of an additional presynaptic mechanism affecting the masseteric reflex. Figure 6A illustrates that the time course of cortically induced inhibition of the masseteric reflex consists of two distinctly separate inhibitory phases that can also be differentiated by pharmacological means (Figure 6B). The first inhibitory phase is identical with the one described by Nakamura et al. (1967) and, on the basis of intracellular studies (Nakamura et al., 1967), can be ascribed to a postsynaptic inhibitory process. The second inhibitory phase can be attributed to a presynaptic inhibitory mechanism affecting trigeminal proprioceptive afferents. Two reasonable arguments may account for this statement: (a) The time course of the second phase of masseteric reflex inhibition (Figure 6A) is identical with the time course of cortically induced changes in presynaptic excitability of masseteric proprioceptive afferents (Sauerland and Mizuno, 1969), and (b) the second phase of masseteric reflex inhibition is effectively abolished by picrotoxin. That picrotoxin possesses the capability of blocking presynaptic inhibition is a well-known fact (Schmidt, 1964; Rudomin, 1966). However, strychnine, which is known to block postsynaptic inhibitory processes, had no effect on the second phase of cortically induced masseteric reflex inhibition. Thus, at least as far as the monosynaptic masseteric reflex of the brain stem is concerned, orbital-cortically evoked inhibitory influences are manifested at the postsynaptic as well as at the presynaptic level. It now appears certain that the orbital cortex is capable of attenuating many neural activities (Sauerland and Mizuno, 1969; Sauerland et al., 1970, 1972) by utilizing presynaptic inhibitory mechanisms.

V. SUMMARY

The role of the brain stem in orbital-cortically evoked changes of somatic reflexes was investigated in the cat. The orbital gyrus and its immediate vicinity constitute the rostral cortical end of a powerful synchronizing and inhibitory system, which was shown to project directly to the ventromedial bulbar reticular formation (chiefly to the nucleus reticularis gigantocellularis) and to the pontine tegmentum (mainly to the nucleus reticularis pontis oralis). As a result of a single-pulse application to the orbital gyrus, short latency direct responses were obtained in the ipsilateral and contralateral pontine and medullary reticular regions. The amplitude of these responses was approximately equal in the medullary inhibitory area, whereas, in the pontine facilitatory region, the size of contralaterally evoked responses was 40–65% less than that of the ipsilaterally evoked potentials.

Orbital-cortically induced inhibition of the monosynaptic masseteric reflex depends upon the integrity of the ventromedial bulbar reticular formation. Transections of the medulla oblongata 2 mm caudal to the obex (P17) and lower had no effect on the efficiency of cortically induced reflex inhibition. More rostral transection (P15–P10) led to ineffectiveness and abolition of the cortically induced inhibitory influence. Moreover, transections at even more rostral levels (P8–P5) caused facilitation of the masseteric reflex following orbital-cortical stimulation. It is assumed that the descending orbitomedullary and orbitopontine fibers synapse with neurons of the reticulospinal tract, and that axon collaterals of these neurons or axons of other reticular cells in the medullary or pontine reticular formation mediate inhibitory or excitatory influences, respectively, to the trigeminal and possibly also to other cranial nerve motor nuclei. A simple schematic drawing (Figure 7) illustrates the role of the brain stem in orbital-cortically induced reflex alteration.

An analysis, utilizing electrophysiological and pharmacological methods, showed that the nature of orbital-cortically induced suppression of the monosynaptic masseteric reflex involves both pre- and postsynaptic inhibitory mechanisms.

ACKNOWLEDGMENTS

This work was supported by US Public Health Service Grants NB 06819 and MH 10083. The authors wish to acknowledge the contributions made by Dr. Yoshio Nakamura and Dr. Noboru Mizuno.

REFERENCES

Brodal, A. (1957). "The Reticular Formation of the Brain Stem. Anatomical Aspects and Functional Correlations." Oliver & Boyd, Edinburgh.

Brutkowski, S. (1965). Functions of prefrontal cortex in animals. *Physiol. Rev.* **45**, 721-746.

Clemente, C. D., and Sterman, M. B. (1967). Basal forebrain mechanisms for internal inhibition and sleep. *Res. Publ., Ass. Res. Nerv. Ment. Dis.* **45** 127-147.

Clemente, C. D., Chase, M. H., Knauss, T. K., Sauerland, E. K., and Sterman, M. B. (1966). Inhibition of a monosynaptic reflex by electrical stimulation of the basal forebrain or the orbital gyrus in the cat. *Experientia* **22**, 844-845.

Hugelin, A., and Bonvallet, M. (1957a). Etude oscillographique d'un réflexe monosynaptique crânien (réflexe masséterin). *J. Physiol. (Paris)* **49**, 210-211.
Hugelin, A., and Bonvallet, M. (1957b). Tonus cortical et contrôle de la facilitation motrice d'origine réticulaire. *J. Physiol. (Paris)* **49**, 1171-1200.
Kaada, B. R. (1951). Somato-motor, autonomic and electrocorticographic responses to electrical stimulation of 'rhinencephalic' and other structures in primates, cat and dog. *Acta Physiol Scand.* **24**, Suppl. 83, 1-285.
King, E. E., Minz, B., and Unna, K. R. (1955). The effect of brain stem reticular formation on the linguomandibular reflex. *J. Comp. Neurol.* **102**, 565-596.
Korn, H., Wendt, R., and Albe-Fessard, D., (1966). Somatic projections to the orbital cortex of the cat. *Electroencephalogr. Clin. Neurophysiol.* **21**, 209-226.
Kuypers, H. G. J. M. (1958). An anatomical analysis of cortico-bulbar connexions to the pons and lower brain stem in the cat. *J. Anat.* **92**, 198-218.
Magoun, H. W., and Rhines, R. (1946). An inhibitory mechanism in the bulbar reticular formation. *J. Neurophysiol.* **9**, 165-171.
Mizuno, N., Sauerland, E. K. and Clemente, C. D. (1968). Projections from the orbital gyrus in the cat. I. To brain stem structures. *J. Comp. Neurol.* **133**, 463-476.
Nakamura, Y., Goldberg, L. J., and Clemente, C. D. (1967). Nature of suppression of the masseteric monosynaptic reflex induced by stimulation of the orbital gyrus of the cat. *Brain Res.* **6**, 184-198.
Newman, P. P., and Wolstencroft, J. H. (1959). Medullary responses to stimulation of orbital cortex. *J. Neurophysiol.* **22**, 516-523.
Niemer, W. T., and Magoun, H. W. (1947). Reticulo-spinal tracts influencing motor activity. *J. Comp. Neurol.* **87**, 367-379.
Penaloza-Rojas, J. H., Elterman, M., and Olmos, N. (1964). Sleep induced by cortical stimulation. *Exp. Neurol.* **10**, 140-147.
Rhines, R., and Magoun, H. W. (1946). Brain stem facilitation of cortical motor response. *J. Neurophysiol.* **9**, 219-229.
Rossi, G. F., and Brodal, A. (1956). Corticofugal fibres to the brain-stem reticular formation. An experimental study in the cat. *J. Anat.* **90**, 42-62.
Rossi, G. F., and Zanchetti, A. (1957). The brain stem reticular formation. *Arch. Ital. Biol.* **95**, 199-435.
Rudomin, P. (1966). Pharmacological evidence for the existence of interneurons mediating primary afferent depolarization in the solitary tract nucleus of the cat. *Brain Res.* **2**, 181-183.
Sauerland, E. K., and Mizuno, N. (1969). Cortically induced presynaptic inhibition of trigeminal proprioceptive afferents. *Brain Res.* **13**, 556-568.
Sauerland, E. K., Knauss, T., and Clemente, C. D. (1966). Inhibition of the monosynaptic masseteric reflex by electrical stimulation of forebrain sites in the cat. *Anat. Rec.* **154**, 415-416.
Sauerland, E. K., Knauss, T., Nakamura, Y., and Clemente, C. D. (1967a). Inhibition of monosynaptic and polysynaptic reflexes and muscle tone by electrical stimulation of the cerebral cortex. *Exp. Neurol.* **17**, 159-171.
Sauerland, E. K., Nakamura, Y., and Clemente, C. D. (1967b). The role of the lower brain stem in cortically induced inhibition of somatic reflexes in the cat. *Brain Res.* **6**, 164-180.
Sauerland, E. K., Mizuno, N., and Harper, R. M. (1970). Presynaptic depolarization of trigeminal cutaneous afferent fibers induced by ethanol. *Exp. Neurol.* **27**, 476-489.
Sauerland, E. K., Velluti, R. A., and Harper, R. M. (1972). Cortically induced changes of presynaptic excitability in higher-order auditory afferents. *Exp. Neurol.* **36**, 79-87.

Scheibel, M. E., and Scheibel, A. B. (1958). Structural substrates for integrative patterns in the brain stem reticular core. *In* "Reticular Formation of the Brain" (H. H. Jasper *et al.*,) (eds.) pp. 31-55. Little, Brown, Boston, Massachusetts.

Schmidt, R. F. (1964). The pharmacology of presynaptic inhibition. *Progr. Brain Res.* **12**, 119-131.

Snider, R. S., and Niemer, W. T. (1961). "A Stereotaxic Atlas of the Cat Brain." Univ. of Chicago Press, Chicago, Illinois.

Sprague, J. M, and Chambers, W. W. (1954). Control of posture by reticular formation and cerebellum in the intact, anesthetized and unanesthetized and in the decerebrated cat. *Amer. J. Physiol.* **176**, 52-64.

Sterman, M. B., and Fairchild, M. D. (1966). Modification of locomotor performance by reticular formation and basal forebrain stimulation in the cat: Evidence for reciprocal systems. *Brain Res.* **2**, 205-217.

Torvik, A., and Brodal, A. (1957). The origin of reticulospinal fibers in the cat. *Anat. Rec.* **128**, 113-135.

Winkler, C., and Potter, A. (1914). "An Anatomical Guide to Experimental Researches on the Cat's Brain." W. Versluys, Amsterdam.

Chapter 11

THE NONSPECIFIC MEDIOTHALAMIC-FRONTOCORTICAL SYSTEM: ITS INFLUENCE ON ELECTROCORTICAL ACTIVITY AND BEHAVIOR*

JAMES E. SKINNER
Physiology Department
Baylor College of Medicine
and
Neurophysiology Department
The Methodist Hospital
Houston, Texas

DONALD B. LINDSLEY
Departments of Psychology and Physiology and
the Brain Research Institute
University of California at Los Angeles
Los Angeles, California

I. INTRODUCTION

The nonspecific systems of the brain are named in contradistinction to the specific sensory ones, which derive their names from Johannes Muller's law of specific nerve energies (Müller, 1838). This law states that each specific sensory system is organized in such a way that it underlies conscious awareness in only a single modality. For example, stimulation of the retina or any other part of the specific visual system by a light flash, electrical shock, or pressure will give rise to conscious awareness in only the visual modality because of the specific organization (specific irritability) of the components of this particular sensory system. The nonspecific brain systems, in contrast, are activated by sensory stimuli in any modality because their components are organized in such a way that they have nonspecific irritabilities. For example, one can be aroused by, reflexively orient toward, or voluntarily draw his attention to any type of sensory stimulus.

The term *unspecific* was used by Lorente de No and others during the late 1940's to designate a type of electrophysiological response that was transmitted to the cortex over unspecific pathways whose origins were unknown (i.e., the

---

*This work was supported, in part, by Grant HE05435 from the Heart Institute, National Institutes of Health, U. S. Public Health Service.

response was not transmitted over the specific projection pathways via the thalamic relay nuclei). The term *diffuse* was coined and used during this same period by several investigators working on the mesencephalic reticular formation (Jasper, Moruzzi, Magoun, and Lindsley) to refer to a particular type of electrocortical response that had a global or widespread cortical distribution (unlike the primary or specific sensory evoked responses). Quite often the terms nonspecific, unspecific, and diffuse have been used interchangeably, but recent observations, discussed below, suggest that perhaps the electrocortical responses are neither unspecific in their projections nor diffuse in their distributions. Thus the term *nonspecific* seems to be more generally suitable to refer to these electrocortical responses since it is not limited by reference to a particular anatomical pathway, but rather is based upon certain qualities of the evoking stimulus.

It is well established that a reticulothalamocortical system plays a significant role in arousal, alerting, and attention (Jasper and Ajmone Marsan, 1952; Lindsley, 1960; Magoun, 1963). High-frequency stimulation of the mesencephalic reticular formation or of the midline nonspecific thalamic nuclei will arouse a sleeping animal or alert a waking animal and cause desynchronization or activation of on-going electrocortical activity (Monnier *et al.*, 1960; Moruzzi and Magoun, 1949). In contrast, low-frequency stimulation of midline thalamic nuclei produces inattention, drowsiness, and sleep and is associated with slow waves and spindle bursts in the electroencephalogram (EEG) (Akert *et al.*, 1952; Akimoto *et al.*, 1956; Hess, 1944, 1954; Monnier *et al.*, 1960). Such synchronized EEG activity in humans is initially accompanied by loss of awareness of and inattention to specific sensory stimuli. Repetitive stimulation of midline nonspecific thalamic nuclei at a rate in the frequency range of spontaneous synchronized activity gives rise to recruiting responses, comprised of incrementing waves of synchronized activity (Jasper, 1949; Morison and Dempsey, 1942; Starzl and Magoun, 1951). Lesion of the mesencephalic reticular formation causes bursts of synchronized activity simultaneously in the thalamus and the cortex (Lindsley *et al.*, 1949), commonly referred to as spindle bursts and characteristically observed during certain stages of sleep. In fact, sleep, except for its more recently described paradoxical phases, is characterized by progressively more and slower synchronized activity. The behavioral states of inattention, drowsiness, and sleep, which are associated with EEG-synchronizing activity mediated by the nonspecific thalamocortical system, have come to be linked with Pavlovian internal inhibition. Such a mechanism appears to be capable of suppressing generally or differentially various functions of the brain. Magoun (1963), in identifying such a thalamocortical system for internal inhibition, states that: "The consequences of the action of this mechanism are the opposite of those of the ascending reticular activating system for internal excitation." In some respects this statement is challenged by the work of

Monnier and associates (1960; Tissot and Monnier, 1959), who have demonstrated that antagonistic functions of activation (desynchronization) and synchronization of electrocortical activity can be elicited by high- or low-frequency stimulation of midline thalamic nuclei, respectively. However, Schlag and Chaillet (1963) have suggested that the desynchronizing effects of medial thalamic stimulation are mediated over thalamoreticular projections sensitive only to the high-frequency stimuli.

One major difference between the mesencephalic reticular and medial thalamic systems is clearly presented when one compares the effects of stimulation and lesion in the two systems. Stimulation of the mesencephalic reticular formation produces cortical activation and behavioral attention, and, conversely, lesions in this region produce cortical deactivation and behavioral inattention (i.e., stimulation and lesion in this region produce complementary effects on both the cortical activity and behavior). On the other hand, stimulation of the medial thalamus produces cortical deactivation and behavioral inattention (Buser *et al.*, 1964), and, conversely, lesions in this structure produce cortical activation (Velasco *et al.*, 1973) but, in marked contrast in the comparison, these lesions seem to produce the same type of behavioral inattention (Skinner and Lindsley, 1967) as that produced by stimulation (i.e., stimulation and lesion in this region do not produce the complementary effects on both cortical activity *and* behavior). Thus, the consequences of action of the medial thalamic system are opposite to those of the mesencephalic reticular system in their effects on cortical activity, but their effects on behavior are not the opposite, a finding that suggests that the types of control of cortical synchronization and activation by the two systems underlie different functions.

Whereas the exact source and destination of the nonspecific thalamocortical system are not known, it is clear that in rats, cats, and monkeys the dorsomedial nucleus and other portions of the midline thalamic nuclear group project via the inferior thalamic peduncle (Manghi *et al.*, 1965; Scheibel and Scheibel, 1967) to the lateroventral and orbital surface of the frontal lobe (Akert, 1964; Auer, 1956; Nauta, 1964; Nauta and Whitlock, 1954). The cortical projections of the nonspecific thalamocortical system appear to be partially congruent with the granular cortex of the prefrontal areas discussed by Akert (1964) and Nauta (1964). The inferior thalamic peduncle seems to be a bidirectional pathway connecting the medial thalamus and the frontal granular cortex (Nauta, 1962, 1964; Scheibel and Scheibel 1967). As we shall see, there are some common features of the behavioral changes resulting from blockade of the inferior thalamic peduncle and those resulting from prefrontal granular cortex ablations. Such commonality supports the view that this nonspecific system in the foreward part of the brain is essentially a thalamocortical one. The frontal granular cortex, however, also appears to have important efferent connections with the claustrum and temporal lobe as well as with the subthalamus, hypothalamus, and brain stem (Nauta, 1962).

The results described in this paper are largely a review of our work on the nonspecific thalamocortical system over the last few years, though some additional results and interpretations are presented. The technique used to investigate this system is primarily that of functional blockade. Ablations, lesions, and reversible cryogenic blockade have been used to block the nonspecific thalamocortical system in the medial thalamus, inferior thalamic peduncle, and frontal granular cortex in the cat. The effects of such blockade have been observed on the synchronous electrical activity of the animal's brain, sensory evoked potentials, slow potentials in the frontal regions, and the behavioral performances of certain complex tasks. The details of the techniques used have been adequately described (Skinner and Lindsley, 1967, 1968, 1971) and summarized in detail elsewhere (Skinner, 1971b), and they will not be described here. The experiments included both acute gallamine-immobilized preparations and freely moving unanesthetized chronic animals. The most common subject in the experiments was a chronic cat preparation that had permanently implanted cryogenic devices that could be cooled in order to block locally the various structures of the nonspecific thalamocortical system and then rewarmed, allowing the structure to recover its function quickly. Such a preparation permits an animal to serve as its own control for the effects of the functional blockade. All of the results obtained in the acute preparations have been confirmed in the healthy, chronic, undrugged preparation where its state of wakefulness, arousal, and attention can be controlled by behavioral techniques.

The results will be presented in sections corresponding to the dependent variable associated with the functional blockade. Section II confirms and elaborates upon previous findings, showing that blockade abolishes spindle bursts (Chow *et al.*, 1959; Lindsley *et al.*, 1949; Velasco and Lindsley, 1965; Velasco *et al.*, 1968; Weinberger *et al.*, 1965) and recruiting responses (Eidelberg *et al.*, 1958; Hanberry *et al.*, 1954; Velasco and Lindsley, 1965; Velasco *et al.*, 1968; Weinberger *et al.*, 1965), and that such blockade also abolishes certain components of other types of synchronous electrocortical activity. Section III shows that blockade of this system at any of its points enhances primary visual and auditory evoked potentials as well as their corresponding association potentials. This effect seems to indicate a widespread reduction of cortical inhibition. Section IV shows that the surface negative slow potential shift in the cortex, a response associated with the expectancy of presentation of a sensory stimulus, is abolished during blockade of the nonspecific thalamocortical system at the level of the inferior thalamic peduncle. Section V illustrates some of the behavioral deficits that occur during this same type of functional blockade.

II. SYNCHRONOUS ACTIVITY

A. Reduction of Electrically Induced Recruiting Responses and Caudate Spindles

Successive ablations of the orbitofrontal granular cortex were made in order to determine the size and placement of the functional blockade most effective in reducing the amplitude of recruiting responses, an effect that Velasco and Lindsley (1965) had found to occur following total orbitofrontal lesion. Figure 1 shows the progressive reduction of recruiting responses following successive ablations. Not until the entire orbitofrontal cortex was ablated did recruiting responses completely disappear. Successive ablations were made bilaterally from either the rostral to the caudal or from the caudal to the rostral parts of the orbitofrontal cortex and, similarly, from the medial to lateral and lateral to medial regions. In all cases, reduction in the amplitude of the recruiting responses were found to be proportional to the amount of tissue ablated, irrespective of location and sequence of lesions. Thus, within the frontal granular cortex of the cat, Lashley's well-known principles of mass action and equipotentiality hold for the recruiting response.

Essentially the same effect on the reduction of recruiting responses was found for electrolytic lesions placed bilaterally in the inferior thalamic peduncle (ITP). Figure 2 shows the placement of 2–3 mm diameter lesions in the critical region of the ITP (dots). These lesions were effective in totally abolishing recruiting responses. Control lesions near this critical region (dark shading) were ineffective. Figure 3 shows that cooling the tips of cryoprobes (Skinner and Lindsley, 1968; Skinner, 1970, 1971b) placed bilaterally in the ITP will abolish recruiting responses elicited by stimulation of one of the midline thalamic nuclei (center median, CM). This location of the functional blockade will not affect augmenting responses recorded by the same electrode in the anterior sigmoid cortex (AS) but elicited by stimulation of a more lateral thalamic site (nucleus ventralis lateralis, VL). This illustrates that the cooling of the cryoprobe tips to 10°C does not produce a functional blockade that extends laterally into the internal capsule where the fibers from the VL are known to be projecting to the frontal cortex (however, more intense cooling of these cryoprobe tips to 0°C will reduce augmenting responses). As shown in Figure 4, similar effects on recruiting and augmenting responses are produced by permanent heat lesions (Skinner, 1971b) placed bilaterally in the ITP in chronic preparations. As will be seen later, however, the slow potential shifts accompanying the augmenting response are affected by blockade in the ITP, but the positive–negative waveform of the augmenting response itself is completely unaffected.

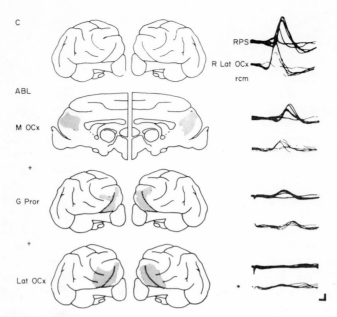

FIG. 1. The effects of successive ablation (ABL) of the orbitofrontal cortex on recruiting responses. Ablations made in three steps beginning with the medial orbitofrontal cortex (M OCx), followed by gyrus proreus (G Pror), and completed with lateral orbitofrontal cortex (Lat OCx) in acute Flaxedilized preparations. At the right, recruiting responses produced by stimulation of nucleus center median (rcm) and recorded on the right posterior sigmoid gyrus (RPS) and the right lateral orbital cortex (R Lat OCx) on the coronal gyrus, before ablation (C = control) and at each successive stage. The asterisk indicates a recording from the contralateral nonspecific midline nucleus center median, which had previously shown recruiting responses. Calibration: 100 µV, 10 msec. (From Skinner and Lindsley, 1967.)

FIG. 2. Effects on recruiting responses and spontaneous spindle bursts of uniform bilateral lesions placed in and around the inferior thalamic peduncle (ITP). Data obtained from 14 acute Flaxedilized preparations. Stippled areas indicate the critical region in which lesions reduced the recruiting responses and spontaneous spindle bursts by 80–100%. Darker oblique-lined areas indicate regions where lesions reduced recruiting responses and spindle bursts by only 0–20%. Maps from Jasper and Ajmone Marsan Atlas (1954). (From Skinner and Lindsley, 1967.) See page 191.

FIG. 3. Superimposed oscilloscope recordings showing the differential effects of cryogenic blockade of the inferior thalamic peduncle (ITP) on recruiting (top) and augmenting (bottom) responses. These recordings were all taken from the same locally anesthetized cat preparation immobilized by gallamine triethiodide (Flaxedil). Recruiting responses elicited by stimulation of the animal's left nucleus centrum medianum (cm) were recorded contralaterally on the right anterior sigmoid gyrus (AS) and are shown in the upper row of traces. Augmenting responses elicited by stimulation of the right nucleus ventralis lateralis (vl) were recorded at the same site as the recruiting responses and are shown in the lower row of traces. The effect of reversible cryogenic blockade is shown by the responses during precooling (Pre-); cooling of ITP (Cool 10°C), 3 min after the onset of cooling; and postcooling (Post-), 3 min after the cessation of cooling. Calibrations: 200 µV and 20 msec. (From J. E. Skinner and D. B. Lindsley, Reversible cryogenic blockade of neural function in the brain of unrestrained animals, *Science* **161**, 595-597 (August 1968). Copyright 1968 by the American Association for the Advancement of Science.) See page 191.

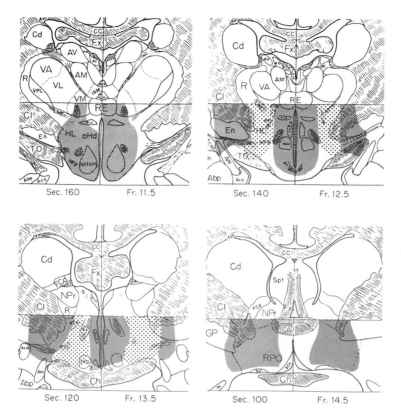

Figure 2

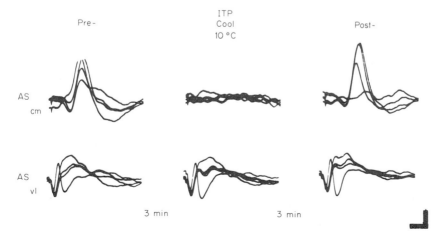

Figure 3

FIG. 4. Blockade of recruiting responses without affecting augmenting responses. Control: Recruiting responses, RR, are elicited by repetitive stimulation of the animal's right nucleus centralis medialis thalamicus (NCM) and are recorded only in the frontal cortical regions (e.g., right anterior sigmoid gyrus, RAS) and not in the more posterior regions (e.g., right anterior suprasylvian gyrus, RASS). Augmenting responses, AR, are elicited by repetitive stimulation of the nucleus ventralis lateralis thalamicus (VL) and are recorded by the same electrodes as the recruiting responses. Les.ITP: Same responses, but after bilateral heat lesions in the region of the inferior thalamic peduncle.

The RR's are abolished or markedly reduced even when the prelesion stimulus intensity is doubled as in the case for the RR's shown. The AR's, however, remain unaffected, except that the base line-to-peak amplitude is reduced in the wave immediately following the last evoked response of the AR train. In all of the experimental preparations of this study, recruiting responses were abolished without affecting augmenting responses. The effect on the RR's was obtained using either the method of permanent lesion as shown in this figure, or reversible cryogenic blockade as shown in Fig. 3. Calibrations: 100 μV and 1 sec. (From Skinner, 1971.) See page 193.

FIG. 5. Caudate spindle abolished by cryogenic blockade of the ITP. Precool: A single shock (6 V, 1 msec duration) to the head of the caudate nucleus produced in the ipsilateral cortex an initial complex surface, positive response followed by a long-duration surface negative wave, upon which a rhythmic spindle burst is superimposed. Cool ITP: During bilateral cryogenic blockade in the ITP sufficient to abolish recruiting responses (5°C tip temperature), the initial surface positive component was enhanced and the rhythmic spindle burst was abolished. Postcool: 3 min after the cessation of cooling, the records returned to the precooled state. RAS: Right anterior sigmoid gyrus. Calibrations: 100 μV and 50 msec. (From Skinner, 1971c.) See page 193.

FIG. 6. Effects of partial ablation of orbitofrontal cortex on recruiting responses and spontaneous spindle bursts. Ablations (shaded area) in the lateral third of the orbital cortex (ABL Lat OCx) in acute Flaxedilized preparations. Recruiting responses evoked by stimulation of the right nucleus centralis medialis (RNCM) and recorded on right anterior (RAS) and posterior (RPS) sigmoid gyri. Spontaneous spindle bursts recorded on the right anterior and posterior signoid and suprasylvian (RSS) gyri and also in the thalamus (RNCM, right nucleus centralis medialis). Postablation records on the right side of the figure taken 3 hr after bilateral aspiration. Calibrations: 100 μV, 10 msec for superimposed CRO traces; 100 μV, 1 sec for ink-writer traces. (From Skinner and Lindsley, 1967.) See page 194, top.

FIG. 7. Effects of bilateral lesions in the critical region of the inferior thalamic peduncle on recruiting responses and spontaneous spindle bursts. The records are all from the same Flaxedil-immobilized preparation, taken before (Control) and 4 hr after lesioning (Les.ITP). Recruiting responses were elicited by stimulation of the right nucleus center median (rcm) and recorded from the ipsilateral anterior sigmoid cortex (RAS) and contralateral thalamus (LCM). Spontaneous spindle bursts were created by large lesions in the mesencephalic reticular formation and recorded in the right anterior sigmoid (RAS), right posterior sigmoid (RPS), right anterior suprasylvian (RSS) cortices, and right centromedial nucleus of the thalamus (RNCM). Calibrations: 100 μV for both CRO and inkwriter records, and 10 msec for CRO and 1 sec for ink-writer traces. See page 194, bottom.

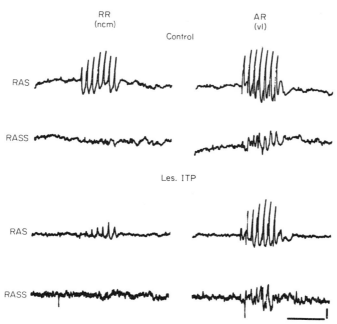

Figure 4

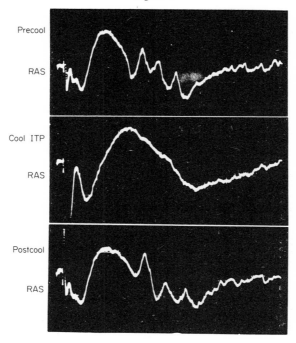

Figure 5

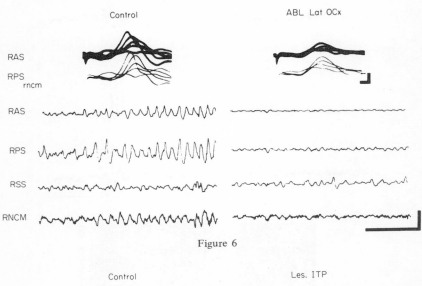

Figure 6

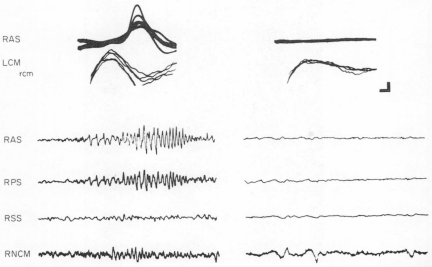

Figure 7

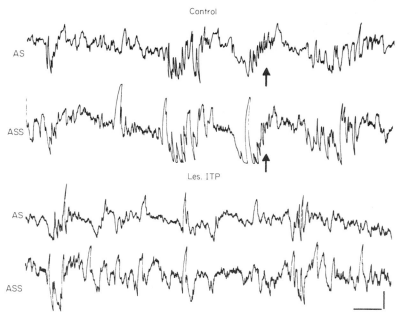

FIG. 8. Sleep spindle component abolished by lesion in the ITP. Control: Records from the anterior sigmoid (AS) and anterior suprasylvian (ASS) gyri were taken during states of drowsiness or during the initial stages of sleep in a 12 hr sleep-deprived animal accustomed to sleeping in the recording chamber. Numerous 12 Hz spindle bursts that have smooth waveforms were recorded on a slow surface negative wave (arrows) interspersed between or combined with lower frequency spindles that had individual waves with inflections in them. Les. ITP: After a bilateral heat lesion in the region of the ITP, the 12 Hz spindles with the smooth waveforms were never observed, but the other spindles containing complex waveforms remained unaffected. Records are from transcortical electrodes in a chronic cat preparation in Figures 8-13. Calibrations: 100 μV and 1 sec. (From Skinner, 1971c.)

Another response, the caudate spindle, resembles the recruiting response in that it is elicited with a lower threshold when the animal is drowsy or has had an injection of sodium pentobarbital. The caudate spindle is the rhythmic afterdischarge that occurs in the frontal cortical regions following a single, brief shock to the head of the caudate nucleus. This response is also abolished following bilateral cryogenic blockade in the inferior thalamic peduncle, and returns on rewarming as illustrated in Figure 5.

B. Reduction in Spontaneous Spindles and Sleep Spindles

When an animal spontaneously goes to sleep, its EEG manifests in the initial drowsy state a synchronous activity that, if recorded monopolarly from the surface of the cortex, is usually 8—12 Hz with a fusiform or spindle shape to the envelope of the waxing and waning potentials that resembles the spontaneous spindles in Figures 6 and 7. If it is recorded transcortically, the EEG manifests a more complex pattern of 8—12 Hz activity combined with slower and larger amplitude waves as seen in the upper two traces in Figure 8. If a large lesion is made in the mesencephalic reticular formation (intercollicular region at stereotaxic level anterior 2.0), the animal will show comatose behavior resembling the state of drowsiness and an EEG pattern that also resembles that obtained during the onset of natural sleep. Figure 6 shows the effects of a partial ablation of the orbitofrontal granular cortex on the recruiting response and spontaneous spindle bursts in an acute immobilized preparation with mesencephalic reticular lesions. Whereas the recruiting responses are markedly reduced, the spontaneous spindles are totally abolished, a result indicating a greater effect of the ablation on the latter response. Figure 7 shows, in the same type of preparation, similar effects on the two types of cortical potentials but from a bilateral lesion of the ITP instead of an ablation of the frontal granular cortex. Figure 8 shows the effects on sleep spindles of permanent bilateral heat lesions in the ITP, sufficient to abolish recruiting responses but not to affect augmenting responses (as in Figure 4). Before the lesions, the sleep-deprived animal showed numerous episodes of 8—12 Hz spindles (arrow) when allowed to sleep but after lesioning showed none of this activity. However, the lower frequency, larger amplitude waves were present when the sleep-deprived animal with the lesions assumed the sleeping posture (recordings in Figure 8 are from transcortical leads). Thus, the 8—12 Hz spontaneous spindle produced by mesencephalic reticular lesions and the 8—12 Hz sleep spindle produced by natural-sleep deprivation are both abolished by blockade in the nonspecific thalamocortical system.

C. Reduction of 8—12 Hz Components of Drug-Induced, Synchronous Activity (Barbiturates, Atropine, and Tetrahydrocannabinol)

Intravenous or intraperitoneal injections of several different types of drugs are known to produce widespread synchronous activity in the cortex. Just how these drugs bring about the synchronous electrocortical activity is not known. Atropine is a known cholinergic blocking agent and the barbiturates seem selectively to inhibit neuronal activity in the mesencephalic reticular formation, but how synchronization of the EEG is produced under the influence of these drugs is not proven. The point is that certain drugs do produce widespread EEG synchronization, certain components of which are abolished by cryogenic blockade in the nonspecific thalamocortical system. Figures 9, 10, and 11 show that the 8—12 Hz components (arrows) recorded by transcortical leads placed in the frontal and parietal regions of the cortex and induced by the influence of barbiturates, atropine, and Δ-1-tetrahydrocannabinol, respectively, are abolished during bilateral cryogenic blockade in the ITP.

D. Reduction of Epileptic Spike Discharges

Since the early pioneering work of Penfield and Jasper on a diffuse or centrencephalic system capable of transmitting epileptic activity from a local focus to widespread regions of the brain, it has been suspected that a diffuse reticulothalamocortical system was involved in the elaboration of the paroxysmal activity. The surgical removal of certain epileptic focusses, these investigators showed, was able to stop the widespread distribution of paroxysmal discharges in many patients and abolish their *grand mal* seizures. The present evidence, however, shows that a small functional blockade is able to prevent the spread of epileptic discharges by blocking the transmitting system rather than removing the focus. Figure 12 shows that a bilateral freeze lesion in the mesencephalic reticular formation is able to produce a transient epileptic focus for several days. Figure 13 demonstrates how this epileptic discharge is altered by bilateral cryogenic blockade in the ITP. The upper trace reveals that the cryogenic blockade previously shown to produce functional blockade of the mesencephalic reticular formation (see Cool MRF 8°C in Figure 12) is not sufficient to block the epileptic focus, suggesting that the freezing of the tissue has produced an epileptogenic process that extends farther from the cryoprobe tips than the actual frozen tissue. (The tissue in the mesencephalon was frozen by lowering the cryoprobe surface temperature to −20°C for 30 sec. This action did not produce a freezing gradient in the tissue for more than a millimeter as measured by a thermocouple, whereas the 8°C cooling has been shown to be effective in producing a functional blockade that extends beyond 1 mm; see Skinner and Lindsley, 1968; Dondey *et al.*, 1962.) However, as shown in the third, from the top trace of Figure 13, cooling the ITP to 10°C bilaterally is effective in abolishing the spike discharge, which is abruptly reduced in amplitude after several minutes of cooling. This blockade is ineffective in abolishing the slow wave that normally follows each spike.

E. Discussion

Bremer (1937) showed that transection of the brain stem at the level of the mesencephalon produces a synchronized EEG pattern in wide regions of the cortex of the cat. Morison and Dempsey (1942) showed that similar patterns of synchronous activity could be produced in the cortex by either low-frequency medial thalamic stimulation or barbiturate injections, although the larger ampli-

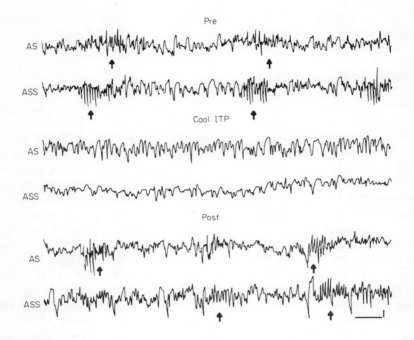

FIG. 9. Barbiturate spindle component abolished by reversible cryogenic blockade in the ITP. Pre: An intravenous injection of sodium pentobarbital (35 mg/kg body weight) produced large amplitude 12 Hz spindles (arrows) interspersed with lower frequency types of synchronization. Cool ITP: Bilateral cryogenic blockade (5°C tip temperature) produced after 3 min a record in which the large amplitude 12 Hz spindles were absent. Post: 3 min after the cessation of cooling, the large amplitude 12 Hz spindles reappeared. AS: Anterior sigmoid gyrus. ASS: Anterior suprasylvian gyrus. Calibrations: 100 μV and 1 sec. (From Skinner, 1971c.)

FIG. 10. Atropine spindle component abolished by cryogenic blockade in the ITP. Pre: Injection of atropine sulfate (1 mg/kg body weight) produced large amplitude 12 Hz spindle bursts (arrows) combined with lower frequency waves. Cool ITP: Bilateral cryogenic blockade in the ITP abolished the 12 Hz component, leaving lower frequency waves of increased duration. Post: 3 min after the cessation of cooling, the 12 Hz spindle components reappeared in the record (arrows). AS: anterior sigmoid gyrus. Calibrations: 100 μV and 1 sec. (From Skinner, 1971c.) See page 199.

FIG. 11. Tetrahydrocannabinol (THC) spindle component abolished by cryogenic blockade in the ITP. Pre: Intraperitoneal injections of 17% Δ-1-THC (16 mg/kg body weight) produced an almost sustained 12 Hz type of synchronous activity with occasional large amplitude spindle bursts (arrow). Cool ITP: 3 min after cooling the cryoprobe tips (5°C), the hypersynchronous activity was markedly reduced in amplitude. Post: 3 min after the cessation of cooling, the records appeared as they did in the precooled state. AS: anterior sigmoid gyrus. Calibrations: 100 μV and 1 sec. (From Skinner, 1971c.) See page 199.

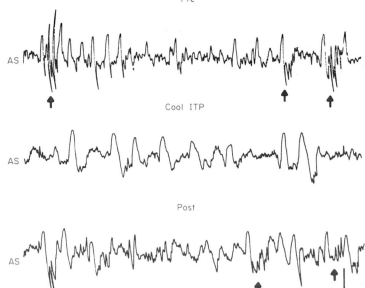

Figure 10

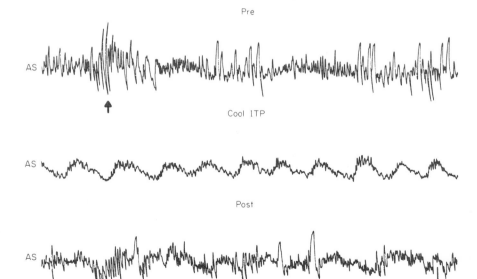

Figure 11

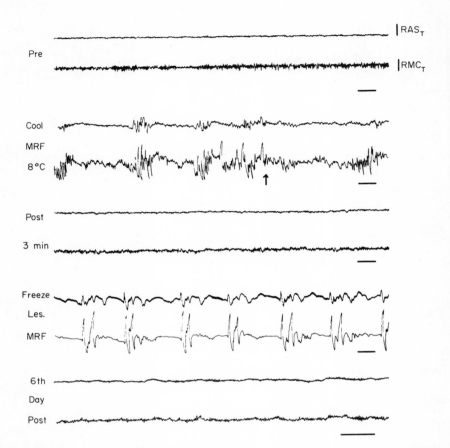

FIG. 12. Transient epileptogenic discharges produced by a focal freeze lesion in the mesencephalic reticular formation. Transcortical recordings were taken from the right anterior sigmoid (RAS_T) and right midcoronal (RMC_T) gyri in the same animal before cooling in the mesencephalic reticular formation (Pre), during bilateral cooling to 8°C of the surface of the cryoprobe tips (Cool MRF 8°C), 3 min after the cessation of cooling (Post 3 min), 4 hr after small freeze lesions produced by tip temperatures of −20°C for 30 sec (Freeze LES MRF), and 6 days after the freeze lesion. Arrow in Cool MRF 8°C indicates tail pinching. Calibrations: 100 μV and 1 sec. (From unpublished data in preparation for publication.)

tude waves tended in both cases to be in the frontal pericruciate regions (anterior and posterior sigmoid gyri).

Stimulation of the medial thalamus (e.g., nucleus centralis medialis or nucleus center median) produces monophasic-negative responses, called *recruiting responses*, in the frontal anterior sigmoid cortex. Stimulation of the more lateral thalamus (e.g., nucleus ventralis lateralis) produces in the same cortical region

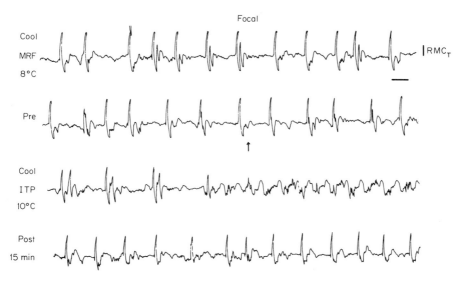

FIG. 13. Effects of reversible cryogenic blockade in the inferior thalamic peduncle on epileptogenic discharges produced by small freeze lesions in the mesencephalic reticular formation. Transcortical recordings are from the midcoronal gyrus (RMC_T). The same effects were observed on synchronous discharges recorded from the anterior suprasylvian, anterior lateral or marginal, and anterior suprasylvian cortices. Cooling the mesencephalic reticular formation to 8°C bilaterally does not abolish the epileptogenic focus (Cool MRF 8°C). However, after rewarming MRF (Pre-) and cooling the inferior thalamic peduncle bilaterally for approximately 10 min (Cool ITP 10°C), the spike discharges are abruptly reduced in all of the recorded regions of the cortex. This blockade, however, does not affect the slow waves following each reduced spike. Following the cessation of cooling (Post 15 min), the spikes return to their normal amplitude in approximately 15 min, which is 12 min after the cooled brain tissue has returned to normal body temperature. Calibrations: 100 μV, 1 sec. (From unpublished data in preparation for publication.)

positive–negative responses called *augmenting responses*. Schlag and Villablanca (1967) found that stimulation of most thalamic nuclei, either midline or lateral, produces simultaneously two types of responses: a positive–negative waveform response recorded in the projection cortex of the particular nucleus stimulated and a monophasic-negative response recorded in the cortex immediately surrounding the projection zone. Jasper (1949) proposed that wave form is an important criterion distinguishing between recruiting and augmenting responses, the former being monophasic negative and the latter positive–negative. Since stimulation of a single thalamic nucleus will produce both types of responses simultaneously, waveform can no longer be used to define the two types of incremental responses. Instead, the exact site of stimulation and the exact pattern of positive–negative and monophasic-negative responses in the cortex should be used. Schlag and Villablanca (1967) suggested dropping the term *recruiting responses*, since electrophysiologically these are indistinct from augmenting responses. However, electrical stimulation of medial and lateral thalamic nuclei produces distinct behavioral reactions, and medial intrinsic thalamic nuclei are anatomically distinct from lateral extrinsic ones, clearly suggesting different underlying functions.

It is proposed here that recruiting responses are those potentials elicited by stimulation of midline thalamic nuclei, and that augmenting responses are those produced by stimulation of lateral thalamic nuclei. Recruiting responses are positive—negative in the orbitofrontal granular cortex, as shown by Schlag and Villablanca (1967), and monophasic negative in the frontal and pericruciate regions. Augmenting responses, depending upon the thalamic nucleus stimulated, are positive—negative in various cortical regions and are monophasic-negative responses in surrounding cortical areas. Stimulation of intralaminar regions of the thalamus, separating the midline and lateral nuclear groups, may stimulate both systems at the same time and erroneously give one the impression, as Jasper (1961) thought, that: "The most medial portions of the recruiting system project forward to the frontal regions, while the lateral portions in the central lateral nucleus project posteriorly." The monophasic negative responses associated with stimulation of lateral thalamic nuclei are not to be confused with those produced by medial thalamic stimulation. The topographical distribution of the monophasic-negative recruiting responses under the present definition is the same as the frontal distribution proposed earlier by Morison and Dempsey (1942) and Starzl and Magoun (1951) and is not the same as the more posteriorly projecting distribution proposed by Jasper (1961).

Spencer and Brookhart (1961a,b) found that the spontaneous spindle waves produced by mesencephalic lesions and recorded from the surface of the frontal cortex were actually composed of two different types, a biphasic positive—negative waveform (type I) and a monophasic-negative wave form (type II). There is also a third type of spindle (mixed), in which inflection points occur in the waveforms because of the interaction of the type I and type II spindles with each other. Laminar analysis of the cortical activity of both the type I spindle and the positive—negative augmenting response indicates very similar profiles, suggesting a common neural mechanism for the genesis of these potentials. In a similar manner, laminar analysis of the type II spindle and the monophasic—negative recruiting responses shows the same results, which suggests that these two potentials may have a common mechanism.

The present results show that cryogenic blockade or permanent lesion in the region of the ITP and, in some experiments, ablation of the frontal granular cortex abolishes an identifiable 8—12 Hz component of electrocortical synchronization elicited by any one of several different methods, without interfering with other synchronous components. In all of the preparations, recruiting responses produced by medial thalamic stimulation were totally abolished, but augmenting responses produced by more lateral thalamic stimulation in the nucleus ventralis lateralis were unaffected. The various synchronous components that remained to be recorded from each electrode in the cortex after the ITP blockade were presumably the type I, or positive–negative, responses of the intact thalamocortical systems projecting directly to the region of the recording electrode, mixed with monophasic-negative, or type II, responses surrounding the projection zones of intact thalamocortical systems, but not projecting directly to the immediate vicinity of the recording electrode. The recorded component that is abolished in all of the cases of synchronization elicited by the various methods is the type II, monophasic-negative, wave form, appearing in the frontal regions where the electrodes were placed, which is associated with activity in the nonspecific thalamocortical system. Since the ITP blockade prevented the direct cortical projection of the responses initiated in the medial thalamic nuclear group, as suggested from the work of Schlag and Villablanca (1967), the positive–negative recruiting responses in the unrecorded orbito-frontal cortex were presumably abolished along with the monophasic-negative responses that were recorded in the surrounding cortices.

Recently Staunton and Sasaki (1971) have shown that cortical responses recorded in parietal cortex and elicited by intralaminar thalamic stimuli are not affected by frontal lobe ablations. Similarly, Robertson and Lynch (1971) have shown that records from the parietal cortex produced by mesencephalic lesions or barbiturate injections also are not abolished by frontal lobe ablations. The authors in both of these studies referred to their recorded cortical responses as classical recruiting responses, spontaneous spindles, and barbiturate spindles. But clearly their responses do not concur with our criteria. The apparently contradictory results obtained by these authors and those obtained by us serve to illustrate the confusion in the literature that results when one defines the cortical responses by *waveform* rather than by the anatomical *system* in which they are evoked. The results of the above authors are, in fact, supportive of our data obtained during ITP blockade (Skinner, 1971a, c) that show cortical responses are not abolished if recorded in posterior cortices or elicited by lateral thalamic stimuli. Clearly, our results are concerned with a mediothalamic-frontocortical system whose monophasic negative surround potentials recorded in the frontal half of the brain are abolished by blockade of the positive–negative projection cortex or projection pathway.

Blockade in the ITP component of the nonspecific thalamocortical system seems to be effective in abolishing the transmission of paroxysmal discharges initiated from a focus in the mesencephalic reticular formation to the frontal and parietal regions of the cortex. Thus, the nonspecific thalamocortical system seems to be a diffuse, centrencephalic-like system that can mediate epileptic discharges from a focus to wide regions of the brain.

III. SENSORY EVOKED POTENTIALS

During experimentally induced states of attention in both man and animals, the evoked potential elicited by the attended stimulus is enhanced, whereas that of the unattended stimulus is unaffected (Galambos et al., 1956; Hearst et al., 1960; Marsh et al., 1961; Galambos and Sheatz, 1962; Garcia-Austt, 1963; Garcia-Austt et al., 1964; Haider et al., 1964; Spong et al., 1965). Thus, a brain mechanism exists for differential enhancement between attended and unattended stimuli.

A. Enhancement of Primary Evoked Potentials

Figure 14 shows a typical orbitofrontal cortical ablation that led to the enhancement of a habituated, unattended, primary visual evoked potential (Figure 15), which was recorded in an immobilized acute preparation and elicited by an electric stimulus to the optic tract.

Similar effects were obtained with cryogenic blockade in the inferior thalamic peduncle (Figure 16). As shown in Figure 17, habituated, unattended, primary evoked potentials elicited by either photic stimulation or stimulation of the optic tract were enhanced in the primary projection cortex, whereas the recruiting response was abolished in the frontal regions.

In control animals, cryoprobes placed near but not in the critical region of the ITP were ineffective in blocking the recruiting response or enhancing the visual evoked potentials (see column C in Figure 26).

Enhancement of the primary visual evoked potential as well as abolition of recruiting responses was obtained during cryogenic blockade in the nucleus

---

FIG. 14. Orbitofrontal cortex ablation in the experiment illustrated by Fig. 15. (A) and (B): mesial surface views of the left and right hemispheres showing ablations that include the orbitofrontal cortex but not the infralimbic cortex. (C): orbital surface views of left and right hemispheres showing ablations of the rostral part of the anterior sigmoid gyrus and of the gyrus proreus. A dot on the animal's right coronal gyrus indicates one of the recording sites (RCor) in Fig. 15. Flaxedil-immobilized preparation.

FIG. 15. Enhancement of primary evoked potentials in the visual cortex and reduction of recruiting responses in frontal regions after ablation of the orbitofrontal granular cortex. The extent of the ablation is shown in Fig. 14. Recruiting responses elicited by stimulation of right nucleus center median (rcm) were recorded from the ipsilateral posterior and anterior sigmoid gyri (RPS, RAS), and primary visual and frontal association evoked responses were elicited by a brief shock to the right optic tract (ot) and recorded, respectively, in the ipsilateral primary visual cortex (RVCx) and orbital gyrus (R Cor). Records were taken previous to the ablation (Pre-ABL) and 3 hr after ablation of the mesial orbital cortex and gyrus proreus (ABL MOCx GPr). Flaxedil-immobilized preparation. Calibrations: 100 μV and 10 msec.

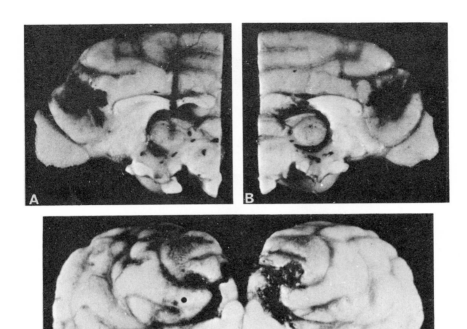

Figure 14

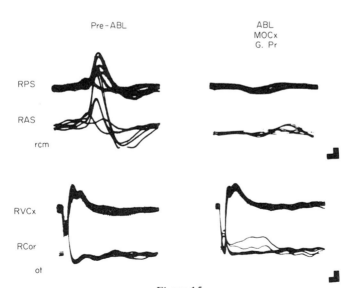

Figure 15

205

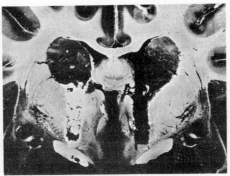
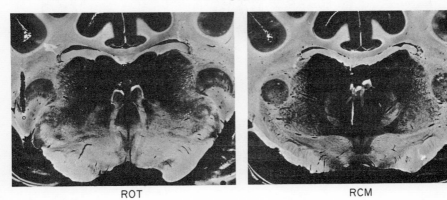

FIG. 16. Cryoprobe and electrode placement in the experiment illustrated by Fig. 17. Bilateral sites of cryoprobes implanted in the critical region of the inferior thalamic peduncle (ITP) and electrode sites for evoking visual and recruiting responses implanted in the right optic tract (ROT) just below the lateral geniculate body and in the right nucleus center median (RCM), respectively, Flaxedil-immobilized preparation.

ventralis anterior of the medial thalamus, illustrating that blockade at any point of the nonspecific thalamocortical system is effective in enhancing unattended visual evoked potentials.

In a chronic animal preparation in which cooling had been performed in the ITP and after the cryoprobe was allowed to rewarm, it was observed that after 1 min the evoked potential was still enhanced and recruiting responses were still blocked (Figure 18). However, after 2 min the evoked potential had begun to reduce in amplitude and the recruiting response recorded at 2.5 min had begun to increase. By the third to fourth minute after rewarming, both responses had returned to their precool amplitudes. This illustrates the close inverse rela-

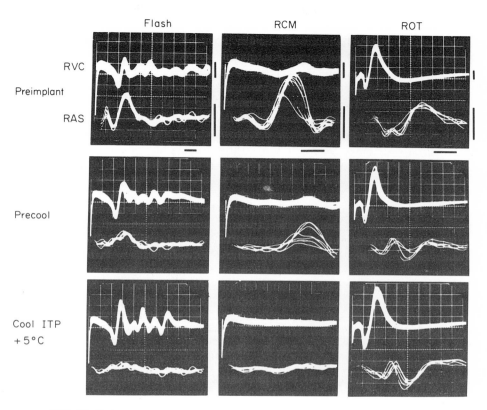

FIG. 17. Enhancement of both photically and electrically evoked visual responses and reduction of recruiting responses following bilateral blockade of the inferior thalamic peduncle. Records are from the right visual cortex (RVC) and the anterior sigmoid gyrus (RAS). Each record shows superimposed traces of evoked potentials produced by either 0.5 Hz photic stimulation (Flash), 8 Hz stimulation of nucleus center median (RCM), or 0.5 Hz stimulation of optic tract (ROT). Preimplant, series recorded before cryoprobes were implanted in the ITP; Precool, series recorded 2 hr after implantation; Cool ITP, series recorded following cooling shortly thereafter. Cat immobilized with Flaxedil. Calibrations: 100 μV and 20 msec. (From Skinner and Lindsley, 1967.)

tionship between the amplitudes of the two responses as well as the proportionality of their amplitudes to the size of the functional blockade as it shrank from its periphery toward the cryoprobe surface during rewarming. Repeated blocking of the ITP always had the same inverse effects on recruiting and visual responses, with no amplitude changes over time when the cooling temperature was held constant. In chronic preparations, cooling was repeated as many as 20 times, and for periods as long as 2 hr, always with the same result.

Figure 19 shows that the cortical enhancement effects occur in the auditory system as well as in the visual. Both the visual and auditory stimuli were well habituated and elicited by stimulating at a constant voltage at the periphery of each sensory system to keep the evoking stimulus at a constant amplitude. When a small functional blockade was made in a chronic preparation in the ITP by cooling to 15°C, primary visual responses were enhanced, but primary auditory responses recorded at the same time were not affected. When the size of the functional blockade was expanded by increasing the cooling to 10°C, both visual and auditory responses were enhanced. The expanded blockade seems to have included fibers related to the auditory system that were previously unaffected.

B. Enhancement at Thalamic and Cortical Levels

Figure 20 shows that the enhancement effects of ITP blockade occurred at both thalamic and cortical levels. Potentials recorded in the visual cortex showed enhancement of different components of the response (numbered in the lower left of Figure 20), depending upon whether the stimulating electrode was in the

---

FIG. 18. Inverse relationship between the amplitudes of recruiting and visual responses during cryogenic blockade of the inferior thalamic peduncle. Visual responses evoked by stimulation of the optic tract (ot) were recorded on visual cortex (VCx) and summed by computer; 30 responses to each superimposed trace evoked during a 1 min period. Three consecutive 1 min recordings were made during each condition. The stimulus artifact has been erased. Recruiting responses recorded on the left anterior sigmoid gyrus (LAS) were evoked by stimulation of the left nucleus centralis medialis (lncm). Temperature in the tissue adjacent to the tip of the cryoprobe is indicated on the left above the EEG traces, and elapsed time after the cessation of cooling is indicated on the right. The top traces show characteristic responses at normal temperature; the second row shows enhancement of the visual cortex response and marked reduction of recruiting responses upon cooling the ITP to 5°C. The third and fourth rows show the effect of progressive rewarming to normal temperature, with return of visual cortex responses and recruiting responses to the control level. Calibrations: 100 μV and 10 msec for superimposed traces, and 1 sec for ink-writer records. (From Skinner and Lindsley, 1967.)

FIG. 19. Independent enhancement of visual and auditory evoked responses during moderate and intense cooling in the inferior thalamic peduncle (ITP). Each superimposed trace represents the sum of 30 evoked responses obtained from the same chronic preparation and recorded in the visual striate cortex (VCx) during optic tract (ot) stimulation and in the auditory middle ectosylvian cortex (ACx) during cochlear nucleus (ch n) stimulation. The records under Pre- were obtained immediately before cryogenic blockade; those under Cool ITP +15°C were obtained during moderate cooling of the ITP; and those under Cool ITP +10°C were obtained during increased cooling. The arrows near the ACx responses are at a constant distance from the baseline and provide visual cues for the readers. Recruiting responses observed with an oscilloscope still showed a small residual response during +15°C cooling but were totally abolished during +10°C cooling, leaving only the stimulus artifact. Calibrations: 200 μV and 10 msec. (From Skinner and Lindsley, 1971.)

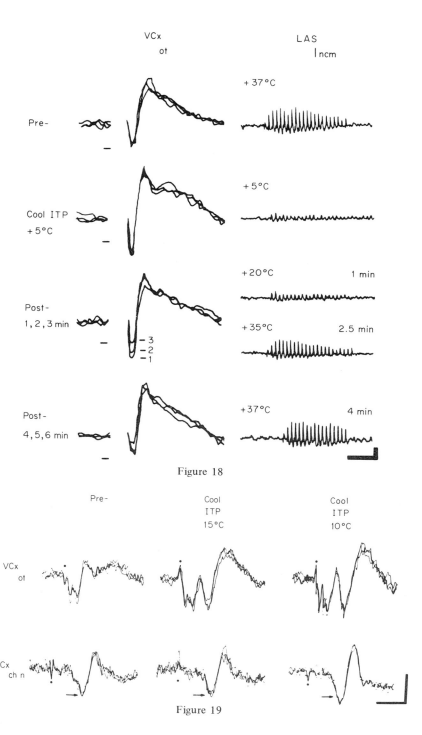

Figure 18

Figure 19

optic tract or in the optic radiations. Stimulation of the pregeniculate optic tract fibers resulted in an increase of all four of the initial positive deflections (Figure 20, OT), whereas stimulation of the postgeniculate fibers in the dorsal part of the lateral geniculate body or in the optic radiations produced an increase in only the third and fourth positive deflections (Figure 20, LG). The current interpretation is that the large first and small second positive deflections recorded from the surface of the cortex are produced by long dipole generators that extend from the geniculate to the cortex and represent the afferent input to the cortex, whereas the third and fourth deflections are produced by small dipole generators located entirely within the cortex and represent the cortical response to the afferent input (Brindley, 1960; Dumont and Dell, 1960; Cordeau et al., 1965). Thus, some influence in the lateral geniculate must have led to enhancement of peaks 1 and 2 during optic tract stimulation, but when this afferent input was held constant by optic radiation stimulation, peaks 3 and 4 were enhanced, which indicates an increased cortical responsiveness.

C. Enhancement of Association and Nonprimary Evoked Potentials

Figure 21 shows that electrical stimulation of the optic radiations near their exit from the dorsal part of the lateral geniculate body results in widespread short-latency cortical potentials. These primary, association, and nonprimary potentials were all enhanced during functional blockade of the nonspecific

---

FIG. 20. Effects of cryogenic blockade of the inferior thalamic peduncle (ITP) on the four initial positive components of the electrically evoked visual cortex potential. All responses are recorded from the visual cortex. Stimulation of the optic tract (OT) in a chronic preparation evokes responses in which all of the components are increased during bilateral cooling to +10°C in the ITP (Cool) as compared to those evoked before cooling (Pre-). Stimulation of the optic radiations near the dorsal part of the lateral geniculate body (LG) evokes responses in a Flaxedil-immobilized preparation in which only the third and fourth components are increased during cooling. See test for explanation. Each response in the upper groups of traces represents the sum of 30 responses averaged on a computer, and those in the lower groups are individual superimposed oscilloscope recordings. Calibrations: 100 μV and 2 msec. (From Skinner and Lindsley, 1971.)

FIG. 21. Short latency cortical potentials evoked by stimulating the optic radiations. Stimulation of the optic radiations near the dorsal part of the lateral geniculate body elicits a primary evoked potential in the visual cortex (VCx), association evoked potentials in the anterior suprasylvian (ASS) and posterior sigmoid (PS), and nonprojection evoked potentials in the auditory cortex (ACx). Responses were evoked before (Pre-), during (Cool ITP +10°C), and 3 min after (Post-) bilateral cooling in the inferior thalamic peduncle. Responses were recorded simultaneously during each of the control and cool conditions, from the same chronic cat and in the same recording session. Recruiting responses were totally abolished during cooling and recovered to the precool level within 3 min after the cessation of cooling. Each trace represents 30 responses summed on a computer. Calibrations: 100 μV and 10 msec. (From Skinner and Lindsley, 1971.)

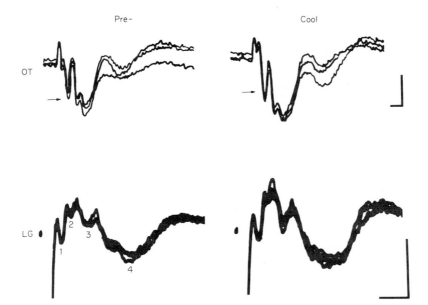

Figure 20

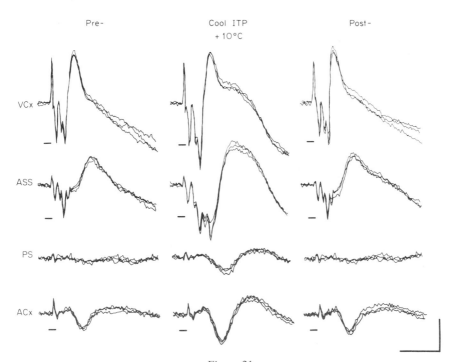

Figure 21

thalamocortical system by cooling in the region of the ITP sufficiently to abolish recruiting responses. Thus, a generalized increase in the cortical responsiveness seems to have occurred during the blockade.

D. Discussion

These results suggest that the nonspecific thalamocortical system normally maintains tonic suppression of evoked potentials elicited by unattended, habituated stimuli at both the level of the thalamus and of the cortex. Blockade of this system in the region on the inferior thalamic peduncle releases the suppression and results in an increased transmission of afferent impulses through the lateral geniculate body and an increased cortical response in specific, association, and nonprojection areas. The sense modalities studied (visual and auditory) may be interpreted as having a discrete anatomical organization in the region of the ITP, because an increase in the size of the blockade increases the number of sense modalities affected. An alternative interpretation is that the auditory system may simply require more intensive blockade in order to be affected.

The differential enhancement of the evoked cortical responses, elicited by the attended and the unattended stimuli and known to occur during attention, seems to be due in part to a maintained suppression of the response of the unattended stimulus rather than to an enhancement of the attended one. Blockade in the nonspecific thalamocortical system releases the tonic suppression of an unattended response and, as we shall see, disrupts the ability of the animal to exhibit attentive behavior. This does not, however, explain the enhancement of the attended stimulus, unless the natural state of the evoked response is the maximum amplitude possible upon which the differential suppression can operate, leaving the attended response disinhibited with all the other unattended responses suppressed. It is possible that either a general or a selective enhancement process is simultaneously operating with the tonic inhibitory process just demonstrated to be mediated by the nonspecific thalamocortical system. The mesencephalic reticular activating system is a likely candidate for an enhancing mechanism because stimulation in the reticular formation accompanying a brief shock to the optic nerve will enhance the evoked response in the visual cortex (Bremer and Stoupel, 1959; Dumont and Dell, 1960).

IV. SLOW POTENTIALS

Directed attention and expectancy, produced by one of several behavioral situations, have been shown to elicit a negative steady potential shift over the frontal and vertex regions of the cortex in man (Walter *et al.*, 1964a, b, 1965;

Walter, 1966). It has been shown in both animals and man that the amplitude of this negative variation of the EEG (called contingent negative variation, or CNV, by Walter et al., 1964a,b) is proportional to the degree of motivation to perform the expectancy task (Irwin et al., 1966; Rebert et al., 1967; Borda, 1970). If the subject is distracted or is inattentive during the interval in which his expectancy is being maintained, the amplitude of the CNV diminishes (Low and McSherry, 1968; McCallum and Walter, 1968; Tecce and Scheff, 1969).

Rotation of the eyeball is known to produce small steady potential shifts that can be recorded from the surface of the cortex. Correlation studies between eye movements and cortical potentials, and investigations in enucleated animals, have shown that the CNV is not an artifact produced by eye movement (Low et al., 1966; Chiorini, 1969; Rebert and Irwin, 1969; Hillyard and Galambos, 1970) but such potentials may contaminate the CNV.

Preceding voluntary movement of a limb, a slow negative cortical potential is generated that terminates with the occurrence of the movement (Kornhuber and Deecke, 1965; Gilden et al., 1966; Deecke et al., 1969). The amplitude of this negative premovement potential is greatest over the rolandic cortex (Vaughan et al., 1968), and a similarly recorded potential can be enhanced in amplitude if the subject is highly motivated to move the limb in response to a signal (McAdam and Seales, 1969). Recent work in monkeys (Borda, 1970) indicates that a negative steady potential shift recorded in the premotor cortex during expectancy is independent of a similar one produced in more frontal regions. The negative steady potential recorded over the motor cortex diminishes or is abolished after repeated presentations of the expectancy trials (Chiorini, 1969), whereas the negative steady potential recorded over the frontal regions persists after overtraining (Borda, 1970). These results suggest that the negative slow potentials recorded over the frontal, premotor, and motor regions are independent. The results to be presented here confirm the independence of negative steady potentials recorded over the frontal cortical regions.

The electrophysiological evidence that best relates the nonspecific thalamocortical system to the one presumed to underlie the generation of the CNV is the result of a study by Goldring and O'Leary (1957), in which it was shown that recruiting responses are superimposed upon a negative steady potential shift. They showed that a brief stimulus pulse to the intralaminar component of the medial thalamus produces a negative slow potential shift in the cortex that outlasts each stimulus by several hundred milliseconds, producing a sustained steady potential shift during repetitive stimulation. It appears from this work that the nonspecific thalamocortical system is capable of generating a sustained slow-wave negative shift in the cortex, a response that resembles the CNV. But as we shall see presently, it is the augmenting response, and not the recruiting response, that is associated with a negative slow potential shift in the frontal regions. This error seems to be caused by the previously discussed confusion about the distinction between recruiting responses and augmenting responses when only the criterion of wave form is considered. What Goldring and O'Leary thought were recruiting responses elicited from the intralaminar thalamic nuclei were most likely augmenting responses of the type discussed by Schlag and Villablanca (1967) in which the surround monophasic-negative responses were recorded.

A. Reduction of Slow Potentials Associated with Augmenting Responses

Direct-coupled monopolar recordings from the surface and depth electrode of each transcortical pair, placed in various regions of the cortex, demonstrated similar potential waveforms for each type of response studied (recruiting, augmenting, and conditioned responses), but the surface and depth records were of reverse polarity. Therefore, the transcortical records of the responses shown throughout Sections IV,A and IV,B are equivalent to monopolar surface records in which negative polarity produces an upward deflection of the inkwriter pens. It was desirable to record transcortically in this experiment to eliminate the slow and steady-state potentials that are produced by movement or position of the eyes and are known to be recorded from monopolar surface electrodes.

Figure 22 shows a representative recruiting response (RR), and augmenting response (AR) produced in the cat by stimulating the NCM and the VL, respectively, and recorded from ipsilateral electrodes placed transcortically across the cortex of the prefrontal (anterior sigmoid, RAS), sensory—motor (coronal, RC), and anterior association (anterior suprasylvian, RASS) cortices. The RR's were recorded only in prefrontal cortical regions, and the AR's were recorded in both prefrontal and sensory-motor cortices. Neither stimulation site produced more posterior responses. The transcortically recorded RR's and AR's were always found to be monophasic-negative or positive—negative potentials, respectively, as were their monopolarly recorded surface responses, confirming Sasaki *et al.* (1970). Because of the difficulty of surgical approach, electrodes were not placed in the orbitocortical region of the orbital gyrus where Schlag and Villablanca (1967) have reported positive—negative waveform responses produced by stimulating the NCM.

Slow or steady potential responses were determined from the level above baseline of the potential preceding each successive response in the 7 or 8 Hz evoked potential train (Figure 22, dotted line). A large amplitude posttrain wave of 200—300 msec duration followed the last response in the train of augmenting responses. Reduction in the frequency of stimulation from 7—8 to 4—5 Hz resulted in an increase in amplitude of the steady potential but not of the poststimulus wave. It is observed in Figure 22 and Figure 23 that no negative steady potential was superimposed upon the RR's, whereas the AR's recorded in the RAS and RC were recorded atop a large surface negative deflection and followed by a large poststimulus wave. In Figure 22 and Figure 23B, the steady potential shifts and poststimulus waves of the AR's are larger in amplitude in the RAS than in the RC, but those in Figure 23A are equally large in both cortical regions. The RR was often seen superimposed upon a small, positive, steady potential shift (Figure 23A), confirming the observations by Goldring and O'Leary (1957) in the cat, in which the nucleus ventralis anterior was stimulated.

Figure 23A shows the effect on the RR's and AR's of bilateral cryogenic blockade in the region of the ITP. Recruiting responses are abolished, but AR's are unaffected. However, the steady potential shift and poststimulus wave associated with the AR are markedly reduced in the RAS and significantly reduced in RC cortices.

Figure 23B shows in another animal the effects on the AR and RR of bilateral cryogenic blockade in the medial half of the anterior limb of the internal capsule. The amplitude of the RR recorded in the RAS is unaffected, whereas the amplitude of the positive–negative response of the AR is significantly reduced. The steady potential shift and poststimulus wave, however, are completely unaffected.

B. Reduction of Conditioned Negative Slow Potentials

After implantation of the various devices in the brain, the animals were allowed a period of at least 4 weeks postoperative recovery. After the recovery period, the animals were first adapted to the apparatus and then exposed to a sensory–sensory, tone–shock, conditioning paradigm. The conditioned stimulus (CS) was a 1000 Hz tone lasting 600 msec, and the unconditioned stimulus (UCS) was a mild electric shock to the neck that also lasted 600 msec. The end of the CS interval and the beginning of the UCS interval were separated by 600 msec, resulting in a 1200 msec interval separating the onset of the CS and UCS. Before introducing the conditioned or unconditioned stimuli, the animals were placed inside the apparatus, which consisted of a small restraining box through which the animal's head protruded, set inside a sound-attenuated chamber containing a speaker and the recording and cryogenic cables. After a week of daily adaptation in which the animals were restrained and their cryogenic devices cooled, they became accustomed to the procedures and were quite cooperative with the experimenter. They would even sleep in the restraining box if undisturbed. After adaptation, the animals were given 100 presentations of the CS alone (habituation to the tone) that occurred at random intertrial intervals of either 15, 30, or 45 sec. For the next 6 days the animals were given 100 conditioning trials per day at the same intertrial intervals. The tone was reinforced by a mild electric shock delivered through uninsulated stainless steel wire sutured through the skin on the back of the neck. The shock was adjusted each day, just before conditioning, to be slightly below the threshold for producing visible neck muscle jerks. On the second day, after 30 trials of conditioning in the noncooled state, the animals were given 30 conditioning trials with bilateral cryogenic blockade in the region of the ITP or medial internal capsule. The animals were then given the remaining trials in the postcooled state for that day to determine whether or not any blocked responses returned after the cessation of cooling. The same daily procedure was followed for the next 4 days until 500 conditioning trials (overtraining) had been presented.

The temperature of the surface of the probe was approximately $-2°C$. This site of temperature recording is in contrast to all of the previous ones in which the thermocouple recorded the temperature in the interface between the surface of the probe and the tissue. The difference in recorded temperatures between the two locations is approximately $10°C$, the surface recording being colder than the interface one.

Figure 24 shows the abolition of the conditioned negative shifts recorded in the AS cortex of three different animals. In all cases, the Precool records were taken early in conditioning, and the cool records were taken immediately afterward. In all cases the return of the conditioned negative shift occurred in the block of 30 trials after cooling had ceased. The relative amplitude enhance-

---

FIG. 22. Recruiting and augmenting responses. Stimulation of the nucleus centralis medialis (NCM) produces recruiting responses, RR, that are recorded only in the frontal cortical region. Stimulation of nucleus ventralis lateralis (VL) produces augmenting responses, AR, that are recorded in frontal cortical regions and sensory—motor areas. Neither response produces posteriorly recorded responses. RAS: right anterior sigmoid (prefrontal cortex); RC: right coronal (sensory—motor) cortex; RASS: right anterior suprasylvian (anterior association) cortex. All stimulating electrodes are ipsilateral to the recording electrodes located on the right side. All records in this and the subsequent three figures are from transcortical electrodes connected to a direct-coupled recording system. Calibrations: $200 \mu V$ and 1 sec. (From Skinner, 1971.)

FIG. 23. Augmenting and recruiting responses during cryogenic blockade of the inferior thalamic peduncle and medial internal capsule. (A) Augmenting responses produced by stimulating the nucleus ventralis lateralis and recorded in the right anterior sigmoid and right coronal gyri (RAS_{vl} and RC_{vl}), and recruiting responses produced by stimulating the nucleus centralis medialis and recorded in the right anterior sigmoid gyrus (RAS_{ncm}) before (Pre), during (Cool ITP), and 3 min after (Post) cooling the tips of the cryoprobes located in the region of the inferior thalamic peduncle. Note the reduction of the steady potential shift and poststimulus wave of the augmenting responses during ITP blockade. (B) Same stimulating and recording sites as in (A), but the bilateral cryoprobes are more laterally placed in the medial part of the anterior limb of the internal capsule (Med. IC). Note that the augmenting response amplitude is decreased in RAS_{vl} during cooling (Cool Med. IC) but that the steady potential shift and poststimulus wave are not affected. All stimulating electrodes are ipsilateral to the recording electrodes on the animal's right side. Because the records from the direct-coupled system were sometimes unbalanced and off scale, not all records shown are from the same stimulus train; however, there was very little variation in the amplitude of the records from stimulus train to stimulus train in these alert chronic preparations. Calibrations: (A) $80 \mu V$ and 1 sec; (B) $200 \mu V$ and 1 sec. (From Skinner, 1971.)

THE NONSPECIFIC THALAMOCORTICAL SYSTEM 217

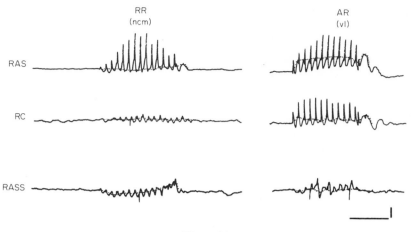

Figure 22

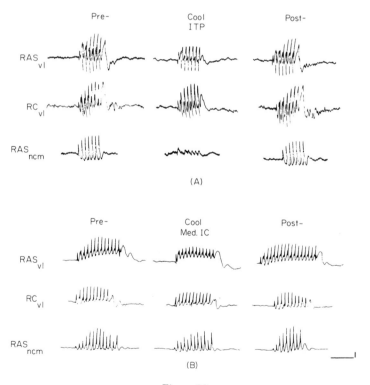

Figure 23

ment of the CS-elicited potential during cooling or noncooling of ITP is not as marked in early conditioning as it is in later overtraining, because in the latter the evoked response during the noncooled state is more reduced in size. In the control animals that had medial internal capsule cryoprobes, cooling had no effect on the negative potentials elicited by the CS.

Figure 25 shows the effect of cryogenic blockade in the region of the ITP on computer-averaged evoked potentials recorded in the AS, C, and AL cortices. The evoking stimulus (CS) was a 600 msec, 1000 Hz tone of moderate intensity that was reinforced 1200 msec after its onset by a mild electric shock. The line marked Hab. indicates the computer-averaged response of the first 30 trials of the presentation of the tone alone. The next line, Cond. (31–60), shows the average response of the second block of 30 trials recorded after the CS was reinforced and a significant change in its evoked response had occurred. Note the increased amplitude of the negative response following the evoked potential in all three cortical regions. The line marked Cool $-2°C$ shows the effect of the ITP blockade, which markedly reduced the amplitude of the negative shift recorded

---

FIG. 24. Abolition of the conditioned negative response in the frontal cortex of three different chronic cat preparations. All recordings are from the anterior sigmoid cortex (as). The top row (Precool) shows 30 computer-averaged responses during early conditioning, and the bottom row (Cool) shows the responses after bilateral cryogenic blockade of the inferior thalamic peduncle sufficient to abolish recruiting responses. The time intervals at the bottom of each column represent the same periods as those described in Figure 25 (CS marks the duration of the conditioned stimulus). Calibrations: Each time interval is 600 msec and the vertical deflection represents 60 μV, except under Cat 3, where it is 150 μV. (From Skinner 1971.)

FIG. 25. Conditioned cortical responses. Three anterior cortical regions, anterior sigmoid (AS), coronal (C), and anterior lateral, i.e., marginal (AL), were recorded simultaneously during several different conditions in a sensory–sensory conditioning paradigm. Indicated at the bottom of the figure underneath each column are three equal time intervals (time moves left to right). The first one marks the base-line period before the presentation of a conditioned stimulus; the middle interval indicates the onset and offset of the 1000 Hz tone used as a conditioned stimulus (CS); and the third interval shows the delay period between the offset of the CS and the onset of the mild shock that served as the reinforcing unconditioned stimulus. The top row (Hab.) shows the three computer-averaged cortical responses to the first 30 presentations of the tone alone. The second row (Cond. 31–60) shows the responses to the second block of 30 presentations of the CS followed by shock. Notice the increased negative response in the AS during the third time interval. The third row (Cool $-2°C$) shows the responses after cooling the inferior thalamic peduncle bilaterally to $-2°C$, which was sufficient to abolish recruiting responses. Note the reduction in AS of the negative response during the third interval. The fourth row (Cond. 531–560) shows the responses after 500 additional trials (overtraining). Note in the AS the reduction of the CS-elicited evoked response and the longer latency to onset of the negative response. Also note the disappearance of the negative responses in C and AL. The last row (Cool $-5°C$) shows the effect of cooling the ITP after overtraining. Calibrations: Each time interval is 600 msec, and the ventrical deflection is 60 μV. (From Skinner, 1971.)

THE NONSPECIFIC THALAMOCORTICAL SYSTEM

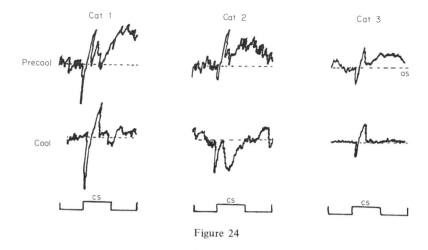

Figure 24

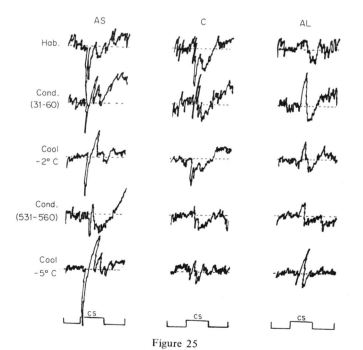

Figure 25

in the AS cortex. The fourth line, Cond. (531–560), shows the computer-averaged response recorded in the noncooled state after 500 trials of overconditioning. Note the reduced amplitude of the CS-elicited evoked potential recorded in the AS, and also note the occurrence of a positive shift following the response and the delayed onset of the negative shift. The negative shifts have dropped out of the recordings in the C and AL after overtraining. The last line in the figure, Cool −5°C, shows the effects of cryogenic blockade on the subsequent block of 30 trials. A slight increase in the degree of cooling results in an increase in the amplitude of the CS-evoked response (compare Cool −2°C and Cool −5°C).

C. Discussion

In contrast to the results of Goldring and O'Leary (1957), recruiting responses were not found superimposed upon a negative steady potential shift but the augmenting responses were. Goldring and O'Leary (1957) may have actually been producing augmenting responses that were superimposed upon negative steady potentials, which they interpreted as recruiting responses because they were recording only in the monophasic-negative surround region of the pattern of evoked potentials. This seems especially likely because they found the steady potential shift only when stimulating in the region of the nucleus paracentralis, which, from the work of Schlag and Villablanca (1967), is known to produce a response elicited by slightly more lateral stimulation in the nucleus ventralis posteromedialis.

It is not known whether the steady potential, defined as the potential above the baseline just preceding each separate evoked response in the repetitive train, is a sustained potential induced during stimulation of the nucleus ventralis lateralis or is an apparently sustained response due to each successive response originating on a long latency component of the preceding one. Increasing the interstimulus interval of the repetitive train (reducing the frequency) caused the steady potential to increase, suggesting that the peak amplitude of the poststimulus wave, defined as the response following the last evoked potential of the repetitive train, is the maximal amplitude of the steady potential. Perhaps the peak of the poststimulus wave is a better index of the amplitude of the steady potential because it does not vary with the frequency of the stimulus train.

An independence between the initial positive–negative components of the augmenting response and the steady potential or poststimulus wave is observed during cryogenic blockade in the region of the ITP or medial internal capsule (Med. IC), as illustrated in Figure 23. Blockade in the ITP (Figure 23A) only slightly reduces the amplitude of the augmenting response in the ipsilateral anterior sigmoid cortex (RAS_{v1}) and not at all in the ipsilateral coronal cortex (RC_{v1}), but markedly reduces the amplitudes of the steady potential and poststimulus wave in both cortical regions, especially in the RAS. Blockade in the Med. IC (Figure 23B) reduces the amplitude of the initial positive—negative components of the recruiting response in the RAS but does not reduce the amplitude of the steady potential or poststimulus wave.

This independence between the VL-elicited augmenting response on the one hand and the steady potential shift and poststimulus wave on the other suggests separate neural mechanisms for their generation, utilizing separate thalamocortical pathways. Only the stimulation of the motor relay nucleus, VL, produces the three responses. Activation of the nonspecific thalamocortical system alone, by stimulation of the NCM, is not sufficient to produce the steady potential shift and poststimulus wave, suggesting that the origin of these responses, like that of the augmenting response, is at the motor relay nucleus. Thus, it appears that the steady potential shift and poststimulus wave are generated by the same thalamic system as the augmenting response but are manifested in the cortex only if the nonspecific thalamocortical system mediating the recruiting response is intact.

During the early stages of conditioning, the conditioned stimulus (CS) evokes a surface negative response in the interval between the CS and the mild shock reinforcement that is recorded in all three anterior cortical areas, the AS, C, and AL cortices. Blockade in the ITP markedly reduces or abolishes this response in the AS cortex but only slightly reduces the magnitude of the response, if any, recorded in the C and the AL. After overtraining (more than 500 conditioning trials), the negative response is not recorded in C or AL but persists in AS. In the latter, the response has a longer latency of onset that rapidly reaches a maximum just before the onset of the expected unconditioned stimulus. These results confirm those of Borda (1970) in showing that after overtraining there is a dichotomy between the surface negative response recorded in the frontal premotor and motor regions. Borda (1970) used a positive-reinforcement conditioning paradigm, but in a tone—shock conditioning situation similar to the one used in the present experiment Chiorini (1969) showed a reduction of the steady potential amplitudes in the sensory-motor cortex after overlearning. These results suggest that the responses recorded in the motor and anterior association cortices are not related to the expected event, because they drop out after overtraining. Since these responses are recorded over the motor cortex, they may be related to the negative premovement potential discussed at the beginning of this section.

The fact that a reversal of potential in the surface negative response occurred when the recording electrode was inserted in the depth underneath the cortical layers suggests that the dipolar generators of the CS-elicited potential are located wholly within the cortex. Intracellular hyperpolarization of cortical neurons has been correlated with long-duration, surface negative responses that occur spontaneously (Jasper and Stefanis, 1965). Recently, Castellucci and Goldring (1970) have reported indirect evidence that suggests that the glial cells also play a role in the electrogenesis of certain long-lasting negative potentials recorded from the surface of the cortex. These data are only suggestive of cortical mechanisms underlying such responses as the CNV or the conditioned negative response reported here, and how these potentials are regulated by subcortical mechanisms is yet another matter. Purpura *et al.* (1966) have shown some interesting synaptic relations between the nonspecific thalamocortical system, the VL motor-projection system, and the anterior cortical regions, which indicates that these subcortical mechanisms may play an important role in regulating events in the anterior regions of the cortex.

V. BEHAVIORAL DEFICITS

An attempt has been made to apply some quantitative tests and measures of behavioral function in order to assess the changes in behavior caused by blockade or interference with the nonspecific thalamoorbitofrontal system believed to regulate electrocortical synchronizing functions and thereby internal inhibition. When large orbitofrontal ablations or electrolytic lesions of the ITP have been produced, it is not only difficult to keep the animals alive, even with intensive care, but there are also symptoms, such as obstinate progression, that interfere with the application of behavioral tests. In contrast, we have found that cryogenic blockade of the ITP does not produce these marked behavioral aberrations and interfering effects. Therefore, in the subsequent sections dealing with the quantitative aspects of behavioral change, only cryogenic blockade has been utilized. Further comment on additional qualitative changes in behavior resulting from orbitofrontal ablations or lesions of ITP will be made at the end of this section.

A. Disruption of Bar Pressing

Concurrently with the assessment of bar-pressing behavior resulting from local reversible cooling of the ITP to $10°C$, recordings of cortical potentials were made in four chronic preparations. In Figure 26, E (experimental condition) shows the effects of local cooling in the critical region of the ITP. The first column (Pre-) shows the visual evoked potentials, recruiting responses, and cumulative bar-pressing record prior to cooling. The second column (Cool) shows these responses after cooling. In addition to the electrophysiological effects already described, bar-pressing behavior was interrupted. The behavior of the animals at this time suggested inattention and uninterest; they tended to turn away from the bar and wander about the experimental chamber, grooming, sniffing, and looking at the one-way mirror. However, there was evidence that the animals could attend, at least briefly, and could press the bar, because it was possible to coax them back to the bar by providing free reinforcements accompanied by the associated click of the apparatus, as they turned away from the bar.

In Figure 26, the control condition (C), with the cryoprobe placed outside the critical region of the ITP although adjacent to it, produced relatively little effect. In five chronic preparations, there was little, if any, change in the sensory evoked response or the recruiting responses, and the animals did not cease bar pressing, even at the time of a distracting stimulus introduced midway in the bar-pressing record (triangle marker). However, the rate of bar pressing was slower.

B. Inability to Perform Single Alternation

Three animals, with cryoprobes implanted in the critical region of the ITP, were trained to press one bar, drink the milk reward, then press alternately the

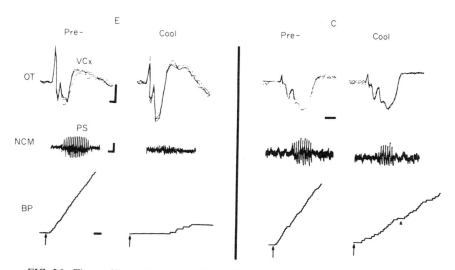

FIG. 26. Three effects of cryogenic blockade of the inferior thalamic peduncle. Stimulation of the optic tract (OT) evoked responses recorded on the striate cortex (VCx). Each of the three superimposed traces represent the mean of 30 averaged responses. Stimulation of the nucleus centralis medialis (NCM) evoked recruiting responses recorded on the ipsilateral posterior sigmoid gyrus (PS). Cumulative response records were recorded for bar pressing (BP) for a milk reward. E, chronic preparations with cryoprobes bilaterally implanted in the critical region of the ITP; C, control preparations with cryoprobes implanted close to, but not in, the critical region of the ITP. Left columns, (Pre-) in E and C, show responses before cooling; right columns (Cool) shows responses during cooling to 10°C. Calibrations: 100 μV for all amplitude markers; 0.5 sec for recruiting response, 5 msec for visual response, and 1 min for bar-press time markers. (From Skinner and Lindsley, 1967.)

other bar, and so on. Incorrect responses were not rewarded, but immediately after making a mistake the animals were allowed to make the correct response. After 3 weeks of daily training, which included approximately 125 trials per day, the animals stabilized at about 75–80% correct alternation performance.

The lowest temperature that did not affect sensory evoked potentials, recruiting responses, and single alternation performance was 25°C (all the temperatures reported in this section were recorded from the probe–tissue interface). The cooling of the ITP to 15°C, the lowest temperature that did not cause bar pressing to cease, affected single alternation performance, partially enhanced visual evoked potentials, and partially reduced recruiting responses.

Figure 27 shows the effects of cooling to 15°C in the ITP on the single alternation task. At cooling to 25°C, as shown by the stippled bars of Figure

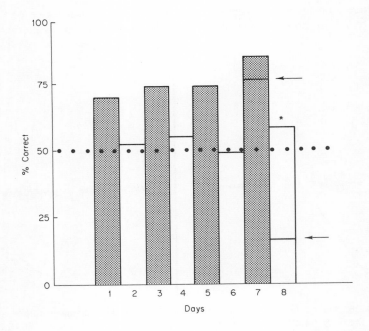

FIG. 27. Deficits in performance of a single alternation task during partial cryogenic blockade of the inferior thalamic peduncle. Each vertical bar represents the mean percent correct scores for the group of animals during cooling to 15°C (clear) or 25°C (shaded). Each of the animals was run with alternate temperatures during 8 consecutive days. The asterisk (day 8) indicates that only the cryoprobe tips were cooled; on the other days, both the tip and the shaft were cooled. The arrows (day 7 and 8) indicate percent correct scores on nonstrategy trials. (From Skinner and Lindsley, 1967.)

27, alternation performance was 70–80% correct; at cooling to 15°C alternation performance was reduced to about 50%, or chance level. Since cooling to 15°C reduced bar-pressing rate more than cooling to 25°C, an analysis of covariance between bar-press rate and percent correct single alternation performance was carried out. This indicated that the decrease in performance was not due to the change in the bar-pressing rate ($p < 0.01$).

During the 8 days of testing, it was noticed that animals, in order to remember the lever last pressed, developed a strategy of leaving their paw outstretched on the correctly pressed bar while drinking, and then removed it and pressed the other bar when it came time to make the next response. Being unable to keep a forelimb outstretched in this fashion continuously, the animals used the strategy in only about one half of the alternation trials, which could account for their 50% correct alternation performance.

If the paw remained outstretched on the correctly pressed bar while the animal was drinking the reward, then the following trial was classified as a strategy trial, and if both paws were underneath the animal, supporting it while it was drinking, then the following trial was classified as a nonstrategy trial. Analyses of trials on days 7 and 8 revealed that, during cooling to 15°C, nonstrategy trials resulted in only 25% correct performance, whereas, during cooling to 25°C, nonstrategy trials resulted in the normal 75% correct performance (in Figure 27, performance on nonstrategy trials is indicated by the arrows). The less than chance performance during cooling to 15°C demonstrated a perseverance of response to the last rewarded bar.

C. Normal Distraction from Bar Pressing

A loud 2 min tape recording of barking dogs played over a speaker in the back of the apparatus box distracted the animals while they were bar pressing for milk. The delay in bar pressing while the animals oriented toward the distracting stimulus was the measure of distractability.

Figure 28 shows the distractability in three animals during eight consecutive days of alternate cooling to 15 and 25°C in the ITP. On the first day, with cooling to 15°C, the novel 2 min distracting stimulus caused the animals to stop bar pressing and orient toward the speaker for about 1 min. On the second day, with cooling to 25°C, the animals stopped bar pressing and oriented toward the speaker for only a few seconds. On the third day, with 15°C cooling, the animals only briefly oriented to the distracting stimulus, showing no increased distractability over the previous day. Another ITP animal was run on all 8 days with cooling to 25°C and showed the same distract–delay values as shown in Figure 28. Thus, cooling in the ITP sufficient to produce the deficit in single alternation performance had no effect on distractability or habituation to a novel stimulus.

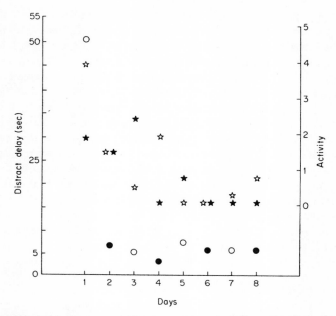

FIG. 28. Effects of cryogenic blockade in inferior thalamic peduncle on distractibility and motor activity. Cryoprobes were cooled to 15°C or 25°C alternately on 8 consecutive days. The same animals as used in the single alternation task were used here. The mean number of seconds the animals stopped bar pressing after the onset of a 2 min recording of barking dogs (distract delay) is indicated by the open circles (15°C) and filled circles (25°C). Measurements of the activity of the same animals were made each day just prior to the distractibility experiments. The mean number of times the animals moved their right hind paws in a 1 min period is indicated by the open stars (15°C) and filled stars (25°C). (From Skinner and Lindsley, 1967.)

D. Normal Motor Activity

Activity was measured by counting the number of times an animal moved his right hind leg during a 1 min period. Measures with cooling to both 15°C and 25°C were taken daily in the same animals as those used in Section V,C, just before they were run in the distractability experiments. There was no difference in activity produced by the different blocking temperatures, and after about 4 days of testing in the apparatus, the animals were quite inactive during both conditions of cooling (Figure 28).

E. Obstinate Progression Produced by Large Orbitofrontal Cortex Ablations and ITP Lesions

In three chronic preparations with large orbitofrontal ablations, and in two chronic preparations with electrolytic lesions of the region of the ITP, behavior was characterized by incessant pacing, violent struggling when forward progress was blocked, and generalized inattention to all stimuli, including novel and unexpected stimuli. Loud hand claps or a bright flash of light appeared to elicit very little reaction, suggesting that the animals were either insensitive to or ignored these stimuli. When momentarily quiet, pinching of the tail did not produce the usual reaction of a normal animal, but instead initiated the pattern of obstinate progression, which continued even when the animal bumped into a wall or forced himself into a corner of the cage. Often the animal would persist in this activity until exhausted. Resistance to forward progress seemed to accentuate the vigorousness of such movements.

EEG recordings during the first week following surgery showed no spontaneous activity in the normal frequency range of 7–8 Hz but instead manifested high-voltage slow waves of 1–3 Hz interspersed with paroxysmal activity. Recruiting responses could not be elicited.

F. Discussion

Early studies of the behavioral changes produced by ablation of the frontal granular cortex interpreted the deficit as a result of the loss of the faculty of attention (Ferrier, 1890; Jacobsen, 1935; Malmo, 1942). In an attempt to clarify the interpretation, recent investigators have explained the nature of the deficit in terms of the "perseveration of central set" (Mishkin, 1964), "inability to inhibit a strong response tendency" (Brutkowski, 1964, 1965), or "increased reflexogenic value of stimuli" (Konorski and Lawicka, 1964), which are all more closely related to the animal's observed behavior than the vaguely defined concept of attention. Whatever the degree to which the description of the behavioral change can be interpreted as a loss in the faculty of attention, it appears that blockade of either the frontal granular cortex or the inferior thalamic peduncle is sufficient to produce the behavioral manifestations. Schulman (1964) has shown that lesions in the nucleus medialis dorsalis will also produce such behavioral changes (i.e., a deficit in the delayed response task). These combined results suggest that the integrity of a thalamocortical system that overlaps with the electrophysiologically defined nonspecific thalamocortical system is necessary for the mediation of these complex behavioral tasks.

Malmo (1942) found that monkeys could perform the delayed response task after prefrontal lobectomy if the immediate environment was darkened to reduce or eliminate extraneous distracting stimuli. He attributed the deficit, manifested during the condition in which the room lights were on, to retroactive inhibition, distractibility, and inattention. Two particular types of tasks are especially disrupted by lesions of the frontal granular cortex: the delayed response task and the discrimination reversal task. In the former, the subject is required to withhold his discriminatory response for a brief time interval after he has received the relevant discrimination cues; the lesioned animal makes errors by persistently responding to certain sensory cues in the test situation (Malmo, 1942; Mishkin and Pribram, 1956; Brush et al., 1961; Mishkin et al., 1962). In the latter task, the subject is required to discontinue his responding to a previously correct discriminative cue and respond instead to the previously incorrect one; the lesioned animal makes errors because it persists in responding to the previously correct cue (Brush et al., 1961; Mishkin et al., 1962). However, the frontal lobe lesioned animals can perform both these tasks correctly if certain strong tendencies to respond to irrelevant aspects of the sensory cues are eliminated. If a tendency to perform a response to a preferred stimulus (usually the first correct one chosen in learning the task) is eliminated by not rewarding this response, then the animal can correctly perform the delayed response task (Konorski and Lawicka, 1964). If a tendency to respond to the previously correct cue is eliminated by substituting a new stimulus in its place at the time of required response reversal, the animal quickly learns to respond to the previously incorrect stimulus (Mishkin, 1964). In summary, the lesioned animal can perform the complex discrimination tasks but is often unable by itself to choose among the various alternative stimuli that control his response behavior. An animal with functional blockade in the inferior thalamic peduncle, as seen in the present results, manifests behavioral deficits that are similar to those produced by lesions in the frontal granular cortex and the nucleus medialis dorsalis thalamicus, thus giving general validity to the statement that the nonspecific thalamocortical system underlies the performance of these tasks.

There have been numerous reports in the literature that frontal lobe lesions produce hyperactivity and hyperdistractability, and an equal number of reports that such lesions do not produce these results. The present results show that large total ablations of the orbitofrontal cortex and large electrolytic lesions in the vicinity of the inferior thalamic peduncle will lead to a form of hyperactivity characterized by a total lack of distractability (obstinate progression). However, cryogenic blockade and small heat lesions, just large enough to abolish recruiting responses, do not produce the hyperactivity and lack of distractability in chronic preparations, suggesting that some parasurgical phenomenon is associated with the methods of ablation and electrolytic lesioning that may account for the divergent results. This parasurgical phenomenon may be related to either an irritated margin of tissue at the border of the functional blockade or an inability to control the localization of the functional blockade because of vascular damage, neither of which seems to be produced by the cooling or heating methods.

VI. CONCLUSIONS

Functional blockade at any point in the nonspecific thalamocortical system renders an animal unable to perform certain types of behavioral tasks (e.g., delayed response, discrimination reversal, response alternation). However, the blockade does not render the subjects totally unable to perform the tasks correctly if certain controls of the experimental situation are provided by the experimenter, such as turning off the room lights to eliminate distracting visual objects, extinguishing strong competing response tendencies, substituting novel stimuli for the response reversal cues, or allowing the animal to develop and use a mnemonic strategy. Whether the provided control is of the stimulus environment or the response choice, or both, is not certain. Perhaps *stimulus and response control* is the better name for the factor in the experimental situation that enables the animals to perform when their nonspecific thalamocortical system is blocked, because both interpretations seem to be supported by the experimental evidence. It may well be that, in the intact animal, these are not independent types of control exerted by the particular mechanism mediating the behavior.

The experimental electrophysiological evidence reported in this review is beginning to give us some insights into the operation of the mechanism underlying the higher order types of behavior that require stimulus selection and response choice.

(1) The nonspecific thalamocortical system projects to the orbitofrontal cortex and mediates a certain type of electrocortical synchronization associated with recruiting responses, caudate spindles, barbiturate spindles, atropine spindles, tetrahydrocannabinol spindles, spontaneous spindle bursts, and natural sleep spindles. Only certain synchronous components are abolished when the nonspecific thalamocortical system is blocked, leaving intact augmenting responses and certain other synchronous components that presumably underlie undisturbed neural functions. The system mediating the particular type of synchronization blocked by cooling in the inferior thalamic peduncle seems to be a diffuse one, reaching out to wide regions of the frontal half of the cortex.

(2) The nonspecific thalamocortical system controls the electrical excitability of at least two sensory systems (visual and auditory), and perhaps all, at both the thalamic and cortical levels. This control maintains a tonic suppression of evoked responses to sensory stimuli in primary, association, and nonprimary cortices. When this suppression is released by blockade in the nonspecific thalamocortical system, unattended habituated stimuli will evoke large amplitude responses, just as attended stimuli will in the intact animal.

(3) The nonspecific thalamocortical system regulates an orbitofrontal projection system that is capable of modifying a surface negative, depth positive electrocortical slow potential elicited by a behaviorally conditioned state of vigilance and expectancy. Blockade in the inferior thalamic peduncle component of the midline nonspecific thalamocortical system will abolish the negative slow potential elicited by the temporal association of two sensory stimuli. This slow potential elicited by behavioral conditioning may be related to the surface negative electrocortical response elicited by repetitive stimulation of the

thalamic motor-relay nucleus, the nucleus ventralis lateralis. This second type of slow potential projects to frontal cortices and is also abolished by blockade in the inferior thalamic peduncle.

How all these data fit together to explain the brain mechanisms underlying such complex behavior as that exhibited in the delayed response, discrimination reversal, and response alternation tasks is for the moment only speculative. Such an explanation must await further experimental evidence specifically relating each of these electrophysiological correlates to the animal's behavior. At this particular time, however, we are beginning to see a common relationship among the mechanisms of internal inhibition as inferred from sleep and other electrocortical synchronization processes, selective attention and perception, and certain other higher order behaviors, all believed to share common mechanisms because they are dependent on the integrity of the midline nonspecific thalamocortical system.

REFERENCES

Akert, K. (1964). Comparative anatomy of frontal cortex and thalamo-frontal connections. *In* "The Frontal Granular Cortex and Behavior" (J. M. Warren and K. Akert, ed.), p. 372. McGraw-Hill, New York.

Akert, K., Koella, W. P., and Hess, J. R., Jr. (1952). Sleep produced by electrical stimulation of the thalamus. *Amer. J. Physiol.* 168; 260-267.

Akimoto, H., Yamaguchi, N., Okabe, K., Nakagawa, T., Nakamura, I., Abe, K., Torri, H., and Masahashi, K. (1956). On sleep induced through electrical stimulation of dog thalamus. *Folia Psychiat. Neurol. Jap.* 10: 117-146.

Auer, J. (1956). Terminal degeneration in the diencephalon after ablation of frontal cortex in the cat. *J. Anat.* 90, 30-41.

Borda, R. P. (1970). The effect of altered drive states on the contingent negative variation (CNV) in rhesus monkeys. *Electroencephalogr. Clin. Neurophysiol.* 29, 173-180.

Bremer, F. (1937). L'activité cérébrale au cours du sommeil et de la narcose. Contribution à l'étude du mécanisme du sommeil. *Bull. Acad. Roy. Med. Belg.* 4, 68-86.

Bremer, F., and Stoupel, N., (1959). Facilitation et inhibition des potentiels évoqués corticaux dans l'éveil cérébral. *Arch. Int. Physiol. Biochem.* 67, 1-35.

Brindley, G. S. (1960). "Physiology of the Retina and the Visual Pathway," pp. 116-120. Arnold, London.

Brush, E. S., Mishkin, M., and Rosvold, H. E. (1961). Effects of object preferences and aversions on discrimination learning in monkeys with frontal lesions. *J. Comp. Physiol. Psychol.* 54, 319-325.

Brutkowski, S. (1964). Prefrontal cortex and drive inhibition. *In* "The Frontal Granular Cortex and Behavior" (J. M. Warren and K. Akert, eds.), pp. 242-270. McGraw-Hill, New York.

Brutkowski, S. (1965). Functions of prefrontal cortex in animals. *Physiol. Rev.* 45, 721-746.

Buser, P., Rougeul, A., and Perret, C. (1964). Caudate and thalamic influences on conditioned motor responses in the cat. *Bol. Inst. Estud. Méd. Biol. (Méx.)* 22, 293-307.

Castellucci, V. F., and Goldring, S. (1970). Contribution to steady potential shifts of slow depolarization in cells presumed to be glia. *Electroencephalogr. Clin. Neurophysiol.* 28, 109-118.

Chiorini, J. R. (1969). Slow potential changes from cat cortex and classical aversive conditioning. *Electroencephalogr. Clin. Neurophysiol* 26, 399-406.
Chow, K. L., Dement, W. C., and Mitchell, S. A., Jr. (1959). Effects of lesions of the rostral thalamus on brain waves and behavior in cats. *Electroencephalogr. Clin. Neurophysiol* 11, 107-120.
Cordeau, J.,P., Walsh, J., and Mahut, H. (1965). Variations dans la transmission des messages sensoriels en fonction des différents états d'éveil et de sommeil. *In* "Aspects anatomo-fonctionnels de la physiologie du sommeil" (M. Jouvet, ed.), pp. 477-508. CNRS, Paris.
Deecke, L., Scheid, P., and Kornhuber, H. H. (1969). Distribution of readiness potential, pre-motion positivity, and motor potential of the human cerebral cortex preceding voluntary finger movements. *Exp. Brain Res.* 7, 158-168.
Dondey, M., Albe-Fessard, D., and LeBeau, J. (1962). Premières applications neurophysiologiques d'une méthod permettant le blocage électif et réversible de structures centrales par réfrigeration localisée. *Electroencephalogr. Clin. Neurophysiol.* 14. 758-763.
Dumont, S., and Dell, P. (1960). Facilitation réticulaire des mécanismes visuels corticaux. *Electroencephalogr. Clin. Neurophysiol.* 12, 769-796.
Eidelberg, E., Hyman, W., and French, J. D. (1958). Pathways for recruiting responses in rabbits. *Acta Neurol. Latinoamer.* 4, 279-287.
Ferrier, D. (1890). "The Croonian Lectures on Cerebral Localization," p. 151. Smith, Elder & Co., London.
Galambos, R., and Sheatz, G. C. (1962). An electroencephalographic study of classical conditioning. *Amer. J. Physiol.* 203, 173-184.
Galambos, R., Sheatz, G., and Vernier, V. G. (1956). Electrophysiological correlates of a conditioned response in cats. *Science* 123, 376-377.
Garcia-Austt, E. (1963). Influence of the states of awareness upon sensory evoked potentials. *Electroencephalogr. Clin. Neurophysiol., Suppl.* 24, 76-89.
Garcia-Austt, E., Bogacz, J., and Vanzulli, A. (1964). Effects of attention and inattention upon visual evoked responses. *Electroencephalogr. Clin. Neurophysiol.* 17, 136-143.
Gilden, L., Vaughan, H. G., Jr., and Costa, L. D. (1966). Summated human EEG potentials with voluntary movement. *Electroencephalogr. Clin. Neurophysiol.* 20, 433-438.
Goldring, S., and O'Leary, J. L. (1957). Cortical D. C. changes incident to midline thalamic stimulation. *Electroencephalogr. Clin. Neurophysiol.* 9, 577-584.
Haider, M., Spong, P., and Lindsley, D. B. (1964). Attention, vigilance, and cortical evoked potentials in humans. *Science* 145, 180-182.
Hanberry, J., Ajmone Marsan, C., and Dilworth, M. (1954). Pathways of nonspecific thalamo-cortical projection system. *Electroencephalogr. Clin. Neurophysiol.* 6, 103-118.
Hearst, E., Beer, B., Sheatz, G., and Galambos, R. (1960). Some electro-physiological correlates of conditioning in the monkey. *Electroencephalogr. Clin. Neurophysiol* 12, 137-152.
Hess, W. R. (1944). Das Schlafsyndrom als Folge dienzephaler Reizung. *Helv. Physiol. Pharmacol. Acta* 2, 304-344.
Hess, W. R. (1954). The diencephalic sleep centre. *In* "Brain Mechanisms and Consciousness" (E. D. Adrian, F. Bremer, and H. H. Jasper, eds.), p. 117. Thomas, Springfield, Illinois.
Hillyard, S. A., and Galambos, R. (1970). Eye movement artifact in the CNV. *Electroencephalogr. Clin. Neurophysiol.* 28, 173-182.
Irwin, D. A., Knott, J. R., McAdam, D. W., and Rebert, C. S. (1966). Motivational determinants of the "contingent negative variation." *Electroencephalogr. Clin. Neurophysiol.* 21, 538-543.
Jacobsen, C. F. (1935). Functions of the frontal association area in primates. *Arch. Neurol. Psychiat.* 33, 558-569.

Jasper, H. H. (1949). Diffuse projection systems: The integrative action of the thalamic reticular system. *Electroencephalogr. Clin. Neurophysiol.* **1**, 405-419.
Jasper, H. H. (1961). Thalamic reticular system. *In* "Electrical Stimulation of the Brain" (D. E. Sheer, ed.), p. 277. Univ. of Texas Press, Austin.
Jasper, H. H., and Ajmone Marsan, C. (1952). Thalamo-cortical integrating mechanisms. *In* "Patterns of Organization in the Central Nervous System" (P. Bard, ed.), p. 493. Williams & Wilkins, Baltimore, Maryland.
Jasper, H. H., and Stefanis, C. (1965). Intracellular oscillatory rhythms in pyramidal tract neurons in the cat. *Electroencephalogr. Clin. Neurophysiol* **18**, 541-553.
Konorski, J., and Lawicka, W. (1964). Analysis of errors by prefontal animals on the delayed-response test. *In* "The Frontal Granular Cortex and Behavior" (J. M. Warren and K. Akert, eds.), pp. 271-294. McGraw-Hill, New York.
Kornhuber, H. H., and Deecke, L. (1965). Hernpotentialänderungen bei Willkürbewegungen und passiven Bewegungen des Menchen: Bereitschaftspontential und reafferente Potentiale. *Pflüegers Arch. Gesamte Physiol Menshen Tiere* **284**, 1-17.
Lindsley, D. B. (1960). Attention, consciousness, sleep and wakefulness. *In* "Handbook of Physiology" (Amer. Physiol. Soc., J. Field, ed.), Sect. 1, Vol. III, p. 1553. Williams & Wilkins, Baltimore, Maryland.
Lindsley, D. B., Bowden, J. W., and Magoun, H. W. (1949). Effect upon the EEG of acute injury to the brain stem activating system. *Electroencephalogr. Clin. Neurophysiol.* **1**, 475-486.
Low, M. D., and McSherry, J. W. (1968). Further observations of psychological factors involved in CNV genesis. *Electroencephalogr. Clin. Neurophysiol.* **25**, 203-207.
Low, M. D., Borda, R. P., Frost, J. D., and Kellaway, P. (1966). Surface-negative, slow-potential shift associated with conditioning in man. *Neurology* **16**, 771-782.
McAdam, D. W., and Seales, D. M. (1969). Bereitschaftpotential enhancement with increased level of motivation. *Electroencephalogr. Clin. Neurophysiol.* **27**, 73-75.
McCallum, W. C., and Walter, W. G. (1968). The effects of attention and distraction on the contingent negative variation in normal and neurotic subjects. *Electroencephalogr. Clin. Neurophysiol.* **25**, 319-329.
Magoun, H. W. (1963). "The Waking Brain," p. 174. Thomas, Springfield, Illinois.
Malmo, R. B. (1942). Interference factors in delayed response in monkeys after removal of frontal lobes. *J. Neurophysiol.* **5**, 295-308.
Manghi, E., Rosina, A., and Mancia, M. (1965). Thalamo-cortical connection as revealed by degeneration study in the cat. *Proc. Int. Congr. Neurol., 8th,* **4**, 15-17.
Marsh, J. T., McCarthy, D. A., Sheatz, G. and Galambos, R. (1961). Amplitude changes in evoked auditory potentials during habituation and conditioning. *Electroencephalogr. Clin. Neurophysiol.* **13**, 224-234.
Mishkin, M. (1964). Perseveration of central sets after frontal lesions in monkeys. *In* "The Frontal Granular Cortex and Behavior" (J. M. Warren and K. Akert, eds.), pp. 219-241. McGraw-Hill, New York.
Mishkin, M., and Pribram, K. H. (1956). Analysis of the effects of frontal lesions in the monkey. II. Variations of delayed response. *J. Comp. Physiol. Psychol.* **49**, 36-40.
Mishkin, M., Prockop, E. S., and Rosvold, H. E. (1962). One-trial object-discrimination learning in monkeys with frontal lesions. *J. Comp. Physiol. Psychol.* **55**, 178-181.
Monnier, M., Kalberer, M., and Krupp, P. (1960). Functional antagonism between diffuse reticular and intralaminary recruiting projections in the medial thalamus. *Exp. Neurol.* **2**, 271-289.
Morison, R. S., and Dempsey, E. W. (1942). A study of thalamo-cortical relations. *Amer. J. Physiol.* **135**, 281-292.

Moruzzi, G., and Magoun, H. W. (1949). Brain stem reticular formation and activation of the EEG. *Electroencephalogr. Clin. Neurophysiol.* 1, 455-473.

Müller, J. (1838). "Handbuch der Physiologiedes Menschen für Vorlesungen." Holscher, Coblenz.

Nauta, W. J. H. (1962). Neural associations of the amygdaloid complex in the monkey. *Brain* 85, 505-520.

Nauta, W. J. H. (1964). Some efferent connections of the prefrontal cortex in the monkey. *In* "The Frontal Granular Cortex and Behavior" (J. M. Warren and K. Akert, eds.), p. 397. McGraw-Hill, New York.

Nauta, W. J. H., and Whitlock, D. G. (1954). An anatomical analysis of the nonspecific thalamic projection system. *In* "Brain Mechanisms and Consciousness" (J. F. Delafresnaye, ed.), p. 81. Thomas, Springfield, Illinois.

Purpura, D. P., Frigyesi, T. L., McMurthry, J. G., and Scarff, T. (1966). Synaptic mechanisms in thalamic regulation of cerebello-cortical projection activity. *In* "The Thalamus" (D. P. Purpura and M. D. Yahr, eds.), pp. 153-172. Columbia Univ. Press, New York.

Rebert, C. S., and Irwin, D. A. (1969). Slow potential changes in cat brain during classical appetitive and aversive conditioning of jaw movements. *Electroencephalogr. Clin. Neurophysiol.* 27, 152-161.

Rebert, C. S., McAdam, D. W., Knott, J. R., and Irwin, D. A. (1967). Slow potential change in human brain related to level of motivation. *J. Comp. Physiol. Psychol.* 63, 20-23.

Robertson, R. T., and Lynch, G. S. (1971). Orbitofrontal modulation of EEG spindles. *Brain Res.* 28, 562-566.

Sasaki, K., Staunton, H. P., and Dieckmann, G. (1970). Characteristic features of augmenting and recruiting responses in the cerebral cortex. *Exp. Neurol.* 26, 369-392.

Scheibel, M. E., and Scheibel, A. B. (1967). Structural organization on nonspecific thalamic nuclei and their projection toward cortex. *Brain Res.* 6, 60-94.

Schlag, J. D., and Chaillet, F. (1963). Thalamic mechanisms involved in cortical desynchronization and recruiting responses. *Electroencephalogr. Clin. Neurophysiol.* 15, 39-62.

Schlag, J., and Villablanca, J. (1967). Cortical incremental responses to thalamic stimulation. *Brain Res.* 6, 119-142.

Schulman, S. (1964). Impaired delayed response from thalamic lesions. *Arch. Neurol. (Chicago)* 11, 477-499.

Skinner, J. E. (1970). A cryoprobe and cryoplate for reversible functional blockade in the brains of chronic animal preparations. *Electroencephalogr. Clin. Neurophysiol.* 29, 204-205.

Skinner, J. E. (1971a). Abolition of a conditioned, surface-negative, cortical potential during cryogenic blockade of the non-specific thalamo-cortical system. *Electroencephalogr. Clin. Neurophysiol.* 31, 197-209.

Skinner, J. E. ((1971b). "Neuroscience: A Laboratory Manual." Saunders, Philadelphia, Pennsylvania.

Skinner, J. E. (1971c). Abolition of several forms of cortical synchronization during blockade in the inferior thalamic peduncle. *Electroencephalogr. Clin. Neurophysiol.* 31, 211-221.

Skinner, J. E., and Lindsley, D. B. (1967). Electrophysiological and behavioral effects of blockade of the nonspecific thalamo-cortical system. *Brain Res.* 6, 95-118.

Skinner, J. E., and Lindsley, D. B. (1968). Reversible cryogenic blockade of neural function in the brain of unrestrained animals. *Science* 161, 595-597.

Skinner, J. E., and Lindsley, D. B. (1971). Enhancement of visual and auditory evoked potentials during blockade of the non-specific thalamo-cortical system. *Electroencephalogr. Clin. Neurophysiol.* 31, 1-6.

Spencer, W. A., and Brookhart, J. M. (1961a). Electrical patterns of augmenting and recruiting waves in depths of sensory-motor cortex of cat. *J. Neurophysiol.* **24**, 26-49.

Spencer, W. A., and Brookhart, J. M. (1961b). A study of spontaneous spindle waves in sensorimotor cortex of cat. *J. Neurophysiol.* **24**, 50-65.

Spong, P., Haider, M., and Lindsley, D. B. (1965). Selective attentiveness and cortical evoked responses to visual and auditory stimuli. *Science* **148**, 395-397.

Starzl, T. E., and Magoun, H. W. (1951). Organization of the diffuse thalamic projection system. *J. Neurophysiol.* **14**, 133-146.

Staunton, H. P., and Sasaki, K. (1971). Recruiting responses not dependent on orbitofrontal cortex. *Brain Res.* **30**, 415-418.

Tecce, J. J., and Scheff, N. M. (1969). Attention reduction and suppressed direct-current potentials in the human brain. *Science* **164**, 331-333.

Tissot, R., and Monnier, M. (1959). Dualité du système thalamique de projection diffuse. *Electroencephalogr. Clin. Neurophysiol.* **11**, 675-686.

Vaughan, H. G., Jr., Costa, L. D., and Ritter, W. (1968). Topography of the human motor potential. *Electroencephalogr. Clin. Neurophysiol.* **25**, 1-10.

Velasco, M., and Lindsley, D. B. (1965). Role of orbital cortex in regulation of thalamo-cortical electrical activity. *Science* **149**, 1375-1377.

Velasco, M., Skinner, J. E., Asaro, K. D., and Lindsley, D. B. (1968). Thalamo-cortical systems regulating spindle bursts and recruiting responses. I. Effect of cortical ablations. *Electroencephalogr. Clin. Neurophysiol.* **25**, 463-470.

Velasco, M., Skinner, J. E., and Lindsley, D. B. (1973). In preparation.

Walter, W. G. (1966). Expectancy waves and intention waves in the human brain and their application to the direct cerebral control of machines. *Electroencephalogr. Clin. Neurophysiol.* **21**, 616P.

Walter, W. G., Cooper, R., Aldridge, V. J., and McCallum, W. C. (1964a). The contingent negative variation: an electrocortical sign of sensori-motor association in man. *Electroencephalogr. Clin. Neurophysiol.* **17**, 340-341.

Walter, W. G., Cooper, R., Aldridge, V. J., McCallum, W. C., and Winter, A. L. (1964b). Contingent negative variation: An electric sign of sensori-motor association and expectancy in the human brain. *Nature (London)* **23**, 380-384.

Walter, W. G., Cooper, R., McCallum, C., and Cohen, J. (1965). The origin and significance of the contingent negative variation or "expectancy wave." *Electroencephalogr. Clin. Neurophysiol.* **18**, 720P.

Weinberger, N. M., Velasco, M., and Lindsley, D. B. (1965). Effects of lesions upon thalamically induced electrocortical desynchronization and recruiting. *Electroencephalogr. Clin. Neurophysiol.* **18**, 369-377.

Part Six

EXPERIMENTALLY BASED MODELS OF FRONTAL LOBE FUNCTION

Chapter 12

STUDY ON THE FUNCTIONAL MECHANISMS OF THE DORSOLATERAL FRONTAL LOBE CORTEX

M. GERBNER

Institute of Psychology
Hungarian Academy of Sciences
Budapest, Hungary

The dorsolateral surface of the frontal lobe contains two structures that have quite opposite functional characteristics. Lesion of one, the prefrontal area, in experimental animals causes anxiety, hyperactivity, and augmentation of the imperative character of stimuli. In addition, disinhibition of conditioned inhibitions, alternating movements between food compartments (in studies where animals have been trained to distinguish between one or another compartments), disturbances in the dynamic stereotypes, and delayed conditioned responses have been described. Impairment of the delayed reaction has been subjected to thorough study by many investigators (Jacobsen *et al.*, 1935; Jacobsen, 1936; Jacobsen and Nissen, 1937; Langworthy and Richter, 1939; Allen, 1940, 1943, 1949; Malmo, 1942; Konorski *et al.*, 1952; Pribram and Mishkin, 1956; French, 1959, 1966; Paillard, 1960; Pribram, 1961, 1966; Pribram *et al.*, 1964; Konorski and Lawicka, 1964; Shustin, 1966; Shumilina, 1966). In man, lesion of the prefrontal area is most commonly associated with restlessness, irresponsibility, freedom from inhibitions, incontinency of urine and stool, ideational apraxia caused by disintegration of the sequence of motor elements, and perseveration. The patient cannot alter the pattern of performing a task once it has been established and is usually incapable of accomplishing more complicated (particularly asymmetrical) programs of action (Harlow, 1848, 1868; Welt, 1888; Brickner, 1932; Jarvie, 1954, 1958; Milner, 1964; Andrew, 1964; Luria, 1966b, 1967; Lebedinsky, 1966).

If the lesion involves not only the prefrontal area but a second structure, the premotor region, the above symptoms are not so distinct and show a certain complexity. Isolated lesions of the premotor area cause retardation of the movements and diminution of spontaneous motor activity in animals, whereas in man, the breakdown of movements, slowing of the rhythm of motor activity, and the abolition of certain motor elements leads to motor apraxia, and the lesion of more lateral regions causes stuttering, groping for words, and other abnormalities of speech (Fulton et al., 1932; Fulton, 1935; Walshe, 1935; Barris, 1937; Richter and Hines, 1938; Aring, 1949; Conrad, 1954; Paillard, 1960; Spirin, 1966; Shkol'nik-Yapros, 1966; Simerniczkaya and Bunatyan, 1966).

In the experiments to be described here, various parts of the frontal lobes of dogs were removed by suction under sterile conditions. Ablation included the following areas: (1) prefrontal cortex, i.e., the outermost layer of the frontal or proreus gyrus to the pole of the frontal lobe orally and to the precruciate sulcus caudally; (2) lateral part of the premotor area, i.e., the lateral third of the anterior sygmoid gyrus caudally to the line drawn about 1.5 mm frontal to the end of the cruciate sulcus; (3) cortex of the medial portion of the premotor area caudally to the line drawn frontal to the cruciate sulcus; (4) the whole prefrontal and premotor areas, i.e., the frontal and anterior sygmoid gyri, and hence, all the regions listed under (1) through (3). The effect of these ablations was studied by three different methods designated hereafter as Methods A, B, and C.

I. METHOD A: FOOD-ACQUIRING MOTOR REACTION

A dog without previous training, when placed on one side of a screen will use its paw to pull pieces of meat from the other side of the screen. The time necessary to pull ten pieces of meat is longer during the first experiment but it becomes successively shorter with practice, and it is almost invariable after five to ten experimental sessions. The four kinds of brain ablation described above were performed successively in five dogs each. The pattern of leg movements did not change after any of the operations.

Prefrontal ablation (1) was followed by no change in the motor response. After removal of the lateral premotor area (2), however, the latency period until the first leg movement and the time lags between the subsequent motor responses were significantly prolonged. Following extirpation of the medial portion of the premotor cortex (3), the above latency and time periods were somewhat longer than prior to ablation. Complete removal of the prefrontal and premotor areas (4) resulted in moderate prolongation of the above time characteristics. It is remarkable that the paw reaction was considerably more delayed when only the lateral parts of the premotor cortex were removed than when much larger areas, i.e., the whole prefrontal and premotor regions, including the lateral parts of the latter, had been extirpated (Figure 1).

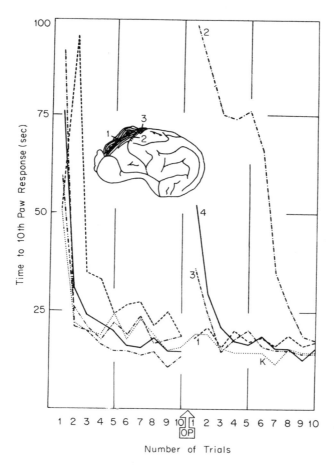

FIG. 1. The effect of frontal lobe ablation on the food-acquiring motor reaction in dogs. OP, ablation. (– – –) Prefrontal ablation (1); (– · –) lateral premotor ablation (2); (– · · –) medial premotor ablation (3); (——) simultaneous ablation of prefrontal and premotor areas (4); (· · ·) control group not subjected to ablation (5).

II. METHOD B: INSTRUMENTAL CONDITIONED REFLEX

Dogs standing quietly in a Pavlov stand in a soundproof cage were taught to place their foreleg on a feeder opposite to them when a sound stimulus was applied. Presentation of food served as reinforcement. After several trials, the motor conditioned reflex was firmly established. Then, sounds of different quality were applied that were not reinforced by food. Initially, these sounds elicited the paw reaction, because of generalization; soon thereafter, however, differential inhibition was established.

Prefrontal ablation (1) in four dogs resulted in disinhibition of the previously established conditonal inhibitions, and the animals also reacted to the differential stimuli. The environmental intersignal reactions observed frequently at the beginning of the experimental trials, which had been already inhibited at the time before the ablation, were disinhibited by the removal of the prefrontal cortex. Upon further training, these inhibitions were reestablished.

Removal of the lateral premotor area (2) resulted in severe defects in the conditioned responses: the animals failed to react to the positive stimuli. In five dogs, abolition of the reaction was only temporary and conditioned reflex activity reappeared successively. In four other dogs, however, the conditioned responses were fully abolished and could be elicited no more. In two dogs, the medial premotor area was removed (3); one dog exhibited disorders in conditioned reflex activity, while the other showed suppression of conditional inhibition; both functions were soon reinstituted. Total extirpation of the premotor and prefrontal areas (4) was undertaken in three dogs; in two of them, ablation resulted in abolition of the conditioned reflexes established previously. Reflex activity reappeared after subsequent trials. In the third dog, conditioned inhibition and the intersignal reactions were initially disinhibited but were successively reestablished on subsequent experiments (Figure 2).

In one dog, two surgical ablations were undertaken. First, the lateral part of the premotor area was removed. The conditioned responses disappeared fully; no reaction could be elicited by the conditional stimulus over a period of 2 months. Then prefrontal ablation was performed in a second operation. The previously lost conditional reflexes now reappeared and were fully restored upon further training. On the other hand, the conditioned inhibition and intersignal reactions, which had not been affected by the first ablation, were initially disinhibited; subsequently they were reestablished (Gerbner and Pásztor, 1965; Figure 3).

III. METHOD C: EXTENDED FORM OF THE FOOD-ACQUIRING REACTION

An explanation was sought for the finding that no change occurred in the food-acquiring motor reaction after prefrontal ablation when disinhibition of the conditional inhibitions was significant. To extend the experimental situation, the

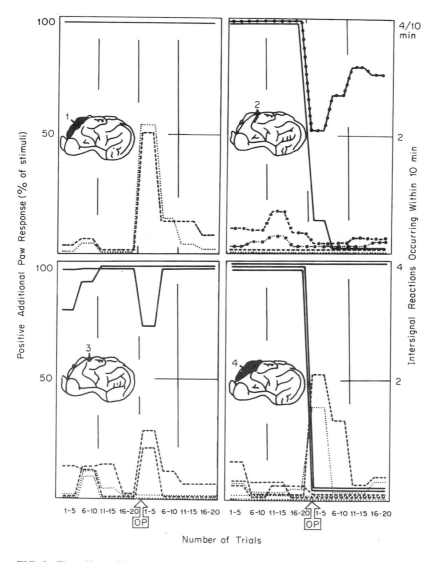

FIG. 2. The effect of frontal lobe ablation on conditional reflexes and inhibitions. OP, ablation. Each point on the abscissa represents the average of five consecutive trials. On the ordinates, in the prefrontal ablations, each point represents the mean obtained in five dogs; in the lateral premotor ablation, the conditioned reflex responses obtained on five dogs are averaged separately (—⋅—), as are the conditioned inhibitions (–●–) and integral reactions (● ⋅ ⋅ ●). The conditioned reflexes (———) and inhibitions (– – –) obtained in the remaining four dogs are also separately indicated, as are the intersignal reactions (⋅ ⋅ ⋅). In the other two types of ablation, each line represents a separate experiment.

241

screen employed in the previous study was separated by a vertical panel. After the dog had pulled one piece of meat through the holes of the screen on one side, the next piece was placed behind the other half of the screen. Thus, the reaction series, initially made up of sequential phases with very short delays

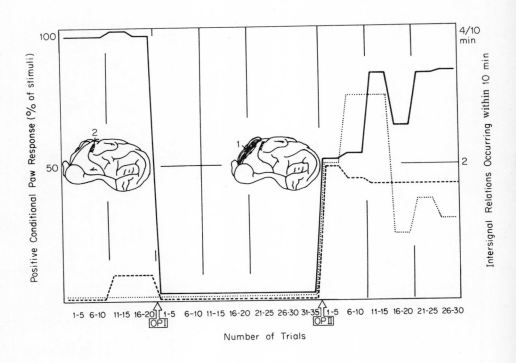

FIG. 3. The effect of prefrontal and superimposed lateral premotor ablation on conditioned reflexes and inhibitions. Symbols, ordinates, and abscissas as in Fig. 2.

between each phase, now included an inhibitory phase, because further responses on the side where the animal had just pulled over the meat were inhibited. The animal was forced to run to the other side of the panel and to initiate a further response there. In five dogs subjected to prefrontal ablation (1), the time lags between the subsequent responses were significantly prolonged. Prior to extirpation, if the animal was offered on one side of the screen a piece of polyethylene similar in size and shape to meat just after he had pulled over the real meat from the other side, the dog would never try to reach for the inedible polyethylene but instead would run over to other side of the panel where he would receive the next piece of real meat. After prefrontal ablation, however, the animals frequently pulled over the polyethylene pieces offered after the meat, and they often ate them. After extirpation of the lateral parts of the premotor area (2), a prolongation of the time lags between the responses, similar to that observed in the simple form of the reaction was recorded (Figure 4).

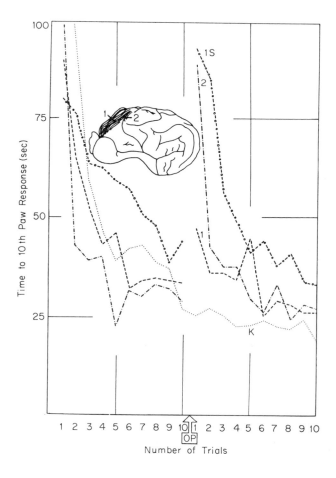

FIG. 4. The effect of frontal lobe ablation on the extended form of the food-acquiring reaction. 1S (– · –) Experiments where polyethylene was offered (see text). Other symbols as in Fig. 1.

IV. RESULTS

Thus, various ablation procedures were followed by well-defined changes in conditioned reflex activity. These were the following:

In the case of prefrontal ablation (1), the conditioned inhibitions were impaired in simple reactions; also, the extended food-acquiring reaction, which included an element of inhibition between the motor phases, was significantly affected. Thus, this type of ablation seems to interfere with inhibitory learned responses.

Removal of the lateral parts of the premotor area (2) seriously affected the conditioned responses and the simple form of the food-acquiring reaction. Thus, the facilitory learned connections were predominantly impaired.

Medial premotor (3), and combined prefrontal and premotor ablations (4) gave rise to transitory impairment of both facilitory and inhibitory connections. The disorders in conditioned reflex activity, however, were significantly less pronounced than in the cases of prefrontal or lateral premotor extirpation, as reflected in the quick restoration of conditioned reflexes upon retraining. All these observations suggest that the facilitory activity of the premotor area and the inhibitory action of the prefrontal region are satisfactorily integrated if the frontal lobe is intact, and the facilitory and inhibitory moments of a learned behavioral response are governed by the complementary and balanced action of these two structures.

V. DISCUSSION

Attempts were made to analyze the observed responses by constructing a model based on the appropriate factors identified in the experimental study. The input represented the stimuli employed, the passive lifting of the animal's paw, and the presentation of food on the one hand, and the ablation of the frontal lobe facilitory and inhibitory regions, on the other. The output current represented the learned paw response of the animal. A relatively simple relationship between these factors reproduced the experimental situation remarkably well. It was assumed that, if the stimulus, the reaction, and the reinforcement coincided, i.e., the presentation of food, indicating the successfulness of the response, the storage of information had shifted toward facilitation, whereas in the absence of reinforcement a shifting toward inhibition had occurred. The probability of the response to the stimulus depended on the relative predominance of facilitory or inhibitory information. If the amount of facilitory information stored by the model was significantly larger than the amount of inhibitory information, a positive reaction was obtained, and vice versa.

The model was constructed in our laboratory. When the circuits representing the stimulus, the reaction, and the reinforcement were closed, appropriate arrow signals appeared on the front of the device, where a schematic diagram of the brain had been drawn. The functioning of the two facilitory and two inhibitory information storages was indicated by lamps in the regions corresponding to the appropriate areas of the frontal lobe. The stimulus employed in the model was in correlation with other positive and inhibitory stimuli too. Therefore, the storages had to contain certain levels of facilitory and inhibitory information. If the stimulus was followed by a response and reinforcement, i.e., if all three switches were turned on, the second facilitory information storage toppled over. This closed the automatic circuit, and on the subsequent presentation of the stimulus, the lamp was switched on through this circuit. If the reaction was not followed by reinforcement, the second facilitory information storage swung back first; if reinforcement was again absent on the next stimulus, the second inhibitory information storage toppled over. This resulted in a gain of the inhibitory input storage, which switched off the automatic current; i.e., the conditioned response was inhibited. Subsequent reinforcement, however, could reestablish the conditional reflex (Figures 5 and 6).

In the course of further experiments, it has been observed that the function of the model is altered by removal of the respective information storage in a manner identical to the way the conditioned reflex activity is altered by ablation of the various parts of the frontal lobe. Moreover, the model simulates the changes in conditioned reflex activity that occur late after the extirpation when the disconnected information storages are switched on again, provided the information is similar to the original information stored by the disconnected elements. The existence of this condition indicates that regions outside the frontal lobe are also involved in the process of learning, and these may substitute for the extirpated regions under appropriate conditions (Szendröy et al., 1966). As seen in Figure 5, showing the function of the model, the following results have been obtained in simulating the ablation experiments. (a) When both inhibitory information storages were disconnected, conditioned inhibitions were disinhibited; the conditioned inhibitions were restored on switching the information storages on again. (b) When both facilitory information storages were disconnected, the previously established conditioned reflexes were abolished. When the storages are switched on again, the amount of facilitory and inhibitory information is equal; hence, the probability of reestablishing a conditional response is 50%. When a conditional response reappears, it can be reinforced, and the response can be fully restored. In the other 50% of cases there is no conditioned response; hence no reinforcement can be applied, and the conditioned reflexes disappear. (c) Switching off of one facilitory and one inhibitory storage, and (d) of two facilitory and two inhibitory storages, results in impairment of the conditional responses in 50% of the cases, and in 50% of the conditional inhibitions; after reinstitution of the circuit, the functions invariably

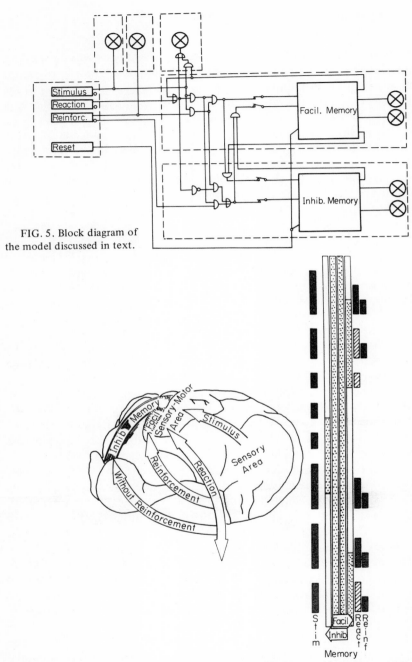

FIG. 5. Block diagram of the model discussed in text.

FIG. 6. Front plane of the model with a schematic representation of the brain (left) and a diagram showing establishment of the conditioned reflexes, their transformation into conditioned inhibitions, and the subsequent reinstitution of the conditioned reflexes.

reappear. Finally, in cases where the disconnection of both facilitory information storages resulted in complete abolition of conditioned reflexes, the subsequent switching off of both inhibitory storages leads to reestablishment of the conditioned reflexes with a probability of 50%. This situation is identical with the experiment where the removal of the facilitory regions of the frontal lobe was followed by permanent abolition of the conditioned responses, and later ablation of the inhibitory areas resulted in reestablishment of the reflexes (Figure 7).

Thus, the results obtained in the ablation experiments could be simulated with an electric model by inhibiting the utilization of information gained in the course of learning. The major characteristics were the following:

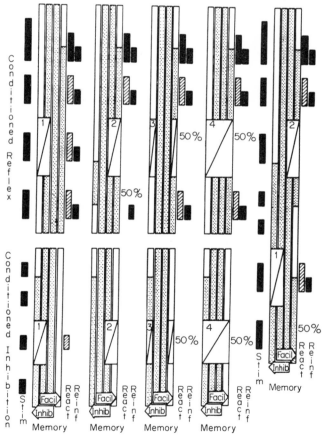

FIG. 7. Changes in the conditioned responses (top) and inhibitions (bottom) after disconnection of the information stores (cancelled) and subsequent switching back on of the storages. The probability for a conditioned response to occur is 50%. Symbols as in Fig. 6.

Prefrontal ablation and the disconnection of the inhibitory storage (1) were followed by no change in the conditioned reflexes, whereas conditional inhibitions were impaired in all experimental animals and on all occasions when the situation was worked out in the model. Conditioned inhibition tended to reappear on subsequent training, in both animal and model.

Lateral premotor ablation and switching off the facilitatory information storages (2) resulted in significant impairment of positive conditioned responses in the animals and in the model as well. Reflex activity was subsequently reestablished in five animals of a total of nine and in half of the experiments carried out on the model.

Ablation of the medial parts of the premotor area and disconnection of one facilitatory and one inhibitory information storage (3), as well as removal of the whole prefrontal and premotor region and disconnection of all four information storages (4), caused impairment of both conditioned reflexes and inhibition in half of the animals and the model experiments. All functions, in the animals and the model experiments as well, were reinstituted on subsequent training (Figure 8).

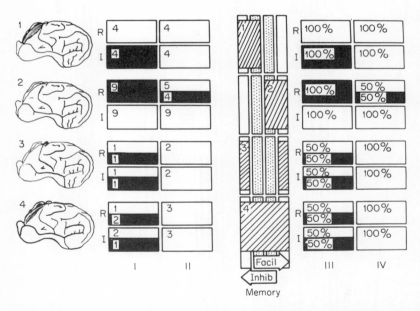

FIG. 8. Derangement of conditioned reflexes (R) and inhibitions (I) immediately after frontal lobe ablation (I), at a later time after frontal lobe ablation (II), on disconnecting the information storages (III) and on switching the latter back on (IV). Hatched area: area of ablation and disconnected information storage, Open block: undisturbed and reinstituted function. Solid block: damaged function. Numbers in rectangles indicate the number of experimental animals or the probability (%) that the model will function.

Thus, the model system reproducing the animal experiments integrates all the stimuli reaching the system in the course of learning, the reactions coincident with them, and the feedback message about the result of the reaction into the facilitory and inhibitory structures. When initiating activity, the system statistically integrates the input stimuli with the information gained previously, and the response that, on the basis of previous experience, seems to be the most adequate in a given situation is brought about. Damage to the frontal lobe disrupts this integrative function of the brain cortex. The concept we suggest on the basis of the present results is in satisfactory agreement with the majority of hypotheses concerning the functional mechanism of the frontal lobe. Also, it allows detailed study, on more concrete grounds, of the various memory hypotheses (Jacobsen et al., 1935; Jacobsen, 1936; Jacobsen and Nissen, 1937; Gross and Weiskrantz, 1964, 1966). Derangement of inhibitory information seems to provide sufficient explanation for the incompleteness of habituation, increase of reactivity to new stimuli, increase of sensitivity to distraction, and insufficient inhibition of reactions given to signals not included in the program (Pribram, 1966; French, 1966; Luria, 1966b). If the frontal lobe is damaged, the equilibrium of facilitory and inhibitory activity is seriously affected, shifting from one element of the behavioral reaction stereotype to another becomes difficult, and pathologic inertia ensues (Luria, 1966a; Luria et al., 1966; Shkol'nik-Yapros, 1966; Spirin, 1966; Lebedinsky, 1966). Secondarily, this may lead to perseveration (Konorski and Lawicka, 1964; Brutkowski, 1966; Spirin, 1966; Gross and Weiskrantz, 1966). The model provides an explanation for the disturbances in integrating previous experiences and recent impressions, and in comparing the initial aim of an action and its result (Bekhterev, 1907; Luria, 1966a). It also explains the impairment of intricate and flexible programs of behavior (Luria, 1966b). The author believes that the synthesis of others' data and the present observations was possible because the model reliably simulated the disturbances in function manifesting themselves in the above symptoms.

REFERENCES

Allen, W. F. (1940). Effect of ablating the frontal lobes, hippocampi, and occipito-temporal (excepting pyriform areas) lobes on positive and negative conditioned reflexes. *Amer. J. Physiol.* **128**, 754-771.

Allen, W. F. (1943). Results of prefrontal lobectomy on acquired and on acquiring correct conditioned differential responses with auditory, general cutaneous and optic stimuli. *Amer. J. Physiol.* **139**, 525-531.

Allen, W. F. (1949). Effect of prefrontal brain lesions on correct conditioned differential responses in dogs. *Amer. J. Physiol.* **159**, 525.

Andrew, J. F. N. (1964). Lesions of the anterior frontal lobes and disturbances of micturition and defaecation. *Brain* **87**, 233-262.

Aring, C. D. (1949). Clinical symptomatology. *In* "The Precentral Motor Cortex" (A. C. Bucy, ed.), pp. 409-423, Univ. of Illinois Press, Urbana.

Barris, R. W. (1937). Cataleptic symptoms following bilateral cortical lesions in cats. *Amer. J. Physiol.* **119**, 213-220.

Bekhterev, V. M. (1907). Osnovy ucheniya o funktsii mozga. Vyp. UP. SPB.

Brickner, B. M. (1932). An interpretation of frontal lobe function based on the study of a case of partial bilateral frontal lobectomy. *Res. Publ., Ass. Res. Nerv. Ment. Dis.* **13**, 259-351.

Brutkowski, S. (1966). O funksional'nykh osobenostyakh tak nazyraemykh "nemykh" zon lobnykh doley zhivotnykh. *In* "Lobnye Doli" (A. R. Luria and E. D. Homskaya, eds.), pp. 100-116. Moscow University Press, Moscow.

Conrad, K. (1954). New problems of aphasia. *Brain* **77**, 491-509.

French, G. M. (1959). A deficit associated with hypermotility in monkeys with lesion of the dorsolateral frontal granular cortex. *J. Comp. Physiol. Psychol.* **52**, 25-28.

French, G. M. (1966). Hyperactivity and impairment of delayed reaction in monkeys after lesions of the frontal lobes. Frontal lobes and regulation of behavior. *Int. Congr. Psychol. 18th, 1966* pp. 50-55.

Fulton, J. F. (1935). A note on the definition of the "motor" and "premotor" areas. *Brain* **58**, 311-316.

Fulton, J. F., Jacobsen, C. F., and Kennard, M. A. (1932). A note concerning the relation of the frontal lobes to posture and forced grasping in monkeys. *Brain* **55**, 524-536.

Gerbner, M., and Pásztor, E. (1965). The role of the frontal lobe in conditioned reflex activity. *Acta Physiol.* **26**, 89-96.

Gross, C. G., and Weiskrantz, L. (1964). Some changes in behavior produced by lateral frontal lesions in the Macaque. *In* "The Frontal Granular Cortex and Behavior" (J. M. Warren and K. Akert. eds.), pp. 74-101, McGraw-Hill, New York.

Gross, C., and Weiskrantz, L. (1966). Posledstviya lateralnykh lobnykh povrezhdeniy u obez'yan. *In* "Lobnye Doli" (A. R. Luria and E. D. Homskaya, eds.), pp. 133-155. Moscow University Press, Moscow.

Harlow, J. M. (1848). Passage of an iron rod through the head. *Boston Med. Surg. J.* **39**, 389.

Harlow, J. M. (1868). Recovery from the passage of an iron bar through the head. *Mass. Med. Soc. Proc.* **2**, 327.

Jacobsen, C. F. (1936). Functions of the frontal association areas in monkeys. *Comp. Psychol. Monogr.* **13**, 1-60.

Jacobsen, C. F., and Nissen, H. W. (1937). Studies of cerebral function in primates. *J. Comp. Physiol. Psychol.* **23**, 101-112.

Jacobsen, C. F., Wolfe, J. B., and Jackson, T. A. (1935). An experimental analysis of the functions of the frontal association areas in primates. *J. Nerv. Ment. Dis.* **82**, 1-14.

Jarvie, H. F. (1954). Frontal lobe wounds causing disinhibition. *J. Neurol. Neurosurg. Psychiat.* **17**, 14-32.

Jarvie, H. F. (1958). The frontal lobes and human behavior. *Lancet* **0**, 365-368.

Konorski, Yu., and Lawicka, W. (1964). Analysis of errors by prefrontal animals on the delayed-response test. *In* "The Frontal Granular Cortex and Behavior" (J. M. Warren and K. Akert, eds.), pp. 271-294. McGraw-Hill, New York.

Konorski, Yu., Stepien, C., Brutkowski, S., Lawicka, W. and Stepien, I. (1952). The effect of the removal of interprojective fields of the cerebral cortex on the higher nervous activity of animals. *Bull. Soc. Scie. Lodz, Cl. IV* **3**, 1-5.

Langworthy, O. R., and Richter, C. P. (1939). Increased spontaneous activity produced by frontal lobe lesion in cats. *Amer. J. Physiol.* **126**, 158-161.

Lebedinsky, V. V. (1966). Vyprolnenie simmetrichnykh i asimmetrichnykh programmu bol'nykh s porazheniem lobnykh doley mozga. *In* "Lobnye Doli" (A. R. Luria and E. D. Homskaya, eds.), pp. 576-603. Moscow University Press, Moscow.

Luria, A. R. (1966a). Lobnye doli i regulyatsiya poverdneniya. *In* "Lobnye Doli" (A. R. Luria and E. D. Homskaya, eds.), pp. 7-37. Moscow University Press, Moscow.

Luria, A. R. (1966b). The frontal lobes and the regulation of behavior. Frontal lobes and regulation of behavior. *Int. Congr. Psychol. 18th, 1966*, pp. 143-151.

Luria, A. R. (1967). "Neuropsychological Analysis of Focal Brain Lesions."

Luria, A. R., Pribram, K. H., and Homskaya, E. D. (1966). Narushenie programmirovaniya dvizhenniy i deystviy pri massivnom porazhenii levoy lobnoy doli. *In* "Lobnye Doli" (A. R. Luria and E. D. Homskaya, eds.), pp. 554-575. Moscow University Press, Moscow.

Malmo, R. B. (1942). Interference factors in delayed response in monkeys after removal of the frontal lobes. *J. Neurophysiol.* **5**, 295-308.

Milner, B. (1964). Some effects of frontal lobectomy in man. *In* "The Frontal Granular Cortex and Behavior" (J. M. Warren and K. Akert, eds.), McGraw-Hill, New York, pp. 313-334.

Paillard, J. (1960). The patterning of skilled movements. *In* "Handbook of Physiology" (Amer. Physiol. Soc., J. Field, ed.), Sect. I. Vol. III, Chapter 67, pp. 1679-1708. Williams & Wilkins, Baltimore, Maryland.

Pribram, K. H. (1961). A further experimental analysis of the behavioral deficit that follows injury to the primate frontal cortex. *Exp. Neurol.* **3**, 432-466.

Pribram, K. H., and Mishkin, M. (1956). Analysis of the effects of frontal lesions on monkeys. III. Object alternation. *J. Comp. Physiol. Psychol.* **49**, 41-45.

Pribram, K. H. (1966). Sovremmennye issledovaniya funktsiy lobnykh doley u obez'yan i cheloveka. *In* "Lobnye Doli" (A. R. Luria and E. D. Homskaya, eds.), pp. 117-132. Moscow University Press, Moscow.

Pribram, K. H., Ahumada, A., Hartog, J., and Ross, L. (1964). A progress report on the neurological processes disturbed by frontal lesions in primates. *In* "The Frontal Granular Cortex and Behavior" (J. M. Warren and K. Akert, ed.), pp. 28-52. McGraw-Hill, New York.

Richter, C. P., and Hines, M. (1938). Increased general activity produced by prefrontal and striatal lesions in monkey. *Brain* **61**, 1-16.

Shkol'nik-Yapros, E. G. (1966). Premotornaya zona kory i sindrom eë porakheniya. *In* "Lobnye Doli" (A. R. Luria and E. D. Homskaya, eds.), pp. 314-355. Moscow University Press, Moscow.

Shumilina, A. I. (1966). Funktsional'nye znachenie lobnykh oblastey golovnogo mozga v uslovna reflektornoy deyatel'nosti sobaki. *In* "Lobnye Doli" (A. R. Luria and E. D. Homskaya, eds.), pp. 61-81. Moscow University Press, Moscow.

Shustin, N. A. (1966). K probleme funktsii lobnykh doley bol'shikh polusharii. *In* "Lobnye Doli" (A. R. Luria and E. D. Homskaya, eds.), pp. 82-99. Moscow University Press, Moscow.

Simernitskaya, E. T., and Bunatyan, B. A. (1966). Narusheniya ritmicheskikh dvizheniy pri opukhol'yakh premotornoy zony bol'shykh polusharii. *In* Lobnye Doli" (A. R. Luria and E. D. Homskaya, eds.), pp. 374-387. Moscow University Press, Moscow.

Spirin, V. G. (1966). Proyavleniya patologicheskoy inertnosti posle operatsiy na perednykh otdelakh bol'shykh polushariy. *In* Lobnye Doli" (A. R. Luria and E. D. Homskaya, eds.), pp. 356-373. Moscow University Press, Moscow.

Szendröy, G., Gerbner, M., and Bakos, S. (1966). Experimental electronic model to simulate the function of the premotor areas of the frontal lobe. *Acta Physiol.* **29**, 434-435.

Walshe, F. M. R. (1935). On the "syndrome of the premotor cortex" Fulton and the definition of the terms "premotor" and "motor" with a consideration of Jackson's views on the cortical representation of movements. *Brain* **58**, 49-80.

Welt, L. (1888). Über Charakterveränderungen des Menschen in Folge der Läsionen des Stirnhirns. *Dent. Arch. Klin. Med.* **42.**

Chapter 13

THE PRIMATE FRONTAL CORTEX AND ALLASSOSTASIS

WALTER GRUENINGER and *JANE GRUENINGER*

Department of Psychology
Stanford University
Stanford, California

I. INTRODUCTION

An apparent paradox was revealed when it was observed that lesions of the frontal cortex produce a marked diminution of the orienting GSR in both humans (Luria and Homskaya, 1964; Luria *et al.*, 1964) and Rhesus monkeys (Kimble *et al.*, 1965). Seemingly in conflict with these results was the common observation that frontal lesions greatly increase, not decrease, behavioral responsiveness to novel stimuli (Pribram, 1961; Brush *et al.*, 1961).

Prior to these observations, it was generally assumed that the autonomic and behavioral indications of orienting were different facets of a common underlying process and that the dorsolateral area of the frontal cortex was associated with a behavioral deficit, whereas autonomic changes were primarily associated with the surrounding cortical areas (Fulton, 1949). This confusing situation raised several questions and prompted the investigations presented in this chapter. What relationship could there be between the delayed response deficit that traditionally typifies the behavioral effect of dorsolateral frontal lesions and the galvanic skin response (GSR) component of the orienting response? Does this cortical area really play some role in the basic homeostatic responses, or were these findings merely an artifact of technique with little deeper significance?

II. PHYSIOLOGICAL EXPERIMENTS

A. Experiment I: EEG Desynchronization and GSR to Tone

1. Comment

Since the early part of this century, electroencephalogram (EEG) desynchronization has been used as a physiological indicator of attention or arousal. To determine the range and limits of the physiological deficit produced by the dorsolateral frontal lesion, it seemed desirable to monitor EEG activity, as well as the GSR, in a replication of the habituation experiment by Kimble, Bagshaw, and Pribram (1965).

2. Method

a. Subjects. The test group originally consisted of eight adolescent rhesus monkeys (numbers 101 through 108). Four animals (numbers 101 through 104) had undergone bilateral removal of the dorsolateral frontal cortex 12 months before the beginning of this experiment. The lesions were made by subpial suction under aseptic conditions and consisted in the removal of the grey matter bounded by the depths of the arcuate sulcus and the anterior tip of the hemisphere. The anterior bank of the arcuate sulcus and the banks and depths of the sulcus principalis were included (Figure 1). Post mortem examination showed animal 103 of this group to have incomplete degeneration of the parvocellular portion of the nucleus medialis dorsalis thalami in the left hemisphere, despite an apparently successful removal of the appropriate cortical area. For this reason, animal 103 will not be considered as part of the frontal group, although the results obtained from this subject will sometimes be included for

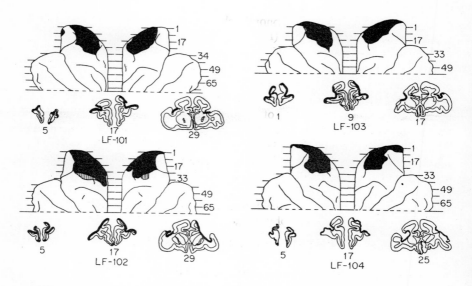

FIG. 1. Reconstruction of lesions. (Reprinted from W.E. Grueninger and K.H. Pribram, Effects of spatial and nonspatial distractors on performance latency of monkeys with frontal lesions, *Journal of Comparative Physiology and Psychology,* 1969, 68 (No.2) 203-209, by permission of the American Psychological Association.)

purposes of comparison. Animals 105 through 108 served as unoperated controls. All subjects had formed a group from the time the operates had recovered from surgery. All had been used together in a number of behavioral experiments as well as in the habituation study of Kimble *et al.* (1965).

b. Apparatus. A Forringer primate chair was used to secure the subjects. Electroencephalogram recordings were made on a four-channel Grass polygraph. Skin resistance was measured by means of a Fels Dermohmmeter and continuous recordings were made with an Esterline-Angus GSR Inkwriter. The GSR electrodes were of zinc–zinc sulfate and were approximately 1 cm in diameter. Fels Electrode Paste (zinc sulfate in agar) coupled these electrodes to the skin. The sound-deadened experimental chamber used was a box approximately 2 ft square by 7 ft high. It was insulated with two layers of acoustic tile and an internal layer of glass wool insulation. A centrifugal fan ventilated the chamber and provided background noise.

c. Procedure. One subject was run per day, and the subject to be run was selected from alternate groups on alternate days. Preparation for the day's run began by removing the animal from its home cage and securing it in the primate chair. The plantar surfaces of the hindpaws were washed with Phisohex and were thoroughly dried. The GSR electrodes were then placed and held in position with elastic adhesive bandages to minimize movement artifact. The subject's scalp was depilated by means of electric clippers and depilatory cream. Small disk-type EEG electrodes were placed over the frontal, parietal, and occipital regions of the left hemisphere. An indifferent electrode was attached to the right ear. All EEG electrodes were secured by a skullcap made from adhesive tape. This sequence was usually accomplished within about 1 hr. When these preparations were complete, the animal was moved into the sound-insulated chamber and was connected to both the Grass polygraph (EEG leads plus signal marker) and to the Fels Dermohmmeter. The skin resistance record was monitored until it appeared that a stable baseline had been reached. When hydration of the GSR electrodes was deemed to be complete and the GSR record appeared stable, the test phase was begun. At a time when the EEG record showed slow waves and the GSR record showed skin resistance to be increasing, or at least to be constant, a tone was delivered through an overhead speaker. On the test presentations this tone was of 2 sec duration and at an intensity of approximately 80dB. As a control, on alternate trials the intensity of the tone was reduced to zero but the criteria for delivering this zero-intensity stimulus remained the same as for the test stimulus. This procedure was continued until no further GSR activity was observed in response to the tone for at least five successive presentations, or for a minimum of 35 tone presentations. These subjects had been previously habituated to tones of 1500 or 1000 Hz in the experiment reported by Kimble *et al.* (1965). The same frequencies were used in this study and each subject was shifted to that tone which had not been used on it in the previous experiment. For purposes of analysis, a GSR had to begin within 5 sec after the termination of the tone, whereas EEG desynchronization had to occur in a window beginning with stimulus onset and ending .5 sec after stimulus offset.

3. Results

Not one orienting GSR was observed in the record of any frontal subject (Table 1). In sharp contrast to this autonomic unresponsiveness were the apparently normal EEG records obtained at the same time from the same subjects. The tone produced desynchronization of the EEG with about equal frequency in the two groups. Subtler techniques might reveal differences (Artemieva and Homskaya, 1973) but a gross examination of the records did find clear desynchronization in the absence of frontal cortex. Curiously, the major difference with regard to the gross EEG activity occurred on the control presentations. The normal subjects responded to the presentations of zero-intensity tone by EEG desynchronization and movement significantly more often than did the experimental animals ($U = 0$, $p = .056$). Ruling out a possible group difference in extrasensory perception, it is possible that the subjects were responding to the faint relay clicks associated with the signal markers. The soundproofing was breached by a 1 ft^2 opening blocked by a single thickness of one-way vision glass for observation purposes. It is also possible that the difference observed was due to a real difference in spontaneous activity between the two groups.

B. Experiment 2: EEG Desynchronization and GSR to Light Flash

1. Comment

The cortical activation described above was in agreement with the behavioral observation that frontal subjects do respond to novelty, although there was no

TABLE 1
EEG Desynchronization and GSR to Tone (First 25 Trials)

		Response to stimulus			
		No response	EEG + GSR	EEG, no GSR	Contaminated
Frontals	101	18	0	6	1
	102	14	0	7	4
	104	12	0	13	0
Normals	105	16	1	4	4
	106	9	10	2	4
	107	14	6	1	4
	108	15	1	5	4
Subtotal frontal	103	7	0	5	13

indication that they were more responsive than the normal subjects. EEG desynchronization is often observed in the absence of a GSR; the reverse seldom, if ever, occurs. This might suggest a threshold response to the novel stimulus that must be exceeded before a GSR can be observed. The responses that the zero-intensity tone elicited from the normal subjects could support the hypothesis that they are more sensitive to the auditory environment than the experimental group. Other investigators have reported a deficit in tasks involving auditory discrimination (Gross and Weiskrantz, 1962). To test the possibility that the frontal unresponsiveness might be due to some sort of relative functional deafness rather than something more general, it seemed desirable to rerun the basic habituation experiment using a nonauditory cue of relatively high intensity.

2. Method

Subjects, apparatus, and procedure were essentially as in the initial study, Section II,A, with a few exceptions.

A light source was attached to the Forringer primate chair approximately 25 cm in front of the subject's head. The intensity of the flash was sufficiently bright to be easily seen through the human eyelid, but did not seem painful to the human volunteer, even with eyelids open and eyes dark adapted. No control presentations of zero intensity were used in this run.

3. Results

Once again, no orienting GSR was observed in the records of the frontally lesioned subjects (Table 2). This intense stimulus often evoked movement from the normal subjects and sometimes even from the frontals. Occasionally, such

TABLE 2
EEG Desynchronization and GSR to Light (First 25 Trails)

		Response to stimulus			
		No response	EEG + GSR	EEG, no GSR	Contaminated
Frontals	101	4	0	20	1
	102	2	0	16	7
	104	0	0	22	3
Normals	105	0	17	3	5
	106	0	10	0	15
	107	1	7	0	17
	108	0	10	0	15
Subtotal frontal	103	0	0	5	20

movement was accompanied by GSR-like changes in skin resistance in both groups. For reasons to be discussed in following sections, such changes were not included in our tabulation of GSR activity. In any case, it is clear that the absence of the orienting GSR in the frontal group cannot be attributed to a specific auditory insensitivity. In this experiment, as in the one preceding, the EEG records of the two groups were quite similar. Despite the similarity in the EEG records, and despite the fact that the light flash was intense enough to evoke movement—39% of the time in the normal subjects and 5.3% of the time from the frontal subjects—not one orienting GSR was recorded from any lesioned subject.

C. Experiment 3: GSR and Corticosteroid Response to Electric Shock

1. Comment

The preceding experiments verified that ablation of the dorsolateral frontal cortex obliterates the orienting GSR. Because the novel stimuli produced grossly normal responses at the cortical level, the lesion must have affected the GSR mechanism itself or the coupling of this mechanism to the rest of the nervous system.

Although all galvanic skin responses share a final common path, the pseudomotor neurons, there are at least three distinct ways in which this final common path can be activated. (1) The GSR can be elicited by stimulation of the hypothalamus even in the decerebrate preparation; (2) it can be elicited by stimulation of the sensory-motor cortex or the pyramidal tract both before and after destruction of the hypothalamus; and (3) it can be elicited from the spinal animal as a nociceptive reflex.

The psychologist usually considers skin resistance changes attributable to the second and third sources above to be bothersome contaminants because they tend to obscure the true GSR. However, these contaminants can be used to determine whether or not the peripheral system remains functional after frontal lesions. Electric shock provides the experimenter with a controllable stimulus for such a test.

Porter (1954) observed an eosinopenia following stimulation of the frontal cortex, and Mason et al. (1961) reported that other portions of the limbic system modulate the release of the adrenal corticosteroids. These results suggested that an examination of the adrenal cortical steroid response, in addition to the GSR, might be fruitful. Dr. Seymor Levine generously offered his time and facilities and made such observations possible.

2. Method

a. Subjects. The subjects were the same as those used previously.

b. Apparatus. The soundproofing on the chamber was improved by the elimination of the one-way vision screen. It was covered with acoustical tiling and glass wool insulation. A constant-current source, which provided up to 5 mA of direct current, was obtained.

c. Procedure. Preparation was as outlined in Section II,A,2,c, but no EEG electrodes were used. A braided copper wire was tied around each of the subject's wrists to serve as the shock electrode. These braids were connected to the current source when the subject was placed in the sound-insulated chamber. The test period lasted 55 min, during which the chamber remained dark, and five shocks were administered. The first shock occurred as soon as the GSR record had stabilized. The others followed at intervals of 20, 10, 15, and 10 min respectively. All shocks were of 3 mA intensity and of 2 sec duration. The skin resistance under the GSR electrodes was recorded continuously during the test period. The run ended when this skin resistance reached a maximum deflection following the fifth shock.

The adrenal steriod response was studied concurrently with the GSR. Three 1 or 2 ml samples of blood were obtained by venapuncture from a superficial leg vein. The first was drawn as soon as the animal was secured in the primate chair, the second immediately before the subject and chair were placed in the test chamber, and the third upon removal of the subject from the sound-insulated box at the end of the GSR measurement period. The first sample was usually obtained within 2 or 3 min after the animal was first approached, the second, ½ hr later, and the third about 1 hr after the second. The usual preparation procedure, with which the animal was quite familiar, occupied the interval between the first and second samples, whereas the five shocks in the dark chamber were administered between the second and third samples. Each subject was run through this procedure in a second session, which followed the first after an interval of 2 weeks, and through the third session, 2 weeks after the second. On these last two sessions, no shock was administered, although everything else remained unchanged. In the second and third sessions, a fourth blood sample was taken 1 hr after the third. During the interval between the third and fourth samples, the animal remained confined in the chair but was removed from the experimental chamber to a relatively quiet room. Steroid analysis was by the fluorometric technique of Glick *et al.* (1964).

3. *Results*

a. Galvanic Skin Response. Under the conditions of this experiment, frontal and normal animals do not differ with respect to the amplitudes of the GSR's produced in response to electric shock. This lack of between-group difference was found whether the statistic compared was absolute resistance change, percentage change, change of the square root of the conductance, or change in log conductance. Whether these statistics were calculated from maximum deflection or from values of resistance 10 sec after shock, no difference was apparent.

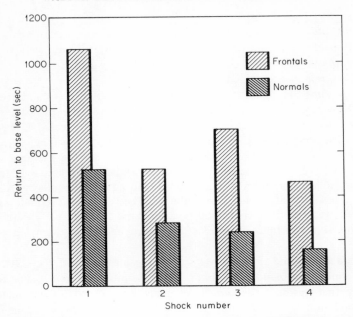

FIG. 2. Time for skin resistance to return to preshock level expressed as group mean return time. (Reprinted from W. E. Grueninger, D. P. Kimble, J. Grueninger, and S. Levine, GSR and corticosteroid response in monkeys with frontal lesions, *Neuropsychologia,* 1965, **3,** 205-216. Copyright 1965 Pergamon Press.)

Figure 2 shows the results of an examination of a second parameter, i.e., the length of time required for the skin resistance to return to its preshock level following the first four of the five shocks. Analysis of variance does show that the two groups differ with respect to this index ($F = 13.18$ with 1 and 5 df, $p < .025$). In fact, the return times for frontally lesioned animals are longer in every instance except one. Examination of the record indicated that even this single instance was due to incomplete hydration of the skin under the GSR electrodes prior to the first shock.

Another statistic that clearly separates the two groups is the total number of GSR-like fluctuations per 50 min session in the experimental chamber. This was the length of time for which uniformly stable GSR records were available for all subjects. All deflections in the resistance record that exceeded 200 Ω and that resembled GSR's in the judgment of the skilled observer were counted. Analysis of variance for the group difference yields an F of 43.73 with 1 and 5 degrees of freedom and a $p < .005$. Even though movement-induced GSR's have been included in this statistic, the frontal animals showed so much less GSR-like activity that there is no overlap between the two groups.

b. Corticosteroids. The mean corticosteroid levels for both groups are graphed in Figure 3. This figure reveals that complete removal of frontal eugranular isocortex does not eliminate the corticosteroid response. The analyses of variance show that the two groups reached essentially the same maximum corticosteroid concentrations ($F = .849$) following overall corticosteroid responses that were surprisingly similar ($F = .020$). Despite these gross similarities, a closer examination shows marked differences between the two groups. These differences can be characterized as an alteration of the pattern of the corticosteroid response and an increased day to day variability of the corticosteroid level in the blood of the subjects with frontal lesions. The average change in the corticosteroid levels between samples one and three was 28.4 mg for the lesioned subjects and 2.76 mg for the normal group. The F ratio for this difference is .02 and one may safely conclude that the average total response to the experimental condition was not altered by the lesion. However, Figure 4 clearly shows that the two groups attained this nearly identical response in quite different ways. This figure graphs the change in adrenal steroid level that occurred in each phase of the experiment. When the results are seen in this form, it is apparent that the major portion of the corticosteroid response in the normal animals occurred during the 30 min interval of the preparatory phase. The electric shock and/or confinement to the dark chamber produced little additional response. However, the lesioned subjects responded in a way that seemed more linear with time, and the major change in steroid levels in this group was recorded during the 60 min confinement. Analysis of variance for the group × interval interaction yields an F ratio of 15.28 with 1 and 5 degrees of freedom ($p < .025$). Figure 4 also shows another rather surprising fact. Whereas the pattern of the corticosteroid response recorded from the frontal group did not differ significantly from day to day, that measured from the normal subjects showed a significant difference of a rather unexpected sort. The greatest corticosteroid response from these controls occurred during the preparatory phase of the first day. The response during this interval was far greater than that recorded from the frontal subjects and nearly double that of the average response recorded from the control group on the following two sessions. Both groups had had considerable experience with the primate chair in the preceding experiments, and the light flash of Experiment 2 was the most noxious stimulus that had been used. The drawing of the first blood sample by an unfamiliar experimenter did provide a cue that this was a different type of experiment. It would seem that the normal monkeys were responding in anticipation of something far worse than the confinement and shocks that they had experienced, because the responses on the succeeding sessions implied a far less potent stressor. Whether or not this was an emotional response to uncertainty, no similar response was recorded from those subjects without frontal eugranular isocortex.

The increased day to day variability observed in the lesioned subjects is suggested in the averages graphed in Figure 3 but is made more obvious in Figure 5, which depicts the initial corticosteroid levels for all subjects for the

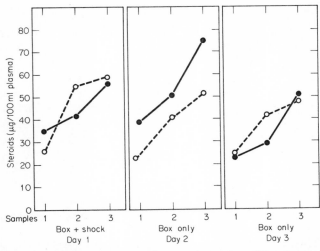

FIG. 3. Blood steroid concentrations under experimental conditions expressed as group means. Open circle, normal group; filled circles, frontal group. (Reprinted from W. E. Grueninger, and S. Levine, GSR and corticosteroid response in monkeys with frontal lesions, *Neuropsychologia*, 1965, **3**, 205-216. Copyright 1965 Pergamon Press.)

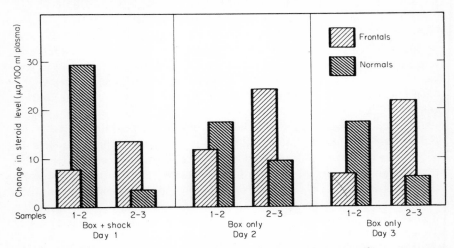

FIG. 4. Change in steroid levels between successive samples expressed as group mean change. (Reprinted from W. E. Grueninger, D. P. Kimble, J. Grüeninger, and S. Levine, GSR and corticosteroid response in monkeys with frontal lesions, *Neuropsychologia*, 1965, **3**, 205-216. Copyright 1965 Pergamon Press.)

three separate sessions. Analysis of variance performed on these initial values informs us that the group difference ($F = 2.33$, df = 1 and 5; $p < .05$) and the group × day interaction ($F = 2.44$, df = 2 and 10; $p < .05$) do not quite reach

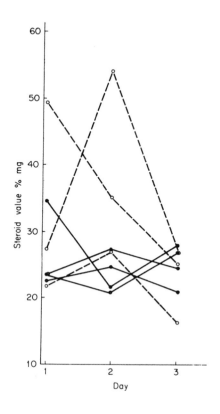

FIG. 5 Initial corticosteroid values for all subjects for the three separate sessions. Open circles, frontal group; filled circles, normal group. (Reprinted from W. E. Grueninger, D. P. Kimble, J. Grueninger, and S. Levine, GSR and corticosteroid response in monkeys with frontal lesions, *Neuropsychologia*, 1965, 3, 205-216. Copyright 1965 Pergamon Press.)

the normally accepted level of significance. Inspection of Figure 5 reveals that the great variability in the lesioned animals drastically reduces the power of the statistical technique. A variance estimate based upon the individual's variability in initial levels yields a value of 141.7 for the experimental group and 17.4 for the controls. A two-tailed Fisher F test shows that these two values differ at the .02 level of probability ($F = 8.143, df = 6$ and 8). The same result is found when using the variance estimate based on the individual's day to day consistency for each given sample ($F = 2.822$ for 18 and 24 degrees of freedom, $p < .05$). However, the variance estimates based upon response variability are 90.66 and 126.37 for the frontal and normal subjects, respectively. The lesion-induced day to day variation does not affect the consistency of the corticosteroid response but does primarily affect the base level from which the response occurred.

Thus, not only is the frontal eugranular isocortex intimately involved in the rapid release of corticosteroids, but it is also involved in stabilizing the base level upon which such a release is superimposed.

D. Experiment 4: Conditioning the GSR

1. Comment

The preceding experiments have shown that a galvanic skin response can be obtained in the absence of the dorsolateral frontal cortex, although it cannot be evoked by mere novelty. Is the frontal cortex responsible for all GSR's that cannot be attributed to movement or to nociceptive reflex? Schwartz (1937), working with cats, found that a unilateral lesion including the medial third of the gyrus poreus and the medial portion of the anterior sigmoid gyrus eliminated the galvanic skin response in the contralateral forepaw except as a segmental reflex. If the dorsolateral frontal cortex in primates is homologous to the medial portion of the gyrus poreus in cats (a general assumption), it would seem probable that only nociceptive stimuli or movement can produce a GSR in frontal subjects. A conditioning paradigm was selected to provide a test of this hypothesis.

2. Method

a. Subjects. Since the object of this experiment was to attempt to obtain an uncontaminated GSR from the lesioned subjects, there was no necessity to include a control group. However, normal subject 105 and subject 103, with its subtotal lesion, were also run.

b. Apparatus. This was essentially the same as that previously used, but the soundproofing was removed, once more, from the one-way vision screen to permit observation.

c. Procedure. Preparation was similar to that of the preceding experiment including the use of the shock electrodes. The box was dimly illuminated to permit visual observation. A 4 sec 1000 Hz tone was presented at least ten times, or until the skin resistance record clearly demonstrated that the subject was giving no response in the absence of movement (phase 1). Next (phase 2), four shocks of 2 mA intensity and .5 sec duration were given unaccompanied by tone, after which at least ten presentations of the tone alone were repeated to check whether pseudoconditioning had occurred with respect to the GSR. Thereafter (phase 3), 50 pairings of tone and shock were administered. The current, initially at 2 mA, was increased to 4 mA for the last 25 pairings. After every fifth pairing, a test presentation of tone alone was inserted. A final series of tone presentations followed the conditioning sequence (phase 4).

3. Results

The results are briefly summarized in Table 3. No tone-evoked GSR was recorded from any of the frontal animals in the absence of movement, although there were many occasions when movement did not occur and a GSR could have been expected. Many local movements occurred without concomitant GSR's.

TABLE 3
Conditioning of the GSR Shows the Number of GSR's Occurring for Each Subject under Each Phase of Conditioning

Subject number	GSR[a]	Tone alone	Elect. shocks	Tone alone	Shock tone pairs 1-10	Two test tones	Shock tone pairs 11-20	Two test tones	Shock tone pairs 21-30	Two test tones	Shock tone pairs 31-40	Two test tones	Shock tone pairs 41-50	Test tones	Tone alone
101	GSR s M	0	0	0	0	0	0	0	0	0	0	0	0	0	b
	GSR c̄ M	1	0	0	0	0	0	0	0	0	0	1	0	0	b
	s GSR s M	9	3	9	6	2	6	2	3	2	0	0	2	0	b
	M s GSR	0	1	1	4	0	4	0	7	0	10	0	9	2	b
102	GSR s M	0	0	0	0	0	0	0	0	0	0	0	0	0	0
	GSR c̄ M	0	3	4	1	1	0	1	3	0	0	0	0	0	0
	s GSR s M	5	6	5	1	1	0	1	0	0	2	0	2	0	1	7
	M s GSR	6	1	1	8	0	10	0	7	2	10	0	10	1	3
104	GSR s M	0	0	0	0	0	0	0	0	0	0	0	0	0	0
	GSR c̄ M	10	4	1	3	1	3	1	2	1	4	1	2	2	17
	s GSR s M	0	0	9	4	1	2	1	7	1	1	1	6	0	14
	M s GSR	0	0	0	3	0	5	0	1	0	5	0	2	0	9
103	GRS s M	5	0	0	0	0	—	—	—	—	—	—	—	—	13
	GSR c̄ M	12	0	26	8	2	—	—	—	—	—	—	—	—	79
	s GRS s M	9	0	1	0	0	—	—	—	—	—	—	—	—	3
	M s GSR	10	4	10	2	0	—	—	—	—	—	—	—	—	35
105	GSR s M	30	—	—	—	—	—	—	—	—	—	—	—	—	—
	GSR c̄ M	18	—	—	—	—	—	—	—	—	—	—	—	—	—
	s GSR s M	34	—	—	—	—	—	—	—	—	—	—	—	—	—
	M s GSR	23	—	—	—	—	—	—	—	—	—	—	—	—	—

[a] GSR s M = GSR present without movement; GSR c̄ M = GSR and movement present; s GRS s M = no GSR nor movement present; M s GSR = movement present without GSR.
[b] Subject became untied.

Furthermore, no conditioning of the galvanic skin response was observed, although behavioral conditioning could clearly be seen in the form of anticipatory movement. As expected, the record of normal subject 105 showed extreme GSR activity in phase 1, making phases 2, 3, and 4 superfluous. Of interest also is the record produced by the subtotally lesioned subject 103. This record is in marked contrast to those of the subjects with complete lesions and is similar to that of normal animal 105.

While the frontal animals did not respond to any tone with an uncontaminated GSR, examination of the records did show spontaneous GSR-like fluctuations unaccompanied by observed movement. Such spontaneous fluctuations occurred at rare intervals throughout the entire procedure and bore no apparent relationship to any tone or shock.

E. Experiment 5: Thermoregulation

1. Comment

The results of the first four experiments raised many questions. How general is the role of the frontal cortex in basic physiological responses? Does it influence all homeostatic systems or some subset, such as those involved in responses to stressors? Is this cortical region interacting specifically with the hypothalamus or is it coupling other parts of the brain to this region? Is the primary effect observed simply a sluggishness of response combined with a decreased sensitivity to external events or internal changes? In an attempt to answer some of these questions, a simple thermoregulatory experiment was designed. The maintenance of body temperature is a basic homeostatic system that would seem relatively far removed from the adrenal steroid response and the GSR. Moreover, it is one of the core systems in which the homeostat's primary receptor is found in the hypothalamus. If the primate frontal cortex interacted with hypothalamic systems to increase their sensitivity and speed their equilibration, the frontal deficit should be observable in an appropriately designed examination of the thermoregulatory process.

2. Method

a. Subjects. The subjects used in this experiment were a new group of ten Rhesus monkeys. Five animals had undergone surgery as described previously in Section II,A,2 approximately 6 months prior to the beginning of this experiment. The remaining five subjects shared a common laboratory experience with the operated animals from the time of surgery, although no control surgery was performed upon them. All subjects had been involved in a number of behavioral experiments and had had previous experience in the Forringer primate chair.

b. Apparatus. In addition to the usual Forringer primate chair, the apparatus included a new soundproofed chamber (Acoustic Industries), the interior of which was 1.2 m deep by 1.2 m wide by 2.15 m high. A four-channel Grass dc polygraph, thermistor probes, heaters and humidifiers, and wet and dry bulb thermometers were also used. A Raytheon Microtherm unit with a 4 in. hemispherical reflector was the source of the thermal challenge. This unit is similar to a microwave oven and heats throughout the soft tissues in the path of the microwave beam which it emits.

c. Procedure. The subject was secured in the Forringer primate chair, its forelimbs and hindlimbs were restrained, and the hair on its anterior chest wall and abdomen was clipped short. The 4 in. diameter hemispherical Microtherm reflector was fastened to the chair in front of the subject, aimed at the thorax and upper abdomen and centered with its front surface 2 in from the lower portion of the sternum. Four thermistor probes were used. One was introduced through the nasal passage and secured with tape so that it hung in the lower portion of the esophagus. One was secured by a thin band of elastic adhesive tape to the fleshy pad of the index finger, and another was secured in the same manner to the contralateral big toe. The fourth probe was used to record temperature in the environmental chamber. Each probe was calibrated prior to each day's experimental run. When the probes and Microtherm unit were in place, the subject and chair were placed in the chamber with temperature controlled at $25°C$. and humidity at saturation. Equilibration to the controlled environment was allowed until recordings showed no appreciable drift for at least 5 min. When this had been achieved, the run began.

The run consisted of four cycles of Microtherm heating and reequilibration. The heating cycles differed as to intensity (power output of the Microtherm unit) and as to duration, as follows: (1) 30% power for 5 min, (2) 60% power for 5 min, (3) 30% power for 10 min, and (4) 60% power for 10 min. The same sequence was used for all subjects. Each heating phase was followed by an equilibration period that continued until the recorded temperatures returned to baseline (the temperature recorded at the onset of the current heating cycle) or until 20 min had elapsed since the end of the heating phase.

3. Results

a. Base Line Data. The two groups were found to be equivalent as far as the initial temperatures were concerned. After the initial equilibration, the following mean temperatures were recorded from the experimental and control groups, respectively: Hand, 32.5 and $30.3°C$; foot, 30.7 and $30.2°C$; core (esophagus), 38.46 and $38.34°C$. By Student's T test none of these differences was significant at even the .10 level of probability. No significant difference was found in overall temperature change from the beginning to the end of the experimental run. Hereafter, all temperatures will be given in terms of the new base temperatures established at the beginning of each heating cycle.

b. Temperature Changes in Extremities and Core. The overall analysis (Table 4) revealed that the hand, foot, and core responses for the two groups were essentially the same. This was true for the total temperature change during the heat-on phases, for the slopes of linear trend (rate of temperature change) for the first 5 min of all heating cycles, for the slopes of linear trend for the first 2 min of all heating cycles, and for the changes that occurred during the first minute of all heating cycles. There may have been a slight difference in the initial recovery rate of the core temperature after heating, but this difference did not reach the .05 level of significance, even if only the most divergent values are selected for a T test (those values obtained 5 min after the start of the recovery phase). In short, the absence of the dorsolateral frontal cortex did not prevent the experimental subjects from maintaining their body temperatures under these experimental conditions.

c. Factor Analysis. This experiment was designed to allow a more sensitive analysis of several possibly pertinent experimental variables, in addition to the general results just summarized. Analyses indicated that variation of Microtherm intensity produced a significant difference in temperature changes after 5 min of exposure, but there was no apparent difference between the two groups of subjects. Likewise, there was a distinct duration effect, but the additional 5 min of heating did not affect the two groups differentially. Analysis for the effect of repetition (learning?) yielded results that are quite suggestive. Because each Microtherm intensity occurred twice in the four heating cycles, it is possible to look for improvement in the performance of the individual subjects. The average change in core temperature recorded from the subjects with frontal lesions was .03°C less during the first 5 min of cycles three and four than it was during the

TABLE 4
Analysis of Variance for Total Heat-On Phases

	Mean change (°C)		
	Hand	Foot	Core
Subjects:			
Frontal	1.405	1.027	.288
Normal	1.800	1.080	.253
Analysis of variance (F values):			
Group	1.20	.040	.052
Condition	1.71	3.43*	13.3*
Group X condition	.398	.141	1.69

*Significant at $p < .05$.

same time period of cycles one and two. When the T test was applied, this change did not differ significantly from zero. In contrast, the normal subjects improved by an average of .10°C. Student's T test indicates that this average differs significantly from the results recorded from the frontals ($p < .02$) and from zero ($p < .02$). This improved performance cannot be attributed to a changed response at the extremities. The analysis could reveal no difference in the temperature changes recorded from hand and foot, although the wide fluctuations normally observed in hand and foot temperatures would tend to obscure any subtle differences that might exist. Equivalent analysis of the 5 min linear trends of core temperature changes tends to corroborate this group difference, although this corroboration is obviously not entirely independent of the overall change recorded on the core temperatures. Examination of the data reveals that the two groups differed only slightly under the condition of 30% Microtherm intensity, whereas the major improvement occurred when the 60% intensity was repeated. In terms of subjective sensation, this 60% level was barely above threshold for the experimenter when his hand was substituted for the usual subject. Even at this higher intensity, the effect was a sensation of warmth from within rather than from some external source.

F. Review of Physiological Results

1. EEG Desynchronization

No abnormalities were apparent in the electroencephalograms of the frontal animals. Desynchronization occurred with comparable frequency and with comparable decrease in amplitude in both groups. This was true whether the stimulus used was a relatively familiar tone or a novel intense light flash.

2. Galvanic Skin Response

Although both tone and light flash were completely incapable of eliciting an uncontaminated galvanic skin response from the frontal animals, a response of apparently normal amplitude was evoked by electric shock. However, when such GSR's occurred, the skin resistance took longer to return to preshock base level in the experimental subjects. The records of skin resistance also showed spontaneous fluctuations, which appeared to be GSR's in both groups. These spontaneous fluctuations were quite rare in the frontal group and were not associable with any other movement of the animal or any external stimulus. The complete failure of the frontal subjects to show conditioning of the GSR is not entirely in accord with the results obtained by Ashby and Basset (1950), and Elithorn et al. (1954) in leukotomized humans. The discrepancy may be the result of species difference or it may a reflection of the different type of lesion— leukotomy versus frontal ablation. However, the fact that the nearly total operation on animal 103 resulted in only a very partial reduction of the GSR activity suggests that GSR conditioning might be accomplished if only a few fibers were spared in the leukotomized subject.

3. Adrenal Corticosteroid Response

Neither the maximum level of circulating corticosteroids nor the overall change in corticosteroid level were affected by frontal ablation under the conditions of these experiments. The normal subjects did show a more rapid rise of level especially during the first experimental session. Also, the initial blood levels of corticosteroids proved more stable from day to day in the normal subjects.

These results extend the findings of Porter (1954), who found an eosinopenia on stimulation of the frontal cortex, and also the seemingly contradictory results reported by Storey et al. (1959), who reported that a corticosteroid response occurred in the total absence of all frontal cortex. As suggested in Section II,C, the frontal cortex would seem to be involved in altering the corticosteroid levels in response to psychological stimuli rather than to the direct noxious stimulation that Storey et al. employed. Mason et al. (1961) reported results that suggest a similar role for the amygdaloid nucleus. They found that removal of this nucleus eliminated the corticosteroid response normally observed in the conditioned avoidance situation only so long as the subject successfully avoided all electric shock.

4. Thermoregulation

The absence of the frontal cortex did not interfere with basic thermoregulatory responses. The time courses for both peripheral and core temperature changes proved equivalent for the two groups. The results did suggest that some adaptation occurred in the normal subjects with repetition of the unusual thermal challenge. No such adaptation was seen in the frontal group. This effect was slight and requires replication.

5. Movement

Hypermotility or hyperreactivity to the environment has been commonly observed in animals with dorsolateral frontal lesions (Isaacs and DeVito, 1958). A peripheral finding in this series of experiments was the observation that movement was most often recorded from the normal subjects. This was most evident when the intense light flash was used as a stimulus. Once restrained, the frontals were relatively quiescent.

6. Summary and Conclusions

The dorsolateral frontal cortex of primates is not necessary for the corticosteroid response, nor is it required for thermoregulation. However, the results reported above indicate that this area normally plays some role in both of these physiological responses. It was suggested in Section II,C,1 that the absence of orienting GSR in the presence of EEG desynchronization implies either malfunction of those elements necessary for a GSR or decoupling of these elements from those areas that usually activate them in the orienting response. Apparently normal GSR's were obtained from the frontal animals when they moved or responded to noxious stimuli or to unknown private events, but never was an uncontaminated GSR elicited directly by tone or light flash.

These results support a tentative conclusion that the primate frontal cortex is involved in coupling at least those core systems tested with areas directly or indirectly involved in processing sensory information. It might be loosely stated that the frontal cortex is necessary for psychological factors to influence these core systems. In his presentation of homeostasis, Cannon (1932) detailed the means whereby the internal state of the body could influence behavior through drive stimuli. The frontal cortex would seem to be involved in a complimentary process whereby the basic homeostatic systems are influenced by external events. In deference to Cannon, perhaps this process should be called allassostasis (*allasso*: Greek, alter or change; *stasis*: Greek, setting) to correspond with homeostasis (*homeo*: Greek, like or similar).* The deficit produced by removal of dorsolateral granular cortex might then be characterized as being an allassostatic deficit.

III. BEHAVIORAL EXPERIMENT

A. Comment

How is this allassostatic deficit related to the overresponsiveness to novelty and the other behavioral alterations commonly associated with frontal lesions? Kimble *et al.* (1965) suggested that the autonomic components of the orienting response are necessary for habituation to occur. Pribram (1969b, 1972) expanded upon this suggestion and emphasized the importance of a stable input from the mass of internal stimuli, which are largely of visceral origin. He hypothesized that lesions of the frontal cortex may disturb habituation by allowing this background to drift. He conjectured that, with a constantly changing back-

---

*Although the term is new, the subjective experience of allassostasis is familiar to everyone. For example: The day has been a full one. Although dinner time has arrived, no thought of food has had a chance to break in upon the hour's events. By chance, you see spareribs barbequing over a charcoal fire. Suddenly your stomach pangs that eating time has long since past. Saliva flows, and thoughts of food stay foremost in your mind. Allassostasis has occurred.

ground, a stimulus would retain its novelty because no single neuronal model could be built. Pribram also cited electrophysiological studies that suggest that the frontal cortex may play a direct role in modifying the sensory input (Spinelli and Pribram, 1967). According to this interpretation, the frontal deficit is largely one of heightened distractibility because the animal without a frontal cortex is unable to habituate to unimportant sensory inputs. This approach would provide a physiological base for the theoretical interpretation of frontal deficit espoused by Malmo (1942) and by Konorski (1961) in terms of increased sensitivity to external inhibition or a decrease in internal inhibition.

Recently, publications by Mishkin (1964), Stamm (1969), and Konorski (1964) have stressed another interpretation of the frontal lobe deficit and have followed through on some results obtained by Mishkin and Pribram (1956). These models would have the frontal cortex primarily involved in processing spatial cues or "spatiokinesthetic transient memory" (Stamm, 1969).

There would seem to be some discrepancy between the Pribram model, which tries to bridge the gap between physiology and behavior, and those just mentioned, which stress the spatial aspects of the behavioral tasks.

In the neuropsychology laboratory at Stanford University, Douglas and Pribram (1969) devised a technique to test the distractibility of amygdalectomized and hippocampectomized monkeys. They had observed that the two groups were differentially sensitive to spatial and nonspatial distractors. Their program for the DADTA III, which generates latency data in the distraction situation, seemed to offer an idea for a basis upon which to design a test of the discrepant models of frontal function.

B. Method

a. Subjects. The animals used in this experiment were those described in Section II,E. However, one member of the normal group had died and was not replaced for this study.

b. Apparatus. The apparatus used for testing was a modification of that described by Pribram (1969a), differing mainly in that stimulus presentation and recording were accomplished by means of a PDP-8 rather than a special purpose computer. The display consisted of a 4 × 4 array of 16 depressable panels upon which stimuli could be projected from the rear. A food cup was located below the display. For purposes of programming, each panel was assigned a letter, as follows:

```
        A  B  C  D

        E  F  G  H

        I  J  K  L

        M  N  O  P
```

The panels were on one of the walls of the testing enclosure, the top two rows at eye level. Illumination was provided by an overhead incandescent light. Subjects were watched by the experimenter through a one-way glass that made up another wall of the enclosure.

c. Procedure. At each session, a few minutes were allowed for adjustment, following which the testing program began. A trial was initiated by the illumination of panel P. Pressure on the panel resulted in the immediate illumination of Panel A. In turn, a press on this panel released a banana pellet and terminated the trial. After a lapse of 10 sec, a new trial would begin, an identical procedure being repeated on all trials. Latencies (to .001 sec) between the presentation of the first stimulus and the first response, and also between responses to the first and second stimuli, were automatically recorded. Only the latter, the inter-response latencies, are considered in this paper. All subjects were originally trained on this response sequence (with no distractors presented) until they met the criterion of a day's run of 50 trials completed within less than 10 min (5-8 days of training).

Similar procedures were used in all distraction tests. In every case subjects were given daily sessions of 50 total trials each, with four distraction trials intermixed among the regular trials. The distraction trials were presented in a pseudorandom fashion, such that they were separated by at least five, but not more than 10, regular trials, and no distraction trial occurred earlier than the eleventh trial of a day's run. The procedure used in a distraction trial was similar to that in regular trials except that the press of Panel P resulted in the appearance of a distracting stimulus simultaneously with the appearance of the stimulus on Panel A. When the distractor was a symbol, the subject could depress the panel upon which it was displayed and these responses were recorded. However, a response to the distractor did not result in a reward nor in any change in the situation. A trial was terminated only when Panel A was pressed, at which time both it and the distracting stimulus disappeared and the reward was delivered. The distraction conditions were given in the order in which they are described.

Condition 1: stimulus varied, location constant. This task was designed to test the subject's sensitivity to variation in a stimulus pattern presented at a specific spatial location and the subject's ability to habituate to distraction in this location despite the variations in stimulus patterns. Eight abstract patterns were used as distractors. On each of the four distraction trials of a day's run one of these patterns appeared in panel location F. No pattern was repeated until all had been used. The sequence was shuffled so that no pattern appeared twice on the same or on successive days. Subjects were tested under this condition for four successive days.

Condition 2: stimulus constant, location varied. This task was designed to test the subject's sensitivity to the distractor's location in space and to test the subject's ability to habituate to or gate out a specific visual stimulus regardless of its location. In this condition, one hitherto unused pattern was presented in each of eight spatial locations. The location used in condition 1 was omitted. The same panel was never used twice on the same or on two successive days. All eight locations were used twice in the 4 day series.

Condition 2^S: stimulus constant, location constant. To ascertain whether the results in condition 2 might have been the result of some unusual property of the stimulus used rather than of the spatial variable under investigation, the same pattern used in condition 2 was presented on four successive distraction trials of the same day in panel location B.

Condition 3: buzzer. A buzzer was used as a distractor instead of the patterns projected on panels. Otherwise, the procedure was the same as in conditions 1 and 2 above.

C. Results

Distraction duration was considered to be the difference between the interresponse latency on a distraction trial and the subject's median interresponse latency on nondistraction trials for that day. The first ten trials of each day were omitted to decrease warmup effect, and the median was chosen instead of the mean to avoid undue influence from occasional unintentional distractions (e.g., subject's dropping of a pellet). The group mean distraction durations recorded under conditions 1, 2, and 3 are presented graphically in Figure 6.

As a check, the raw interresponse latencies on distraction trials were also analyzed. The findings were essentially the same as those presented in the figure. The overall average interresponse latencies on nondistraction trials were .537, .463, .441, .857, and .583 sec for the five subjects with frontal lesions and .565, .975, .528, and .366 sec for the four normal controls. Thus, the median interresponse latencies of the two groups did not differ in the absence of distractors. ($U = 10$, maximum possible overlap for groups of this size.)

The results are analyzed as they specifically apply to four views of the deficit produced by this lesion, i.e., increased distractibility (Pribram, 1961, 1973), failure to habituate (Luria *et al.*, 1964), difficulty in responding to spatial cues (Mishkin *et al.*, 1966) and inability to inhibit responses (Brutkowski *et al.*, 1963). In the interests of clarity, the results will be discussed as they are presented.

1. Distractibility

The analyses of variance summarized in Table 5 indicate that the mean distraction durations recorded from the frontally ablated group were significantly higher than those of the normal group under all experimental conditions. When the subject's mean distraction durations for each condition were ranked, there was only a single overlap, which occurred in condition 1.

The increase in response time that the mean distraction duration represents could have been brought about by either or both of two underlying factors. The

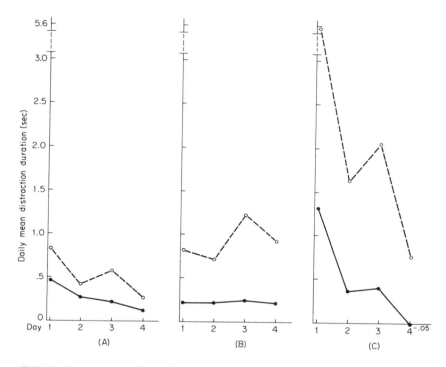

FIG. 6. Group mean distraction duration. Open circles, frontal group; filled circles, normal group. (Reprinted from W. E. Grueninger and K. H. Pribram, Effects of spatial and nonspatial distractors on performance latency of monkeys with frontal lesions, *Journal of Comparative Physiology and Psychology,* 1969, **68,** (No.2), 203-209, by permission of the American Psychological Association.)

frontally lesioned subjects may have been distracted by a greater proportion of the distractors, or they may have recovered more slowly whenever they were distracted. The proportion of latencies on the distraction trials that were greater than a subject's median normal trial latency provides some insight into this problem. If the distractors had no effect whatever, we would expect this statistic to have an average value of .50, whereas any delay the distractor might produce should bias this indicator toward 1.00. Through examination of this proportion (Figure 7) we find that under condition 1 the two groups are essentially identical in the consistency of their distraction. A Mann-Whitney U Test for overall group difference yields a U value of 9.5, where 10 would be the value obtained for the maximum theoretically possible overlapping of the two sets of scores. This means that under this condition all of the difference observed in the mean distraction durations must have come from a slower recovery on the part of the frontally lesioned animals. The results from the first 2 days of condition 2 confirm this finding ($U = 10$). In the third condition (Figure 7C) a different

TABLE 5
Statistical Analysis of Distraction Duration

	F value	Degrees of freedom	Significance
Condition 1: Stimulus varied, location constant			
Group	8.332	1 and 7	$p < .05$
Day	4.483	3 and 21	$p < .05$
Group × day	.309	3 and 22	NS*
Condition 2: Stimulus constant, location varied			
Group	5.912	1 and 7	$p < .05$
Day	.679	3 and 21	NS
Group × day	.463	3 and 21	NS
Location	2.434	7 and 49	$p < .05$
Group × location	1.351	7 and 49	NS
Condition 3: Buzzer			
Group	47.471	1 and 6	$p < .001$
Day	6.683	3 and 18	$p < .01$
Group × day	1.596	3 and 18	NS

*Not significant at the $p < .05$ level.

situation prevailed. The loud buzzer placed at the back of the cage proved a very strong distractor. However, with experience, the control animals, at least, were no longer distracted so much as galvanized by the occurrence of this stimulus. The frontally lesioned subjects seemed more prone to orient toward the back of the cage when the buzzer sounded, whereas the control animals often ducked or lunged more vigorously than usual at the second task stimulus. The end effect on the normal group, at least, would appear somewhat equivalent to stimulus intensity dynamism. For this group, the stimulus still evoked a response but the response was directed toward the rewarded task instead of the distractor.

2. Behavioral Habituation

Analysis of variance confirms the apparent decrease in distraction durations through the course of conditions 1 and 3. Because the group × day interaction in the two conditions was negligible, it is apparent that the frontally ablated group showed behavioral habituation at the same rate as the normal animals. Of course, the higher initial values of the frontally ablated animals would result in their requiring a greater number of days for the mean distraction duration to reach some specified level.

A comparison of the graph of distraction duration and of the proportion of distraction trials greater than the subjects' median latency for condition 1 (stimulus varied, location constant) proved interesting. It shows that the apparent habituation observed in both groups is not due to a decrease in the

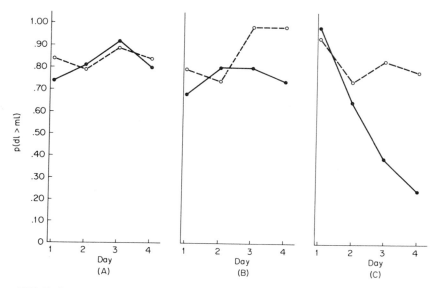

FIG. 7. Proportion of distraction trials with latencies greater than the subjects' median latency. Open circles, frontal group; filled circle, normal group. (A) Condition 1; (B) condition 2;(C) condition 3. (Reprinted from W.E. Grueninger and K.H. Pribram, Effects of spatial and nonspatial distractors on performance latency of monkeys with frontal lesions, *Journal of Comparative Physiology and Psychology,* 1969, **68** (No.2), 203-209, by permission of the American Psychological Association.)

probability that a given distractor will produce an effect. The habituation would seem to be the result of faster recovery rather than gating out or decreasing sensitivity to the distractors. This increase in behavioral efficiency, as opposed to desensitization or gating of the input, was very evident in condition 3 (buzzer). In this condition, both groups showed a significant decrease in distraction duration. The normal subjects' distraction durations and proportions of distraction trial latencies greater than median both sank below the level expected if no distractor were present. With respect to the latter statistic, Figure 7C clearly shows the sharp contrast between the trends of the two groups ($U = 0$, $p = .036$). The absence of a consistent trend in the scores of the frontally ablated monkeys strongly suggests that these subjects would never show the paradoxical negative distraction observed in the control subjects in the presence of this strong distractor.

3. Spatial Deficit

It was predicted that if the frontally ablated animals were deficient in registering spatial information, they would have much greater difficulty habituating under the first condition (stimulus varied, location constant) than would the normal controls. This should have been the case because the distractors differed widely in pattern but were all consistent in location. However, the rate of behavioral habituation proved to be the same for both groups. If the animals with lesions were less able to process and record the location of the distractor, they might also have shown relatively rapid habituation under condition 2 (stimulus constant, location varied). On the contrary, the frontally operated monkeys were even more disrupted throughout this condition than in the first. Ranking the subjects on the basis of the difference between their distraction durations under conditions 1 and 2 showed no overlap between groups ($U = 0$, $p = .016$). Although neither group habituated under condition 2, Figure 7B shows that repetition actually heightened the distractor's effectiveness upon the subjects with lesions. Under condition 2^S, where the stimulus used in condition 2 was presented only in location B, the mean distraction duration of the frontal operates approximated the low level of the normal group by the fourth presentation. Thus, although the subjects with dorsolateral frontal lesions proved capable of learning to minimize the disruptive effect of varied patterns appearing in one specific spatial location, the distraction produced by a single pattern could be maintained for the animals by simply shifting its spatial location. It might be said that with regard to distraction, the frontally ablated animals were more, not less, sensitive to spatial location than were the normal controls.

In order to examine our results in terms of the spatial aspects of the cognitive field, the results of condition 2 were analyzed with respect to location instead of experimental day. The mean distraction durations obtained for the various locations used are presented in Table 6. The location effect was significant, as might be expected, but the value of the F ratio for the group by location interaction was only 1.351. Thus, there would appear to be no difference between the two groups with respect to which locations proved most effective. The frontally operated monkeys were not at all immune to the distractors that were at greater distances from the task stimuli.

In summary, under our experimental conditions, the frontal lesion increased, not decreased, the sensitivity of the animal to a distractor's location. This was true regardless of where the distractor appeared.

4. Inhibitory Deficit

Although the pressing of the distractor panel was never rewarded, a record was kept of all such presses. With four distraction trials per day, subjects with frontal lesions averaged 1.6, .2, .6, and 0 presses for the 4 days of condition 1 and .8, 1.2, 1.8, and .8 presses for the 4 days of condition 2. The only press recorded from a normal subject occurred on day 1 of condition 1, making the group average for that day .25 presses. In contrast, if we combine both conditions we find that no frontally ablated subject pressed less than three distractors. This yields a U of zero and $p = .016$.

TABLE 6
Panel Matrix Presenting Mean Distraction Duration
as a Function of Distractor Location[a]

Task stimulus 2	**.595** .217	**1.524** .207	**.345** .179
.976 .315		**.500** .116	**.745** .203
	1.497 .318	**1.209** .290	
			Task stimulus 1

[a]Boldface = frontal; lightface = normal.

The analysis might suggest that the increased distraction durations recorded for the frontally ablated animals were the result of time wasted in the pressing of unrewarded stimuli. However, this is not the case. When the subject's overall averages were calculated, omitting all trials on which a press occurred, there was still only one overlap between the two groups ($U = 1, p$ .032). Therefore, not all of the frontally ablated animals' increases in distraction duration can be attributed to the time wasted by actually pressing unrewarded panels.

In condition 3, it was observed that the increased distraction duration of the frontally ablated group seemed largely due to their persistent orientation toward the buzzer when it sounded. In contrast, the normal subjects eliminated such disruptive responses and actually showed an increased response speed under the influence of the distractor.

D. Discussion

The results clearly show that the frontal animal stays distracted longer than the normal, but can, within limits, behaviorally habituate at a normal rate. The lesioned animals do build neuronal models of their environment that include spatial information. It is interesting to note that it is the duration of the distraction effect that clearly separates frontal from normal. Only in certain circumstances is there a group difference in the probability that the distracting stimulus will be sampled. Furthermore, the behavioral habituation observed in both groups under these experimental conditions is not so much an ignoring of the distractor as it is a decrease in the time spent in behavior directed toward the distractor and away from the task.

There is nothing in these results that would support an interpretation of the frontal deficit in terms of short-term spatiokinetic memory. On the other hand, the fragment of the Pribram (1973) model invoked in Section III,A as a possible bridge between the physiological and behavioral realms is not, in itself, sufficient to explain the data. If difficulty with habituation were the sole basis for the frontal deficit, one might expect to observe this effect primarily as a renewing of the distractors' effectiveness from day to day, or possibly even from trial to trial. However, no difference was observed in the behavioral habituation curves of the two groups under any of the three conditions. The distraction durations recorded from the frontal animals showed the same relative lengthening from the first to the last day a given distractor was used.

As Pribram (1969b) indicated, the influence of the dorsolateral frontal cortex is not limited to the core homeostatic systems. He presented evidence that this cortical area affects activity in the visual system as far peripherally as the optic tract. In the experiments reported here, some of the effects observed under conditions 2 and 3 can be interpreted as a lesion-produced inability to suppress irrelevant distracting stimuli. In addition, the results from all conditions show that a normal frontal cortex helps to limit or suppress competing responses directed toward the distractor. The forcefulness of this response suppression was made obvious under condition 3.

In the traditional switchboard model of the brain, this control over both input and output might seem dualistic and confusing. However, when the brain is considered in servomechanistic terms, no such problem arises. The *Plan* (Miller et al., 1960) is such a servomechanistic conceptualization that incorporates the input and output aspects of a goal-directed sequence.

In the light of the above considerations, a simple model can be suggested. The dorsolateral frontal cortex functions to bring about and maintain a relative facilitation in those neuronal elements appropriate to the Plan that is in the process of execution, regardless of minor interruptions. From this assumption, the behavioral deficits usually observed following frontal lesions can be predicted.

Insofar as all observable behavior is a product of motor (spatiokinetic) activity, the frontal cortex might be said to participate in spatiokinetic storage. However, it is the anticipated future that is being preserved (i.e., the yet to be executed Plan), and not a dying image of the spatial past.

Such an interpretation is in close agreement with that set forth by Pribram (Miller *et al.*, 1960; see also Pribram *et al.*, 1964), when he suggested that the frontal cortex might function as a " '...working memory,' where various Plans could be temporarily stored (or perhaps regenerated) while awaiting execution."

IV. SPECULATIVE SYNTHESIS

In Section II it was tentatively concluded that the dorsolateral frontal cortex is involved in allassostasis, i.e., in biasing the core homeostatic systems on the basis of the inputs from the exteroreceptive processing systems. Section III concluded with the hypothesis that the frontal cortex functions as a working memory for the storage of soon to be executed Plans. Because ultimate simplicity is an aim of scientific understanding, this section will be devoted to the presentation of a model that attempts to reconcile this apparent disparity.

A. Biological Interface

Bishop (1959), in his investigations on peripheral sensory nerves, found evidence for the existence of at least three evolutionary layers in the nervous system of mammals. Of these, the unmyelinated C fibers are the most primitive, the small myelinated (gamma and delta) fibers are more recent, and the beta group of large myelinated fibers represents the most recent evolutionary addition to the peripheral sensory system. Successive evolutionary stages demonstrate a marked increase in the speed of transmission and the detail of the information transmitted. In addition, new central terminals have evolved to receive their primary projections. Primitive C fibers convey their information directly to the phylogenetically ancient neuropil in the brain stem, whereas the largest, most recent, fibers project to the newly evolved neocortex.

Pribram (1960b) drew attention to the fact that the basic receptors of the core homeostatic systems lie close to the walls of the third and fourth ventricles in the brain stem, where they are embedded in a reticulum of small neurons typified by unmyelinated fibers and multisynaptic conduction paths (the neuropil). He suggested that such a region is ideal for graded responses rather than all-or-none digital transmission. Pribram hypothesized that the frontolimbic system acts to bias the homeostatic systems by producing graded responses in the vicinity of these receptors or in the vicinity of those neuronal elements that these receptors affect. The results presented in Section II support this allassostatic function of the frontolimbic system.

The modern high-speed digital computer provides a clear instance of the difficulties that can arise in the interaction of systems that operate at widely different speeds. If the central processing unit of a modern computer were to operate a teletype by simply firing an otherwise appropriate string of type commands at its maximum possible rate, very little would be accomplished. A central processing unit is ordinarily capable of issuing commands about 10,000 times as fast as a standard teletype can execute them. This problem is commonly solved in an interposed device called the *interface*. This device usually makes it possible for the central processor's command to be stored temporarily until execution has occurred. By contrast, transmitting information from a low-speed device to a high-speed device is comparatively simple. It is merely necessary for the high-speed device to sample the input at least as often as this input shows significant change.

In this framework, it is easy to understand how the core receptors responsible for homeostasis can bias the activity in the primitive nerve network, as well as in the more recent evolutionary layers. Likewise, it is easy to understand how the neuropil can transmit its biasing graded responses to the neocortex via the small fiber system that terminates on the apical dendrites of cortical cells. However, a problem is encountered when the flickering evanescence of the neocortical sensory systems is called upon to exert a lasting and significant influence on the plodding elements of the core systems. As Pribram has suggested, it could well be the frontolimbic system, including the dorsolateral frontal cortex, that serves as a biological interface and mediates allassostasis (Pribram, 1960b).

B. Buffer Register or Working Memory

In the case of the teletype, a specified type command is stored in a buffer register until it has been executed. This register can be described as an electronic short term memory unit. It is set by the high-speed output circuits of the computer's central processing unit and is cleared upon execution of the stored command. Such a temporary storage mechanism should be a useful part of the biological interface under consideration. Without it, the dispositional state of the organism would be largely under the control of the core receptors and the activity of the neuropil itself. Only transient and limited modulations might be expected from the activity in the evolutionarily new sensory processing systems. However, with a biological buffer register, a significant flicker from the exteroceptive processing systems should be sufficient to produce and maintain an allassostatic readjustment until an appropriate response or responses alters the situation and resets the system. One might expect such a register to maintain both dispositional state and behavior sequence intact, despite organismically insignificant interruptions. The results presented in Section II implicated the dorsolateral frontal cortex in allassostasis, whereas those in Section III evoked the notion of a working memory. The frontal cortex may not play only one role; however, it would seem that all the behavioral and physiological data could be explained if this cortical area played a necessary part in such a biological buffer register.

Perhaps this model might be clarified by a simple example. A small mammal is about to emerge from the shadows of dense undergrowth into the sunlit glare of a clearing. Through the leaves, it catches a momentary glimpse of a well-known and feared predator. Thanks to the biological buffer register, the momentary flicker of sensory information is sufficient to produce and maintain the allassostatic response, which prepares it for maximal flight. It runs unhesitatingly to a nearby hiding place, ignoring the tempting berries it happens to pass en route. Safe and out of sight, its heart rate slows, its blood pressure returns to normal, and finally it relaxes its vigil and starts to groom. Picture the same animal without the biological buffer register to rely upon. The momentary sensory flicker would produce no lasting shift in the homeostatic systems. Habit might make it turn and start to run to the hiding place, but the sight of the berries would distract it and it would begin to eat, remaining in plain sight, awaiting the predator's pleasure. In this model, the frontal cortex, as part of the biological buffer register, maintains the allassostatic readjustment as well as the related behavioral Plan until the appropriate physiological or behavioral goal has been achieved.

C. Neural Mechanisms

It should be obvious that the analogy between the computer's buffer register and the system described above is, at best, at the rough level of function rather than of design. Because the function itself is still speculation, any statement concerning the neural mechanisms involved is mere conjecture. However, some fascinating experiments by J. S. Stamm and his collaborators do seem pertinent.

Stamm (1969), and Stamm and Rosen (1972, 1973) found that electrical stimulation of the prefrontal cortex surrounding the sulcus principalis disrupted performance on a delayed response task only when such stimulation occurred during or shortly after the onset of the delay period, or (with certain electrode placements) after the response had occurred. Stimulation during the early portion of cue presentation, the later portion of the delay period, the response period, or the intertrial interval proved almost entirely ineffective. Stamm and Rosen also recorded steady potentials from the area and found surface negativity peaking at the onset of the delay period and at the time the response was made. Needless to say, these results lend strong support to the model advanced here. The surface negativity indicates that something special is happening at just those times when one would expect the biological buffer register to be reset; i.e., immediately after sensory processing and upon completion of the goal-directed sequence. Furthermore, disruption proved possible only when such resetting should have been in progress.

When cessation of stimulation coincided with the onset of the response period, performance was markedly disturbed. Such was not the case if stimulation began at this time or simply continued into the response period. This might appear somewhat perplexing; however, a momentary decrease or complete pause in background activity is commonly observed immediately following stimulation of neural tissue. This may be due to hyperpolarization or perhaps to recurrent inhibition. The effect is brief and normal background activity soon reappears. Since stimulation-induced activity at the onset of the response period does not impair performance, it must be such a momentary pause in activity that does. This indicates that continued output from the frontal cortex is necessary to insure the initiation of a correct response in this task.

If this interpretation is correct, the next question to be answered is whether the output from the frontal cortex is patterned on a unit by unit basis or is nonspecific. The fact that disruption can be caused by stimulation during the first few seconds of the delay period implies that some pattern of activity is being set up somewhere at this time. If such a pattern is being set up in the frontal cortex itself, it would have to become extremely stable to resist disruption under the conditions of Stamm's experiments. Alternately, it is possible that the output from the frontal cortex is not patterned. The frontal efferents may exert their influence in a general fashion by blocking disrupting inputs to the regions that they serve or, possibly, by acting to stabilize the then current pattern of graded responses by adjusting local feedback loops. Either process might be mediated by the shift in the balance of inhibitory processes that Pribram has suggested. The actual answer must await further experimentation.

D. Experimental Results in the Light of the Model

1. Galvanic Skin Response

If one assumes that the GSR represents a shift in the settings of the core homeostatic systems, the results observed seem relatively understandable. The exteroreceptive inputs, both auditory and visual, were not capable of producing a suitable change in those core systems, and, therefore, produced no galvanic skin response. However, apparently normal GSR's were observed in response to inputs that probably affected the neuropil directly (nocioceptive stimuli transmitted by C fibers).

2. Corticosteroid Response

The corticosteroid response to confinement in the chair seemed to be the same in both groups. The difference observed can be interpreted as an abnormal response on the part of the normal subjects at the start of the first session. In this view, they were frightened by the drawing of the first blood sample and by the presence of the new experimenter, which implied a changed experimental situation. Since this stimulus was temporary compared to the slow corticosteroid response, the subjects without a frontal cortex did not show a similar reaction. Furthermore, in its function of helping to maintain the appropriate setting or adjustment until it is no longer appropriate, the frontal cortex apparently helps to stabilize the corticosteroid level found at the beginning of the experimental sessions. This frontally enhanced stability of the core systems probably also helps to explain the variability of performance observed in monkeys with frontal lesions during overtraining on behavioral tasks (Pribram et al., 1966).

3. Thermoregulation

Since the heat input from the Microtherm unit provided the organism with little in the way of exteroreceptive input, no real difference should be expected between the two groups. In fact, little, if any, difference was observed.

4. Behavioral Latencies and Distractibility

As mentioned earlier, a computer's high-speed central processing unit need only sample low-speed inputs as often as a significant change occurs in these inputs. Such sampling is sometimes under program control, but commonly it involves the use of an interrupt. The interrupt is an electronic equivalent of a tap on the shoulder indicating that some peripheral device should be sampled. The orienting response to novelty would seem to be a similar phenomenon. When the exteroreceptive input does not agree with expectation (the neuronal model built up from previous experience), the organismic equivalent of an interrupt occurs. Desynchronization of the EEG might be an indication of such an occurrence. It would seem reasonable that such a disparate input should have to be sampled until its significance to the organism could be estimated. Upon the first several occasions, one might expect a defensive allassostatic shift due to the anxiety of uncertaintly. Thereafter, such an interrupt need only involve sufficient time and response for the input to be sampled and identified. If the disrupting input were of greater significance to the organism than the on-going task, doubtless, an allassostatic response would occur and the GSR could be observed.

Accepting the above assumptions, the interpretation of the results in Section III is fairly straightforward. In the normal subjects, the frontal cortex functions to maintain the core homeostatic settings and also the prepotency of the Plan, which has been interrupted by the novel input. It acts as a finger on the page or a "working memory," as Pribram has called it. Unless the interrupting stimulus is found to be of greater significance, the normal subject can immediately return to the interrupted task with all dispositional states still intact. In contrast, the frontal subject experiences the same interrupt but has no finger on the page and must sample all external stimuli to find one with significance (the lighted panel that will be rewarded with a pellet). In addition, by maintaining the facilitation of the interrupted Plan, the frontal cortex probably provides a constant tug on the sleeve that helps to shorten any unimportant response sequence initiated by the interrupt. The behavioral habituation observed probably represents the complement of a learning curve in which the novel stimulus is being categorized as an event of no significance. With practice, the recognition of the stimulus requires less processing time.

The relative facilitation of one Plan implies a relative suppression of all others. This suppression includes the suppression of irrelevant inputs (as in the last days of conditions 2 and 3) as well as the suppression of irrelevant or competing outputs (as in all three conditions). The strength of the stabilizing control available to the normal animal, but not to the frontal decorticate, was strikingly evident when the loud buzzer provided the distraction in condition 3.

5. Spatial Deficit

Mishkin (1964) and Mishkin et al. (1966) have found that lesions of the dorsolateral frontal cortex do not impair a visual version of the delayed alternation task as much as do lesions of the orbital surface. Instead, they find that the dorsolateral lesion produces a more marked deficiency in spatial delayed alternation tasks. It would not seem at all surprising if there should be other buffer registers in the vicinity of the area discussed above. It would also seem reasonable to expect that the dorsolateral area might be limited in its influence. Its connections would seem to be suited to action upon spatiomotor activity through its relationships with the head of the caudate nucleus. It might even be possible to isolate different cortical regions responsible for the maintenance of shifts in predispositions toward behavioral responses in certain regions in space as opposed to shifts in core dispositional states. Only further experimentation can tell (see Stamm and Rosen, 1973).

E. Previous Theoretical Interpretations in the New Light

In the light of the new model presented above, it can be seen that the many previous interpretations of prefrontal function were not erroneous and conflicting hypotheses but were specific manifestations of the more general underlying mechanism. As Jacobsen et al. (1935) originally suggested, the prefrontal cortex does participate in immediate memory, in that it functions to preserve the effect of an input. However, it is not involved in sensory processing or in the actual storage of sensory information. Several investigators have proposed that the dorsolateral frontal cortex enhances the vividness (Thompson, 1964; Pribram, 1950) or the reflexogenic strength (Konorski, 1964) of the cue; this is exactly what the biological buffer register accomplishes when it allows a momentary input to exert a continuing influence over behavior and/or homeostasis. Many investigators (e.g., Malmo, 1942; Milner, 1964; Pribram, 1969b, 1973) have underlined the importance of interference factors in the frontal deficit; as outlined above, the biological buffer register functions to maintain the relative facilitation of the Plan being executed, thus suppressing irrelevant inputs as well as suppressing possible competing responses or response tendencies (Mishkin et al., 1962; Mishkin, 1964; Milner, 1964). By allowing relevant inputs to both set and reset the system, the mechanism set forth above does parse the temporal flow (Pribram and Tubbs, 1967), and by maintaining stability until resetting occurs it makes it possible for the organism to compare the outcomes of its actions with its expectancy or intention (Pribram, 1960a, b, 1969b).

Thus, the dorsolateral frontal cortex, as a necessary part of the biological buffer register, performs a single function that has previously been described in many apparently discrepant ways.

F. Recapitulation

It has often been reported that lesions of the dorsolateral frontal cortex in primates increase behavioral responsiveness to novelty. Surprisingly, experimenters at the neuropsychology laboratory at Stanford University discovered that this lesion greatly diminishes the galvanic skin response component of the orienting response. This seeming paradox prompted a series of experiments that yielded the following results.

(1) Habituation tests revealed that the lesion does not grossly alter EEG desynchronization in response to tone or flash, although it totally eliminates the GSR to these stimuli.

(2) Even after 50 tone–shock pairings, no GSR to tone was recorded from the experimental group.

(3) Electric shock evoked skin resistance changes of normal amplitude but of increased duration from the experimental subjects.

(4) The corticosteroid response was not eliminated by the lesion but sensitivity to what might be termed psychological factors appeared to have been diminished or eliminated.

(5) When a Microtherm heating unit provided a thermal challenge, subjects with lesions evinced apparently normal thermoregulatory responses.

(6) Distractibility was studied by recording latencies in a behavioral task where visual distractors that did or did not vary in location and also an auditory distractor were occasionally inserted into an on-going two step sequential task. The lesion lengthened the duration of distractor-directed responses without necessarily increasing the probability that the distractor would be sampled. Distractors that varied in spatial location did affect subjects with lesions more frequently than they did controls.

In the preceding pages it has been speculated that the frontal eugranular isocortex evolved as part of a biological interface that couples the flickering evanescence of the neocortical information processing systems to the plodding neural elements of the more primitive layers of the brain. The dorsolateral frontal cortex in particular has been portrayed as part of a buffer register, which stabilizes allassostatic readjustments until appropriate internal or external responses have been completed. If anything at all is stored within the substance of the frontal cortex, even temporarily, it is not the dying image of a spatial past, but a pattern of efferent activity that promotes the completion of any Plan being executed.

In this model, the frontal cortex is somewhat analogous to a music stand that preserves the pages of a score unriffled by a summer breeze. The music stand merely holds the place, it neither writes nor reads the notes upon the page.

ACKNOWLEDGMENT

The research reported here was supported by NIMH Grant MH 12970.

REFERENCES

Artemieva, E. Yu., and Homskaya, E. D. (1973). Changes in the asymmetry of EEG waves in different functional states in normal subjects and in patients with lesions of the frontal lobes. *In* "Psychophysiology of the Frontal Lobes" (K. H. Pribram and A. R. Luria, eds.), pp. 53-70. Academic Press, New York.

Ashby, R. W., and Basset, M. (1950). The effect of prefrontal leucotomy on the psychogalvanic response. *J. Ment. Sci.* **96**, 458.

Bishop, G. H. (1959). The relation between nerve fiber size and sensory modality: Phylogenetic implications of the afferent innervation of cortex. *J. Nerv. Ment. Dis.* **128**, 89-114.

Brush, E. S., Mishkin, M., and Rosvold, H. E. (1961). Effects of object preferences and aversions on discrimination learning in monkeys with frontal lesions. *J. Comp. Physiol. Psychol.* **54**, 319-325.

Brutkowski, S., Mishkin, M., and Rosvold, H. E. (1963). Positive and inhibitory motor CRs in monkeys after ablation of orbital and dorsolateral surface of the frontal cortex, *In* "Central and Peripheral Mechanisms of Motor Functions" (E. Gutman, ed.). Czechoslovakia, Academy of Sciences.

Cannon, W. B. (1932). "Wisdom of the Body." Norton, New York.

Douglas, R. J., and Pribram, K. H. (1969). Distraction and habituation in monkeys with limbic lesions. *J. Comp. Physiol. Psychol.* **69**, 473-480.

Elithorn, A., Piercy, M. F., and Crosskey, M. A. (1954). Autonomic changes after unilateral leucotomy. *J. Neurol., Neurosurg. Psychiat. [N. S.]* **17**, 139.

Fulton, J. F. (1949). "Functional Localization in Relation to Frontal Lobotomy." Oxford Univ. Press, London and New York.

Glick, D., Redlich, D., and Levine, S. E. (1964). Fluorometric determination of corticosterones and cortisol in 0.02-0.05 ml. plasma or submiligram samples of adrenal tissue. *Endocrinology* **74**, 653-654.

Gross, C. G., and Weiskrantz, L. (1962). Evidence of dissociation between impairment of auditory discrimination and delayed response in frontal monkeys. *Exp. Neurol.* **5**, 453-476.

Isaacs, W., and DeVito, J. L. (1958). Effect of sensory stimulation on the activity of normal and prefrontal lobectomized monkeys. *J. Comp. Physiol. Psychol.* **51**, 172-174.

Jacobsen, C. F., Wolfe, J. B., and Jackson, T. A. (1935). An experimental analysis of the functions of the frontal association areas in primates. *J. Nerv. Ment. Dis.* **82**, 1-14.

Kimble, D. P., Bagshaw, M. H., and Pribram, K. H. (1965). The GSR of monkeys during orienting and habituation after selective partial ablations of the cingulate and frontal cortex. *Neuropsychologia* **3**, 121-125.

Konorski, J. (1961). Disinhibition of inhibitory CRs after prefrontal lesions in dogs. *In* "Brain Mechanisms and Learning" (J. F. Delafresnaye, ed.), Blackwell, Oxford, 567-593.

Konorski, J. (1964). Integrative Activity of the Brain." Univ. of Chicago Press, Chicago, illinois.

Luria, A. R., and Homskaya, E. D. (1964). Disturbances in the regulative role of speech with frontal lobe lesions. *In* "The Frontal Granular Cortex and Behavior" (J. Warren and K. Akert, eds.). McGraw-Hill, New York.

Luria, A. R., Pribram, K. H., and Homskaya, E. D. (1964). An experimental analysis of the behavioral disturbance produced by a left frontal arachnoidal endothelioma (meningioma). *Neuropsychologia* **2**, 257-280

Malmo, R. B. (1942). Interference factors in delayed response in monkeys after removal of frontal lobes. *J. Neurophysiol.* **5**, 295-308.

Mason, J. W., Nauta, W. J. H., Brady, J. B., Robinson, J. A., and Sachar, E. J. (1961). The role of limbic system structures in the regulation of ACTH secretion. *Acta Neuroveg.* **23**, 4-14.

Miller, G. A., Galanter, E. H. and Pribram, K. H. (1960). "Plans and the Structure of Behavior." Holt, New York.

Milner, B. (1964). Some effects of frontal lobectomy in man. *In* "The Frontal Granular Cortex and Behavior" (J. M. Warren and K. Akert, eds.), p. 313. McGraw-Hill, New York.

Mishkin, M. (1964). Perseveration of central sets after frontal lesions in monkeys. *In* "The Frontal Granular Cortex and Behavior" (J. M. Warren and K. Akert, eds.), p. 219. McGraw-Hill, New York.

Mishkin, M., and Pribram, K. H. (1956). Analysis of the effects of frontal lesions in the monkey. II. Variations of delayed response. *J. Comp. Physiol. Psychol.* **49**, 36-40.

Mishkin, M., Prockop, E. S., and Rosvold, H. E. (1962). One trial object discrimination learning in monkeys with frontal lesions. *J. Comp. Physiol. Psychol.* **55**, 178-181.

Mishkin, M., Vest, B., Waxler, M., and Rosvold, H. E. (1966). Arc-examination of the effects of frontal lesions on object alternation. *Proc. Int. Congr. Psychol. 18th, 1966* p. 43.

Porter, R. W. (1954). The central nervous system and stress induced eosinopenia. *Recent Progr. Hor. Res.* **10**, 1-18.

Pribram, K. H. (1950). Some physical and pharmacological factors affecting delayed response performance of baboons following frontal lobotomy. *J. Neurophysiol.* **13**, 373-382.

Pribram, K. H. (1960a). The intrinsic systems of the forebrain. "Handbook of Physiology" (Amer. Physiol. Soc., J. Field, ed.), Sect. I, Vol. II, pp. 1323-1344. Williams & Wilkins, Baltimore, Maryland.

Pribram, K. H. (1960b). A review of theory in physiological psychology. *Annu. Rev. Psychol.* **11**, 1-40.

Pribram, K. H. (1961). A further experimental analysis of the behavioral deficit that follows injury to the primate frontal cortex. *Exp. Neurol.* 3, 432-466.

Pribram, K. H. (1969a). DADTA III: An on-line computerized system for the experimental analysis of behavior. *Perceptual and Motor Skills* 29, 599-608.

Pribram, K. H. (1969b). The primate frontal cortex. *Neuropsychologia* 7, 259-266.

Pribram, K. H. (1973). The primate frontal cortex—Executive of the brain. *In* "Psychophysiology of the Frontal Lobes" (K. H. Pribram and A. R. Luria, eds.), pp. 293-314. Academic Press, New York.

Pribram, K. H., and Tubbs, W. E. (1967). Short-term memory, parsing, and the primate frontal cortex. *Science* 165, 1765-1767.

Pribram, K. H., Ahumada, A., Hartog, J., and Ross, L. (1964). A progress report on the neurological processes disturbed by frontal lesions in primates. *In* "The Frontal Granular Cortex and Behavior" (J. M. Warren and K. Akert, ed.), p. 28. McGraw-Hill, New York.

Pribram, K. H., Konrad, K., and Gainsburg, D. (1966). Frontal lesions and behavioral instability. *J. Comp. Physiol. Psychol.* 62, 123-124.

Schwartz, H. G. (1937). Effect of experimental lesions of the cortex on the "psychogalvanic reflex" in cat. *Arch. Neurol. Psychiat.* 38, 308.

Spinelli, D. N., and Pribram, K. H. (1967). Changes in visual recovery function and unit activity produced by frontal cortex stimulation. *Electrocenphalogr. Clin. Neurophysiol.* 22, 143-149.

Stamm, J. S. (1969). Electric stimulation of monkeys' prefrontal cortex during delayed response performance. *J. Comp. Physiol. Psychol.* 67, 535-546.

Stamm, J.S., and Rosen, S.C. (1972). Electrical stimulation and steady potential shifts in prefrontal cortex during delayed response performance by monkeys. *Acta Biol. Exp. (Warsaw)* (in press).

Stamm, J.S., and Rosen, S.C. (1973). The locus and crucial time of implication of prefrontal cortex in the delayed response task. In preparation.

Storey, J., Melby, J. C., Egdahl, R. H., and French, L. A. (1959). Adrenal cortical function following stepwise removal of the brain in the dog. *Amer. J. Physiol.* 196, 583.

Thompson, B. (1964). A note on cortical and subcortical injuries and avoidance learning by rats. *In* "The Frontal Granular Cortex and Behavior" (J. M. Warren and K. Akert, eds.), p. 60. McGraw-Hill, New York.

Part Seven

CONCLUSION

Chapter 14

THE PRIMATE FRONTAL CORTEX — EXECUTIVE OF THE BRAIN

K. H. PRIBRAM

Department of Psychology
Stanford University
Stanford, California

I. INTRODUCTION

The gross anatomical conglomerate anterior to the central fissure is given, in primate brains, a certain distinctiveness by the development of a bony orbit and is thus labeled the frontal lobe. Its posterolateral extent receives projections from the ventral thalamic nucleus and is covered with an agranular or dysgranular cortex. It constitutes the classical precentral motor cortex from which movements can be elicited at low thresholds by electrical excitation.

The remainder of the lobe, the subject of this essay, although somewhat homogeneous in its phylogenetic derivation and in its function, is made up of functionally discrete parts. A large portion of its medial extent derives projections from the anterior thalamic nuclei (Pribram and Fulton, 1954). Within this region, three subregions can be discerned, a supracallosal, a precallosal, and a subcallosal, each receiving projections from one of the anterior group of thalamic nuclei. Near the callosum the cortex covering this medial frontal region is transitional, junctional in type, and dysgranular; in keeping with these characteristics, it is found to give rise to movements when electrically stimulated (Kaada *et al.,* 1949) and so (with the orbitoinsulotemporal cortex, see next paragraph) has been labeled a mediobasal motor cortex (Pribram, 1961a). The precallosal portion of the region, further forward, is homotypical and eugranular in architecture, as is most of the remainder of the frontal lobe. This remainder, comprising the polar and lateral reaches of the lobe, receives projections from the major microcellular portion of the medial thalamic nucleus. The function of this cortex is the substance of this chapter.

Before proceeding to this discussion, however, we shall note another area of dysgranular transitional cortex important to the picture of the frontal cortex as a whole. This area lies on the posterior portion of the orbital surface, sandwiched between the orbital extensions of the medial and lateral cortices. This posterior orbital cortex derives its projections (Pribram et al., 1953) from the midline of the thalamus (the midline magnocellular portion of the nucleus medialis dorsalis) and is heavily and reciprocally connected with the adjacent anterior insular, periamygdaloid, and temporal polar cortex (Fulton et al., 1949).

In summary, apart from the classical precentral motor cortex, the primate frontal lobe can be divided into three major parts: (1) the medial, defined by its projections from the anterior nuclear group; (2) the dorsolateral (including the pole), defined by its projections from the microcellular portion of the medial thalamic nucleus; and (3) the posterior orbital, deriving projections from the midline of the thalamus and so heavily connected to the adjacent anterior insula, temporal pole and periamygdaloid cortex that these structures have often been considered together as a unit (Pribram and Bagshaw, 1953).

These anatomical considerations are important because investigators using behavioral techniques for the analysis of brain function have often ignored them, with consequent confusion in experimental results and their interpretations. Of special importance here are recent experiments in which large dorsolateral frontal resections have been compared with resections of the orbital cortex (Mishkin, 1964; Stamm and Rosen, 1973; Mahut, 1964; Pinsker, 1966). As noted, the orbital surface includes three different divisions: medial, lateral, and posterior. Differences in results obtained in various investigations might well be expected when different amounts of cortex from each of these categories have been included in the lesion.

Nonetheless, attempts to treat the frontal tissue selectively have proved to be most worthwhile. In order to ascertain for myself what the reported results might mean when comparison was made between lesions anatomically appropriate to each of the major divisions (excluding the precentral motor cortex), I prepared separate groups of monkeys with medial frontal, dorsolateral frontal, and orbitoinsulotemporal resections. With respect to alternation behavior (see Section V), at least, the effect of the dorsolateral lesion could be distinguished from that of the other lesions (Pribram et al., 1966a; Figure 1). This finding reinforced my view that the effect on behavior reported by other investigators resulted from invasion into medial and orbitoinsulotemporal portions of the frontal lobe—portions which have in recent years been included under the rubric "limbic systems" because of their position at the inner edge of the cerebral hemisphere, their allo- or juxtallocortical cytoarchitectonics, and their functional relatedness (Pribram, 1958, 1961a).

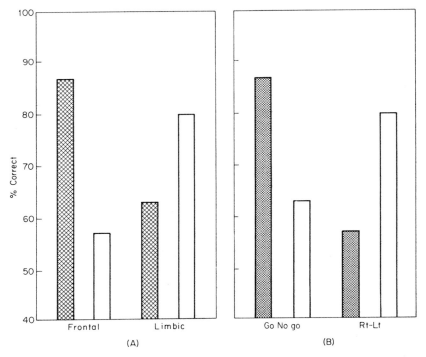

FIG. 1. Comparison of the effect of frontal and limbic lesions on (A) go—no go (crosshatched bars) and right—left alternations (open bars). Comparison on the basis of lesion locus. (B) Comparison on the basis of task. Stippled bars represent the frontal group, open ones the limbic group.

II. SPATIAL AND TEMPORAL ORGANIZATION

Despite this clarification, another, perhaps more serious problem of interpretation was aggravated by the results of the alternation experiments (Mishkin and Pribram, 1955, 1956; Pribram et al., 1966b). Alternation behavior is disrupted by lesions anywhere in the frontolimbic formations (and also in some parts of the basal ganglia) but not by lesions in the sensory-motor projection systems or in the cortex associated with these systems. Its temporal organization is the characteristic of the alternation task. This leads to an hypothesis that the frontolimbic formations might be directly involved in providing the temporal structure necessary to the proper execution of all behavior.

The hypothesis was put to test and found to be only partially supported. The sequential organization of behavior (pressing without repetition a series of

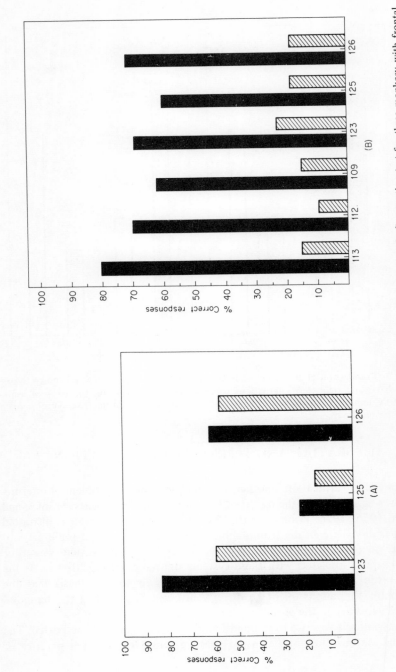

FIG. 2. (A) Percentage of correct responses in the pre- (solid) and postoperative (hatched) retention test for three monkeys with frontal lesions on the externally ordered sequence G–R. (B) Percentage of correct responses in the pre- (solid) and postoperative (hatched) retention test for frontal and temporal groups on the internally ordered sequence G–O–"4."

spatially randomized panels displaying distinct symbols) is indeed disrupted by frontal lesions, but only if that organization must be supplied by the monkey. (Figure 4). When external cues to organization are supplied by the procedure (as when reinforcement is contingent on pressing, on each trial, the identical order of symbols), frontally lesioned monkeys perform as do their unoperated controls (Pinto-Hamuy and Linck, 1965; Figure 2).

Another serious doubt about interpreting the frontal syndrome in purely temporal terms was raised by the finding that frontal lesions produce greater disturbance in spatial than in nonspatial and go–no go problems (Pribram and Mishkin, 1956; Pribram *et al.*, 1952). This led to the inference that the dorsolateral frontal cortex is somehow essential to the processing of spatial cues (Mishkin, 1964; Figure 3).

That this simple interpretation must be modified has been shown by a subsequent study (Pohl, 1970). In this experiment, two different spatial tasks were devised: one to test for responses based on external, the other on internal spatial cues. In the external cue task, the monkey had to respond to the location nearest a signal object, which varied from side to side. In the other (Figure 4), the monkey had to respond repetitiously to one side. Once these criteria had been attained, the problem was reversed: now reward went to responses away

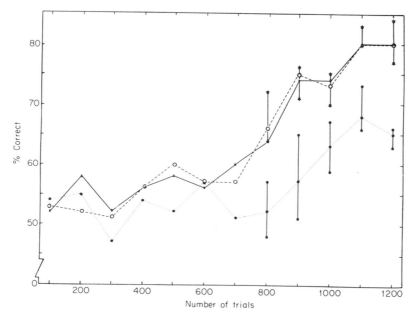

FIG. 3. Performance graph for monkeys in the nonspatial object alternation experiment. (*) Normals; (○) temporals; (●) frontals. Vertical lines show range of data.

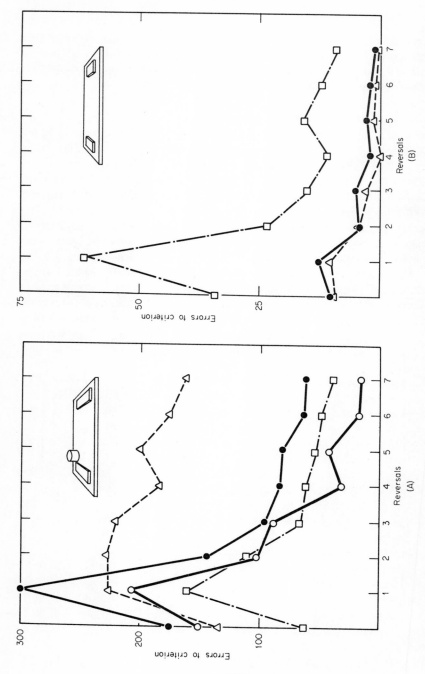

FIG. 4 (A) Performance of unoperated controls and three lesion groups on the landmark reversal task. (●) Temporal; (△) parietal; (□) frontal; (○) unoperated. (B) Performance of lesion groups on the place reversal task. Symbols as in Fig. 3A.

from the signal and, in the second test, to responses made to the side other than before. Monkeys with parietal lobe lesions failed the externally (but not the internally) cued task, whereas the frontally lesioned monkeys were severely hampered in their internally (but not the externally) cued performance.

III. DISTRACTION: PROACTIVE AND RETROACTIVE INHIBITION

Thus, both the temporal and spatial organization of behavior are disrupted by frontal lesions when, and only when, that organization is incompletely specified by environmental contingencies. In such situations, the organization of behavior becomes dependent upon internal processes. What might these processes be and by what mechanisms are they effected?

A clue to the answers to these questions comes from the results of the following experiment (Grueninger and Pribram, 1969). Monkeys were trained to respond quickly in succession to two cues. Then another cue, a distractor, was introduced simultaneously with the second cue and the consequent change in latency of response to the succession of cues was recorded. Two types of distractors were used, spatial and nonspatial. Monkeys with dorsolateral frontal resections were distracted more than unoperated controls by spatial interpolations and, moreover, they reacted more to spatial than to nonspatial distractors. Thus, an explanation for the observation that frontal lesions affect spatial delay problems more than nonspatial, can be suggested: dorsolateral frontal lesions do not preclude the processing of spatial information; instead, such lesions make the organism more sensitive to distraction, especially by spatial cues, a condition not unlike that found in unoperated but naive monkeys (Figure 5).

This result makes plausible the hypothesis that shifts in the spatial dimension in the spatial alternation problem act as distractors, interfering with adequate performance on the task, whereas shifts in the nonspatial dimensions are less distracting. The greater difficulty experienced by frontally lesioned monkeys on spatial tasks would thus be accounted for.

In earlier experiments (Malmo, 1942), monkeys were run under a variety of conditions designed to minimize interference. An explanation in terms of retroactive and proactive inhibition was set forth and became the classical way to account for the impaired performance of frontally lesioned animals on the spatial delayed response problems, the other task typically failed by such animals. The interference hypothesis is at present, as it has been for three decades, the most viable and useful in explaining the effects of resection of the dorsolateral frontal cortex of primates. This would suggest the hypothesis that under ordinary conditions the frontal cortex functions to inhibit interference.

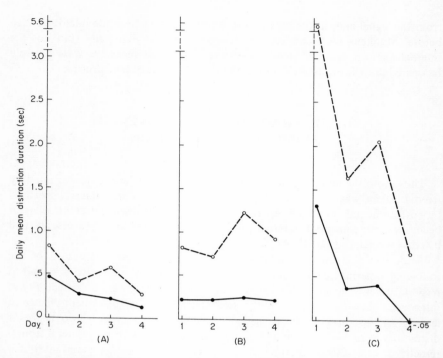

FIG. 5. Daily mean distraction duration (mean distraction trial latency minus median latency) for (A) condition 1: stimulus varied, location constant; (B) condition 2: location varied, stimulus constant; and (C) condition 3: buzzer. (○) Frontals; (●) normals.

The remainder of this chapter will be devoted to detailing the results of experiments aimed at obtaining an understanding of the mechanisms of inhibition of interference by the frontal cortex. A synthesis will be attempted in terms of a model. Perforce, this chapter will concentrate on the functions of the dorsolateral frontal cortex, leaving those of the limbic (medial and orbitoinsulotemporal) formations for another occasion. However, from what has already been reviewed, it should be clear that the point of view entertained here is that all three of the anterior frontal regions are concerned in this function.

IV. CONTROL OF INPUT

Two methods for investigating the mechanisms of frontal influence on the inhibitory organization of brain processes were explored. One consisted of continuing the studies of the effects of frontal resection on behavior; the other employed psychophysiological and neurophysiological techniques to record the effects of frontal resections on the processing of input to the brain. The experiments on input control stemmed from a series of studies using the delayed response and alternation tasks that showed that reaction to cue variables rather than response contingencies were affected by frontal resection (Mishkin and Pribram, 1955, 1956; Pribram and Mishkin, 1956). In these studies, the procedure was changed from a go right–go left situation to a go–no go task. This change resulted in a surprising improvement in the performance of the delay tasks by frontally lesioned monkeys, which could be attributed either to the change in response choice or to the change in cue (bait versus bare hand in delayed response; having just responded or withheld a response in the alternation task). Further experimentation showed that, in fact, it was the change in the cues, not in the response choice, that made the difference. When, in the delayed response task the original two-cup situation was retained and the monkeys cued to the left cup by a peanut and to the right cup by a bare hand displayed between the two cups, the monkeys performed as well as in the go–no go procedure. Also, when responses had to be alternated between two objects, irrespective of their placement, alternation performance improved. The results of these experiments supported the interference hypothesis and so will be taken up first. [It should be noted that these data also gave rise to the spatial hypothesis as an explanation for the effects of frontal damage, an hypothesis that continued to be supported by other results until found partially wanting by the results of Pohl's (1970) experiments, described earlier.]

Because in an earlier experiment monkeys with frontal lesions were found to react behaviorally with alacrity to novelty even under conditions when control subjects did not (Pribram, 1960), an investigation of the effect of such lesions on the orienting reaction seemed in order. Sokolov (1960) had just described the physiological measures of orienting in detail; of these, the galvanic skin response (GSR) was the simplest to use and so was applied to the study in frontal patients (Luria and Homskaya, 1964; Luria et al., 1964) and monkeys (Kimble et al., 1965). Much to our surprise, this and other psychophysiological measures failed to confirm the behavioral observations of distractibility, i.e., hyperreactivity to novelty; just the opposite was found. Psychophysiological measures of orienting did not occur at all (Kimble et al., 1965), or if they did, they were sluggish in their appearance and disappearance (Grueninger et al., 1965). Could it be that the absence of the psychophysiological components of orienting is directly related to the hyperreactivity to novelty and continued distractibility? Are these psychophysiological reactions perhaps indicative of a mechanism necessary for behavioral habituation to take place?

Further analysis suggested that this was indeed the case. The orienting reactions described by Sokolov (1960) could be classified into at least two categories: those involved in sampling the situation and those necessary to register it in awareness and memory. Registration appears to be necessary for behavioral habituation to occur (Kimble et al., 1965; Bagshaw and Benzies, 1968). Frontal lesions, by interfering with registration, make the organism susceptible to changes in input to which normal subjects had become accustomed, i.e., habituated (Figure 6).

Next, it became imperative to check whether one could obtain any direct neurophysiological evidence for frontal control over input processing. This evidence was obtained in the following manner (Spinelli and Pribram, 1967). Small gross electrodes were implanted in the visual system and records were made in the fully awake monkey faced with paired bright flashes of light. Computer summation (averaging) techniques were used to enhance the reliably repetitive aspects of the neuroelectric responses evoked by the flashes. The amplitude of the pair of responses was measured and that of the second response expressed as a percent amplitude of the first. During the experiment, the

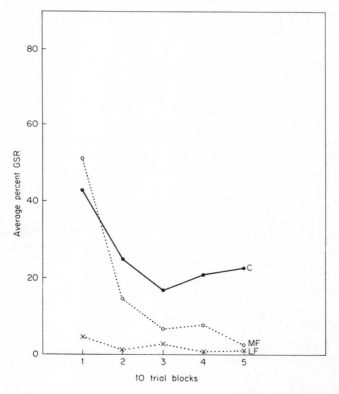

FIG. 6. Curves for percent GSR response to the first 50 presentations of the original stimulus on the second run for the normal (CH), medial frontal (MF), and lateral frontal (LF) groups.

interflash interval separating each pair of flashes was systematically varied and the percent amplitude was plotted as a function of the interflash interval. This provided a recovery function that indicated the percent of the visual channel available for processing the second of the pair of flashes when it occurred.

A stable recovery function served as a base line for testing the possible effects of electrical stimulation of the frontal cortex. As shown in Figure 7, an effect was obtained, recovery was markedly enhanced by the frontal lobe excitation.

To test the validity of this result, microelectrodes were inserted into various levels of the visual channel and visual receptive fields were plotted. Now, the effect of electrical excitation of the frontal cortex on stable (for over an hour)

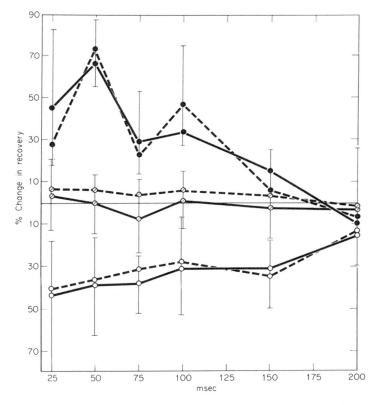

FIG. 7. The change in recovery of a response to the second of a pair of flashes compared with prestimulation recovery function. Control stimulations were performed on the parietal cortex. Records were made immediately after the onset of stimulation and weekly for several months. The response curves obtained immediately after onset (——) and after 1 month (– – –) are presented. Vertical bars represent variability of the records obtained in each group of four monkeys. (○) Temporal lobe stimulation; (●) frontal lobe stimulation; (⊘) control stimulation.

receptive fields was explored and again a dramatic result was obtained (Figure 8). There can be no question that by an as yet unknown pathway the dorsolateral frontal cortex can exert an influence on visual input processing.

The simplest way to conceptualize this effect of the frontal cortex on the input channel in psychological terms is to suggest that it produces some alteration in the attentive process (Gerbrandt *et al.*, 1970). Unfortunately, the term *attention* is commonly used in a variety of ways. However, in frontal lobe function, some precision in its use has been provided by the results of the experiments that showed that registration was impaired in subjects with frontal lesions. The term *registration* thus implys a focusing of attention, an *assimilation*—to use Piaget's term—of the situation by the organism.

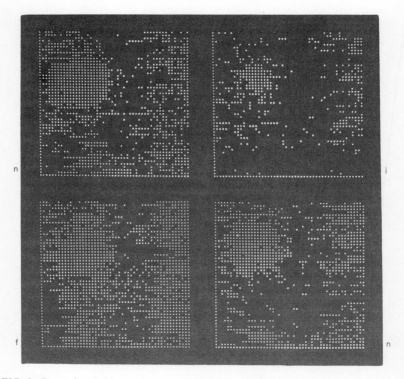

FIG. 8. Receptive field maps from a lateral geniculate unit. (n) Top left; control; (i) mapped while inferotemporal cortex was being stimulated; (f) mapped during frontal cortex stimulation; (n) bottom right, final control. A third control was taken between the (i) and the (f) maps and was not included because it was not significantly different from the first and the last. Note that inferotemporal stimulation (i) decreases the size of the "on" center; frontal cortex stimulation (f), although not really changing the circular part of the receptive field, brings out another region below it. The level of activity shown is three standard deviations above the normal background for this unit.

V. THE FRONTAL CORTEX AND BEHAVIOR

To return now to the other major method for investigating inhibition of interference among the brain's processing mechanisms: the continuing experimental analysis of behavioral disturbance produced by frontal lesions. What might be the effects on behavior of a deficiency in registration, an impaired process of assimilation? William James noted that, "what holds attention determines action." And as already mentioned, frontally lesioned monkeys continue to react to cues as if they were novel long after control subjects have behaviorally habituated to them.

Interestingly and somewhat surprisingly, however, the deficiency appears even more dramatically in a different aspect of behavior determination: monkeys with frontal resections do not process the consequences of their behavior as do normal animals (Pribram, 1960). Reinforcement, whether reward, punishment (English and Rosvold, 1956; Pribram, 1961b), or error (Pribram, 1961b), is processed sluggishly by the operated subjects. In a sense, this deficit is most obviously apparent in classical conditioning situations, where even reinforcement by punishment is severely affected (Rosvold and Szwarcbart, 1964). This inability to stably maintain reward-guided behavior is manifest in the two-choice discrimination (Pribram *et al.*, 1966a), and becomes critical in a multiple-choice situation in which monkeys are expected to reach a criterion of five consecutive errorless trials. In this task, frontally lesioned subjects will repeatedly make three and even four correct responses and then make an error, thus delaying the attainment of the criterion (Pribram, 1961b). In fact, the failure to perform the classical, spatial, delayed alternation task (Figure 9) has been shown by W. A. Wilson (1962) to depend for its solution on just this same insensitivity to reinforcing stimuli: frontally lesioned monkeys have difficulty in remembering the position of the preceding reinforcer. Moreover, even when the successive form of alternation is presented, a form which, as noted earlier, frontally lesioned monkeys can learn to perform, learning is characterized by an extraordinary number of errors (Pribram, 1960; Figure 10).

These results raise questions about the relationship of attention, at least in its registrational aspects, to reinforcement. Must reinforcers be registered in awareness or only in memory in order to guide behavior? Just what is the connection between awareness and memory? Does attention invariably imply awareness? Is one function of reinforcers to attract attention or is behavior guided by them without such an intervening step in most instances? The answers to these questions are not at present available but should be forthcoming as a result of experiments in which responses are observed in animals and man using the elegant techniques developed for eyeball photography accomplished while the subject is performing a task (Mackworth, 1967, 1968). When man is the subject, the relationship between attention and awareness should be subject to check by asking for a verbal statement describing awareness and correlating this with the evidence from the eye camera and perhaps even with some concomitant neuroelectric measurements.

VI. THE MODEL

I feel reasonably sure that the dorsolateral frontal cortex, like the limbic formations of the forebrain (including the medial and orbital frontal cortex), are concerned in the inhibition of interference among brain events. With respect to lesions of the frontal cortex, this involvement becomes manifest on the input side as a difficulty in attention, a difficulty in registering novelty so that habituation, or assimilation, fails to take place. On the output side, the feedback to actions from their outcomes is impaired and reinforcers become relatively ineffective.

The intact frontal brain tissue must help to accomplish registration and reinforcement by some not too complicated mechanism. What could be its nature? In order to obtain some clue, I turn, as I so often have in the past, to the analogy of those hardware brains, especially computers, that so effectively mimic many of the functions ordinarily carried on by the wetware in our heads (Miller *et al.*, 1960; Pribram *et al.*, 1964; Pribram, 1971). Mechanical as well as biological thinking machines continually face the simultaneous demands of a variety of inputs and outcomes. These could easily interfere with one another and with any of the central operations being carried on at the moment by the computer. To prevent this, some noticing order must govern the acceptance of first this, then that, product of the input–output devices. In its simplest form, each of these devices is fitted with a marker or flag, which decrees that while busy with its productions, the computer temporarily shuts off the paths to and from other

---

FIG. 9. Graph showing the differences in the number of repetitive errors made by groups of monkeys in a go–no go type of delayed reaction experiment. Especially during the initial trials, frontally operated animals repeatedly return to the food well after exposure to the nonrewarded predelay cue. Note, however, this variation of the delay problem is mastered easily by the frontally operated group. The 12 rhesus monkeys used in the multiple object experiment had served as subjects some 2 years earlier in the delayed response experiment portrayed here. (———) Normals; (– – –) temporals; (· · ·) frontals.

FIG. 10. Graph of the average number of trials to criterion attainment taken in the multiple object experiment by each of the groups in each of the situations after search was completed, i.e., after the first correct response. Note the difference between the curves for the controls and for the frontally operated group, significant at the .05 level by analysis of variance ($F = 8.19$ for 2 and 6 df) according to McNemar's (1955) procedure performed on normalized (by square root transformation) raw scores. (———) Normals; (– – –) temporals; (· · ·) frontals.

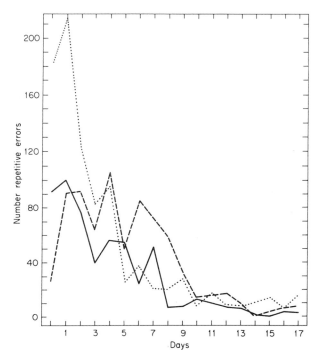

Figure 9

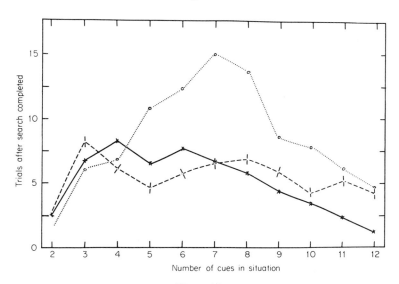

Figure 10

devices. In more complicated forms, only part of the computer may be thus preempted, or a program can be used to regulate the flow of information. Simple flexible noticing order programs have been used for years for this purpose; more recently these have burgeoned into full scale executive routines that effect the timesharing of large multiple user machines.

The essence of flexible noticing orders is their multiply recursive nature. Instead of the branching hierarchy that characterizes programs used to make discriminations and to compute, flexible noticing orders must in some way keep track of the various routines that involve the computer. Which routine has precedence depends upon an order that can be flexibly rearranged according to other flexible programs. What is allowed to occur at any moment, therefore, is weighted on the basis of a number of simultaneously noticed events. What occurs becomes dependent upon the context of these noticed events. The structure of programs with such characteristics is called *context sensitive* or *context dependent*, whereas the structure of hierarchical programs is said to be *context free*, because the occurrence of a particular item in the program is completely specified by the hierarchy of the routine in which it occurs. As already described, frontal lesions have their effect on behaviors that are incompletely specified by the environmental situation (schedules, routines) in which the behavior takes place. It is therefore plausible to hypothesize that the frontal cortex is especially concerned in structuring context-dependent behaviors.

To put this in slightly different terms, in a simple sensory discrimination (whether a simultaneous or a successive procedure is used), the cue–response–reinforcement contingencies remain invariant across trials. The close coupling of appropriate behavior to this invariance makes it free of determinants at other levels. In the delayed response and alternation tasks, on the other hand, these contingencies vary from trial to trial and the subject must take note of these variations so that second-order invariances can be extracted. These second-order invariances provide a context in which appropriate behavior is generated. It is in this sense that the behavior becomes context dependent.

On an earlier occasion, I had already compared the functions of the frontal cortex to that of a flexible noticing order, a primitive executive program (Pribram *et al.*, 1964). I suggested then that Ukhtomski's (1926) "dominant focus" might provide the neurological mechanism by which flexibility in noticing order might be achieved. It remains here to bring this model up to date by alterations and specifications made possible by the neurophysiological and neurobehavioral results that have accrued since the proposal was made.

Two findings are of special significance. One concerns the recovery function data reported in this manuscript. The other was provided by the discovery by Grey Walter (1964) of a contingent negative electrical variation (CNV) originating in the front part of the brain whenever an organism is preparing to perform (i.e., during the foreperiod) in a reaction-time experiment. This rather extensive neuroelectric phenomenon certainly behaves as a temporary dominant focus. Could it be a signal that the brain is busy and act as an interference prevention device blocking input, much as does the marker or flag in mechanical computer systems? Partial answers to this question might come from experiments that study the relationship between this electrical brain event and those that are involved in the registrational aspects of the orienting reaction, like the experiments by Lacey (1969) on the foreperiod (readiness to respond) phenomenon. As noted elsewhere in this volume (Chapter 7), such experiments have shown the frontal cortex to be especially active in the production of the CNV in tasks where behavior has not as yet become completely dependent upon environmental contingencies—when a marker or flag is especially necessary to prevent distracting interference (Donchin et al., 1971).

The importance of the recovery function data for frontal lobe function is somewhat more complicated to present because a neuronal minimodel of the recovery cycle phenomenon must first be constructed. This minimodel has been detailed on several occasions (Pribram, 1968, 1971): in essence it is based on the inhibitory interactions that occur in the input channels. Two reciprocally related forms of inhibitory phenomena are recognized: those in which the activity of a neuron decreases that of its neighbors and those that result in diminishing the activity of the neuron itself. These two types of reciprocal afferent inhibition have been compared to Pavlov's *external* and *internal* forms, respectively. The suggestion was made that the usual function of the frontal cortex is to weight the balance of these reciprocal processes toward internal inhibition, i.e., self diminishing neural activity (Figure 11). This suggestion was based on the recovery function data, which showed that recovery of cells in the visual system excited by flashes was enhanced; i.e., cells returned to their preexcited state more rapidly.

The model was further developed (Pribram, 1971) to suggest that the orienting reaction and its habituation depended upon the activation of these reciprocal afferent inhibitory mechanisms: orienting on the contrast-enhancing effect of the inhibition of a neighbor's activity; habituation on the subsequent diminishing of this effect caused by the progressive diminution in the number of the initiating impulses.

VII. TEST OF THE MODEL

The model is thus operationally spelled out at several levels. Neuronally, the effect of frontal excitation on afferent inhibitory mechanisms can be checked. At a somewhat grosser level, the effects of frontal excitation and resection on foreperiod and orienting responses can be investigated using eye movement

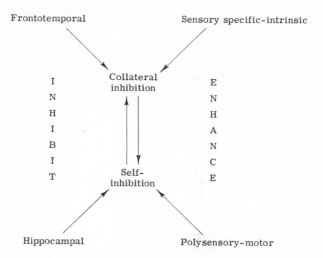

FIG. 11. A model of the inhibitory interactions taking place within the afferent channels. Collateral and recurrent afferent inhibitions are bucked against one another, forming a primary couplet of neural inhibition within afferent channels. Four forebrain mechanisms are assumed to provide efferent control on the primary couplet: Two of these, frontotemporal and sensory specific-intrinsic (which included the inferotemporal cortex), influence the couplet by regulating collateral inhibition; two others, hippocampal and polysensory-motor, regulate recurrent inhibition. The sensory specific-intrinsic and polysensory-motor association cortical systems exert their control by enhancing, whereas the frontotemporal and hippocampal systems exert control by inhibiting afferent neural inhibition.

recordings, the contingent negative variation, and other brain wave manifestations, as well as the more usual psychophysiological indicators (GSR, heart and respiratory rate changes, and alterations in blood flow). Finally, at the behavioral level, an attempt can be made to compensate for the deficiency produced by frontal lesions by providing the organism with a frontal prosthesis, as it were. Using such an external substitute for its frontal cortex, a lesioned monkey should be able to perform as well as an unoperated control in the delayed alternation situation. Perhaps by providing the operated monkey with context, a marker or flag, by building into the task a simple executive program, problems previously failed could in this fashion be solved.

Such an experiment was performed (Pribram and Tubbs, 1967). The usual equal interval that separates the alternation trials was modified by interposing a 15 sec interval between each R–L couplet (Figure 12). Thus if the classical 5 sec delayed alternation problem can be represented as R5–L5–R5–L5–5R–5L–5..., the variation here presented would read: R5L–15–R5L–15–R5L–15–R5L....

Monkeys who had failed to adequately perform the classical task after 1000 trials performed remarkably well on the variation within 250 trials, much as did

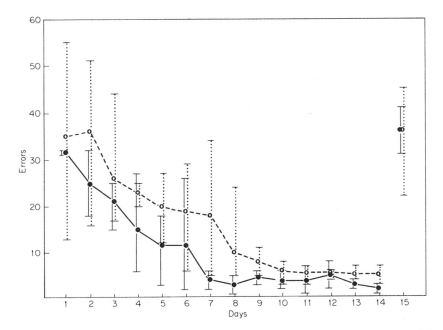

FIG. 12. Graph of the average number of errors made by monkeys having ablations of the frontal cortex (○) and by their controls (●). Bars indicate ranges of errors made. Records of the number of errors made on return to the classical 5 sec alternation task for day 15 are shown.

their normal controls. This dramatic improvement is not caused by changing the problem into a successive discrimination: interposing a red light or a loud buzzer as a marker for every other trial failed to produce such dramatic improvement in performance of either the control or the operated monkeys, although both control groups eventually (after 1000–1500 trials) learned to perform these tasks (Tubbs, 1969). Meanwhile, any return to the 15 sec interposition was immediately effective in restoring adequate performance in both operated and unoperated groups.

These results suggest that for the frontally lesioned primate, the alternation task and perhaps many other situations appear much as would this printed page were there no spaces between the words and no punctuation at the ends of phrases, sentences, and paragraphs.

Spaces serve as markers or flags to externally organize context-dependent input. The letters of the alphabet make meaningful organizations when arranged in different orders. This flexible order of arrangement is tolerated provided the length of the arrangement is limited (the magic number is 7 ± 2; Miller, 1956).

The finding that, when context is furnished by markers, frontally lesioned primates can so readily perform a task that had been their nemesis for decades supports the hypotheses that performance of this task is an instance of context-dependent behavior and that the function of the frontal cortex is, in the absence of sufficiently simple environmental structure, to internally organize a context upon which behavior must depend in such situations. In short, the frontal cortex appears critically involved in implementing executive programs when these are necessary to maintain brain organization in the face of insufficient redundancy in input processing and in the outcomes of behavior.

REFERENCES

Bagshaw, M. H., and Benzies, S. (1968). Multiple measures of the orienting reaction and their dissociation after amygdalectomy in monkeys. *Exp. Neurol.* **20**, 175-187.

Donchin, E., Otto, D. A., Gerbrandt, L. K., and Pribram, K. H. (1971). While a monkey waits: Electrocortical events recorded during the foreperiod of a reaction time study. *Electroencephalogr. Clin. Neurophysiol.* **31**, 115-127.

English, J. M., and Rosvold, H. E. (1956). The effect of prefrontal lobotomy in rhesus monkeys on delayed-response performance motivated by pain-shock. *J. Comp. Physiol. Psychol.* **3**, 286-292.

Fulton, J. F., Pribram, K. H., Stevenson, J. A. F., and Wall, P. D. (1949). Interrelations between orbital gyrus, insula, temporal tip and anterior cingulate. *Trans. Amer. Neurol. Assoc.* 175

Gerbrandt, L. K., Spinelli, D. N., and Pribram, K. H. (1970). The interaction of visual attention and temporal cortex stimulation on electrical activity evoked in the striate cortex. *Electroencephalogr. Clin. Neurophysiol.* **29**, 146-155.

Grueninger, W. E., and Pribram, K. H. (1969). The effects of spatial and nonspatial distractors on performance latency of monkeys with frontal lesions. *J. Comp. Physiol. Psychol.* **68**, 203-209.

Grueninger, W. E., Kimble, D. P., Grueninger, J., and Levine, S. E. (1965). GSR and corticosteroid response in monkeys with frontal ablations. *Neuropsychologia* **3**, 205-216.

Kaada, B. R., Pribram, K. H., and Epstein, J. A. (1949). Respiratory and vascular responses in monkeys from temporal pole, insula, orbital surface and cingulate gyrus. A preliminary report. *J. Neurophysiol.* **12**, 347-356.

Kimble, D. P., Bagshaw, M. H., and Pribram, K. H. (1965). The GSR of monkeys during orienting and habituation after selective partial ablations of the cingulate and frontal cortex. *Neuropsychologia* **3**, 121-128.

Lacey, J. I. (1969). *In* "Readiness to Remember" (D. P. Kimble, ed). Gordon & Breach, New York.

Luria, A. R., and Homskaya, E. D. (1964). Disturbance in the regulation role of speech with frontal lobe lesions. *In* "The Frontal Granular Cortex and Behavior" (J. M. Warren and K. Akert, eds.), pp. 353-371. McGraw-Hill, New York.

Luria, A. R., Pribram, K. H., and Homskaya, E. D. (1964). An experimental analysis of the behavioral disturbance produced by a left frontal arachnoidal endothellome (meningioma). *Neuropsychologia* **2**, 257-280.

Mackworth, N. H. (1967). A stand camera for line-of-sight recording. *Percept. Psychophys.* **2**, 119-127.

Mackworth, N. H. (1968). The wide-angle reflection eye camera for visual choice and pupil size. *Percept. Psychophys.* **3**, 32-34.

Mahut, H. (1964). Effects of subcortical electrical stimulation on discrimination learning in cats. *J. Comp. Physiol. Psychol.* **58**, 390-395.

Malmo, R. B. (1942). Interference factors in delayed response in monkeys after removal of frontal lobes. *J. Neurophysiol.* **5**, 295-308.

Miller, G. A. (1956). The magical seven, plus or minus two, or some limits on our capacity for processing information. *Psychol. Rev.* **63**, 81-97.

Miller, G. A., Galanter, E. H., and Pribram, K. H. (1960). "Plans and the Structure of Behavior." Holt, New York.

Mishkin, M. (1964). Perseveration of central sets after frontal lesions in monkeys. *In* "The Frontal Granular Cortex and Behavior" (J. M. Warren and K. Akert, eds.), pp. 219-241. McGraw-Hill, New York.

Mishkin, M., and Pribram, K. H. (1955). Analysis of the effects of frontal lesions in monkeys. I. Variations of delayed alternation. *J. Comp. Physiol. Psychol.* **48**, 492-495.

Mishkin, M., and Pribram, K. H. (1956). Analysis of the effects of frontal lesions in monkeys. II. Variations of delayed response. *J. Comp. Physiol. Psychol.* **49**, 36-40.

Pinsker, H. M. (1966). Behavioral effects of dorsolateral, lateral orbital and medial orbital lesions in frontal granular cortex of rhesus monkeys. Ph.D. Thesis, University of California, Berkeley.

Pinto-Hamuy, T., and Linck, P. (1965). Effect of frontal lesion on performance of sequential tasks by monkeys. *Exp. Neurol.* **12**, 96-107.

Pohl, W. G. (1970). Dissociation of spatial discrimination deficits following frontal and parietal lesions in monkeys. Ph.D. Thesis, George Washington University, Washington, D.C.

Pribram, K. H. (1958). Comparative neurology and the evolution of behavior. *In* "Evolution and Behavior" (G. G. Simpson, ed.), pp. 140-164. Yale Univ. Press, New Haven, Connecticut.

Pribram, K. H. (1960). The intrinsic systems of the forebrain. *In* "Handbook of Physiology" Washington, (Amer. Physiol. Soc., J. Field, ed.), Sect. 1, Vol. II. Williams & Wilkins, Baltimore, Maryland.

Pribram, K. H. (1961a). Limbic systems. *In* "Electrical Stimulation of the Brain" (D. E. Sheer, ed.), pp. 311-320. Univ. of Texas Press, Austin.

Pribram, K. H. (1961b). A further experimental analysis of the behavioral deficit that follows injury to the primate frontal cortex. *Exp. Neurol.* **3**, 432-466.

Pribram, K. H. (1968). The primate frontal lobe. *In* "Systems Approach in Studies of the Brain Functional Organization" (A. I. Shumilina, ed.). Moscow.

Pribram, K. H. (1971). "Languages of the Brain." Prentice-Hall, New York.

Pribram, K. H., and Bagshaw, M. H. (1953). Further analysis of the temporal lobe syndrome utilizing fronto-temporal ablations. *J. Comp. Neurol.* **99**, 347-375.

Pribram, K. H., and Fulton, J. F. (1954). An experimental critique of the effects of anterior cingulate ablations in monkeys. *Brain* **77**, 34-44.

Pribram, K. H., and Mishkin, M. (1956). Analysis of the effects of frontal lesions in monkey. III. Object alternation. *J. Comp. Physiol. Psychol.* **49**, 41-45.

Pribram, K. H., and Tubbs, W. E. (1967). Short-term memory, parsing and the primate frontal cortex. *Science* **156**, 1765-1767.

Pribram, K. H., Mishkin, M., Rosvold, H. E., and Kaplan, S. J. (1952). Effects on delayed-response performance of lesions of dorsolateral and ventromedial frontal cortex of baboons. *J. Comp. Physiol. Psychol.* **45**, 565-575.

Pribram, K. H., Chow, K. L., and Semmes, J. (1953). Limit and organization of the cortical projection from the medial thalamic nucleus in monkeys. *J. Comp. Neurol.* **98**, 433-448.

Pribram, K. H., Ahumada, A., Hartog, J., and Roos, L. (1964). A progress report on the neurological processes disturbed by frontal lesions in primates. *In* "The Frontal Granular

Cortex and Behavior" (J. M. Warren and K. Akert, eds.), pp. 28-55. McGraw-Hill, New York.
Pribram, K. H., Konrad, K., and Gainsburg, D. (1966a). Frontal lesions and behavioral instability. *J. Comp. Physiol. Psychol.* **62**, 123-124.
Pribram, K. H., Lim, H., Poppen, R., and Bagshaw, M. H. (1966b). Limbic lesions and the temporal structure of redundancy. *J. Comp. Physiol. Psychol.* **61**, 368-373.
Rosvold, H. E., and Szwarcbart, M. K. (1964). Neural structures involved in delayed-response performance. *In* "The Frontal Granular Cortex and Behavior" (J. M. Warren and K. Akert, eds.), pp. 1-15. McGraw-Hill, New York.
Sokolov, E. N. (1960). Neuronal models and the orienting reflex. *In* "The Central Nervous System and Behavior" (M. A. B. Brazier, ed.), pp. 187-276. Josiah Macy, Jr. Found., New York.
Spinelli, D. N., and Pribram, K. H. (1967). Changes in visual recovery function and unit activity produced by frontal cortex stimulation. *Electroencephalogr. Clin. Neurophysiol.* **22**, 143-149.
Stamm, J. S., and Rosen, S. C. (1973). The locus and crucial time of implication of prefrontal cortex in the delayed response task. *In* "Psychophysiology of the Frontal Lobes" (K. H. Pribram and A. R. Luria, eds.)., pp. 139-153. Academic Press, New York.
Tubbs, W. T. (1969). Primate frontal lesions and the temporal structure of behavior. *Behav. Sci.* **14**, 347-356.
Ukhtomski, A. A. (1926). Concerning the condition of excitation in dominance. *Nov. Refleksol. Fizol. Nerv. Syst.* **2**, 3-15; *Psych. Abstr.* p. 2388 (1927).
Walter, W. G. (1964). Slow potential waves in the human brain associated with expectancy, attention and decision. *Arch. Psychiat. Nervenkr.* **206**, 309-322.
Wilson, W. A., Jr. (1962). Alternation in normal and frontal monkeys as a function of response and outcome of the previous trial. *J. Comp. Physiol. Psychol.* **55**, 701-704.

AUTHOR INDEX

Numbers in italics refer to the pages on which the complete references are listed.

A

Abe, K., 186, *230*
Adamovich, V.A., 40, *50*
Agafonov, V.G., 102, *106*
Ahumada, A., 152, *153*, 237, *251*, 281, *290*, 306, 308, *313*
Ajmone Marsan, C., 186, 188, 190, *231*, *232*
Akert, K., 157, *165*, 186, 187, *230*
Akimoto, H., 186, *230*
Aladzhalova, N.A., 69, *69*, 110, *122*
Albe-Fessard, D., 180, *183*, 197, *231*
Aldridge, V.-J., 126, 133, *138*, 212, 213, *234*
Alexander, G.E., 150, *153*, 163, 164, *165*
Allen, W.F., 237, *249*
Andrew, J. F. N., 237, *249*
Anokhin, P.K., 29, *50*, 102, *106*
Aring, C.D., 238, *250*
Artemieva, E.Yu., 8, 9, *24*, *25*, 55, 59, *70*, 71, *86*, 256, *288*
Asaro, K.D., 188, *234*
Ashby, R.W., 269, *288*
Aslanov, A.S., 29, *51*
Auer, J., 187, *230*

B

Bagchi, B., 72, *88*
Bagshaw, M.H., 135, *138*, 253, 254, 255, 271, *289*, 294, 295, 301, 302, *312*, *313*
Bakos, S., 245, *252*
Baranovskaya, O.P., 8, *24*, 71, *87*
Barlow, J., 53, *70*
Barris, R.W., 238, *250*
Basset, M., 269, *288*
Beer, B., 204, *231*
Bekhterev, V.M., 4, *24*, 249, *250*
Bekhtereva, N.P., 72, *87*
Benzies, S., 302, *312*
Berger, H., 53, *70*, 102, *106*
Bishop, G.H., 53, *70*, 281, *288*
Black, S., 120, *122*
Blinkov, S.M., 23, *25*
Bogacz, J., 204, *231*
Bonvallet, M., 102, *107*, 168, 169, 180, *183*
Borda, R.P., 126, 137, *137*, *138*, 213, 221, *230*, *232*
Borissova, T.P., 103, *106*
Bossom, J., 135, 136, *138*
Bowden, J.W., 186, 188, *232*
Brady, J.B., 258, 270, *289*
Bremer, F., 197, 212, *230*
Brickner, B.M., 237, *250*
Brindley, G.S., 210, *230*
Brodal, A., 177, 178, 179, *182*, *183*, *184*
Brookhart, J.M., 202, *234*
Brush, E.S., 228, *230*, 253, *288*
Brutkowski, S., 139, *152*, 167, *182*, 227, *230*, 237, 249, *250*, 274, *288*
Bujas, L., 92, *106*
Bunatyan, B.A., 238, *251*

AUTHOR INDEX

Buser, P., 187, *230*
Butorin, V. I., 103, *107*

C

Cannon, W.B., 271, *288*
Caspers, H., 117, *122*
Castellucci, V.G., 221, *230*
Chaillet, F., 187, *233*
Chambers, W.W., 180, *184*
Chang, H.T., 53, *70*
Chase, M.H., 167, 180, *182*
Chiorini, J.R., 213, 221, *231*
Chow, K.L., 188, *231*, 294, *313*
Clark, W., 93, *107*
Clemente, C.D., 167, *168,* 169, *170, 172, 177,* 178, 180, 181, *182, 183*
Cohen, J., 136, *137,* 212, *234*
Cohen, S.M., 151, *153*
Coldstone, M., 117, *122*
Conrad, K., 238, *250*
Cooper, R., 109, 110, *122,* 126, 133, *138,* 212, 213, *234*
Cordeau, J.P., 210, *231*
Costa, L.D., 135, *138,* 213, *231, 234*
Critchley, M., 23, *25*
Crosskey, M.A., 269, *288*
Crow, H.H., 109, *122*
Crow, H.J., 109, *122*

D

Danilova, N.N., 72, 84, *87, 88*
Davis, P.A., 84, *87,* 102, *107*
Deecke, L., 135, 136, *137, 138,* 213, *231, 232*
de Lange, J.V.N., 92, *107*
Dell, P., 102, *107,* 210, 212, *231*
Dement, W.C., 188, *231*
Dempsey, E.W., 186, 197, 202, *232*
Denissova, A.M., 103, *107*
De Vito, J.L., 270, *289*
Dieckmann, G., 214, *233*
Dietsch, G., 72, *87*
Dilworth, M., 188, *231*
Donchin, E., 126, 127, 128, 131, 132, 134, 135, 137, *137,* 309, *312*
Dondey, M., 197, *231*

Dongier, M., 85, *87*
Dongier, S., 85, *87*
Douglas, R.J., 279, *288*
Dovey, F.J., 72, *88*
Drift, J.H., 72, *87*
Dumont, S., 210, 212, *231*
Dzugayeva, S.B., 98, *107*

E

Egdahl, R.H., 270, *290*
Eidelberg, E., 188, *231*
Elithorn, A., 269, *288*
Elterman, M., 167, *183*
English, J.M., 305, *312*
Epstein, J.A., 293, *312*

F

Fairchild, M.D., 180, *184*
Farley, B., 93, *107*
Fatler, T.O., 10, *24*
Ferrier, D., 227, *231*
Filippycheva, N.A., 10, *24,* 71, *87*
Fishkopf, L., 93, *107*
French, G.M., 237, 249, *250*
French, J.D., 188, *231*
French, L.A., 270, *290*
Frigyesi, T.L., 222, *233*
Frost, J.D., 125, *138,* 213, *232*
Frustorfer, H., 137, *138*
Fulton, J.F., 238, *250,* 253, *288,* 293, 294, *312, 313*
Fuster, J.M., 150, *153,* 158, 163, 164, *165*

G

Gadzhiev, S.G., 15, *24*
Gainsburg, D., 285, *290,* 294, 305, *314*
Galambos, R., 204, 213, *231, 232*
Galanter, E.H., 280, 281, *289,* 306, *313*
Gamberg, A.L., 103, *107*
Garcia-Austt, E., 204, *231*
Gastaut, H., 85, *87*
Gavrilova, N.A., 29, *51*
Genkin, A.A., 8, *24,* 54, *70*
Gerbner, M., 240, 245, *250, 251, 252*
Gerbrandt, L.K., 126, 127, 128, 131, 132, 134, 135, 136, 137, *137,* 304, 309, *312*

Gibbs, F.A., 72, 74, *87*
Gilden, L., 213, *231*
Glass, A., 91, *106*
Glick, D., 259, *288*
Goff, W.R., 136, *137*
Goldberg, L.J., 180, 181, *183*
Goldring, S., 213, 215, 220, 221, *230, 231*
Golikov, N.V., 35, 40, *50*
Grass, A.M., 72, *87*
Grindel, O.M., 72, 84, 85, *87*
Gross, C.G., 135, 136, *138,* 139, 140, *153,* 249, *250,* 257, *289*
Grossman, R.G., 162, *165*
Grueninger, J., *260, 262, 263,* 301, *312*
Grueninger, W.E., *254, 260, 262, 263, 275, 277,* 299, 301, 312

H

Haider, M., 204, *231, 234*
Hanberry, J., 188, *231*
Harlow, J.M., 237, *250*
Harper, R.M., 169, 181, *183*
Hartog, J., 152, *153,* 237, *251,* 281, *290,* 306, 308, *313*
Harvey, E., 53, *70*
Hearst, E., 204, *231*
Hess, J.R., Jr., 186, *230*
Hess, W.R., 186, *231*
Hillyard, S.A., 213, *231*
Hines, M., 238, *251*
Hobart, G., 53, *70*
Homskaya, E.D., 6, 8, 9, 13, 23, *24, 25,* 29, *50,* 55, 62, 69, *70,* 71, *86, 87,* 135, *138,* 249, *251,* 253, 256, 274, *288, 289,* 301, *312*
Hugelin, A., 168, 169, 180, *183*
Hyman, W., 188, *231*

I

Ilyanok, V.A., 72, 74, 84, *87*
Irwin, D.A., 213, *231, 233*
Isaacs, W., 270, *289*

J

Jackson, T.A., 237, 249, *250,* 286, *289*
Jacobsen, C.F., 125, *138,* 139, *153,* 227, *231,* 237, 238, 249, *250,* 286, *289*

Jarvic, H.F., 237, *250*
Jarvilehto, T., 137, *138*
Jasper, H. H., 91, *106,* 186, 190, 201, 202, 221, *232*

K

Kaada, B.R., 167, *183,* 293, *312*
Kalberer, M., 186, 187, *232*
Kamenskaya, V.M., 103, *107*
Kaplan, S.J., 297, *313*
Karpov, B.A., 16, 18, *24, 25*
Kellaway, P., 126, *138,* 213, *232*
Kennard, M.A., 238, *250*
Kimble, D.P., 135, *138,* 253, 254, 255, *260, 262, 263,* 271, *289,* 301, 302, *312*
Kimwich, H., 102, *107*
King. E.E., 180, *183*
Kiryakov, K., 92, *107*
Knauss, T., 167, 169, 180, *182, 183*
Knott, J.R., 213, *231,* 233
Koella, W.P., 186, *230*
Kogan, A.B., 85, 86, *88*
Konorski, J., 227, 228, *232,* 272, 286, *289*
Konorski, Yu., 237, 249, *250*
Konovalov, Yu. W., 69, *70*
Konrad, K., 285, *290,* 294, 305, *314*
Kooi, K., 72, *88*
Korn, H., 180, *183*
Kornhuber, H.H., 135, 136, *137, 138,* 213, *231,* 232
Kornmüller, A.E., 32, *50*
Kovner, R., 152, *153*
Krokovica, 92, *106*
Kruger, L., 136, *138*
Krupp, P., 186, 187, *232*
Kubell, Z., 102, *107*
Kuypers, H.G.J.M., 178, *183*

L

Lacey, J.I., 309, *312*
Langworthy, O.R., 237, *251*
Lawicka, W., 227, 228, *232,* 237, 249, *250*
Langworthy, O.R. 237, 249, *250*
LeBeau, J., 197, *231*
Lebedinsky, V.V., 12, 20, *25,* 237, 249, *251*
Leifer, L., 135, *137*

AUTHOR INDEX

Levine, S.E., 259, *260, 262, 263, 288,*
  301, *312*
Lim, H., 295, *314*
Linck, P., 297, *313*
Lindsley, D.B., 83, *88,* 91, *106,* 186, 187,
  188, 189, 190, 192, 197, 204, 207, 208,
  210, 223, 224, 226, *231, 232, 233, 234*
Livanov, M.N., 29, *32, 51,* 72, *88*
Loomis, A., 53, *70*
Low, M.D., 126, *138,* 213, *232*
Luria, A.R., 3, 6, 7, 13, 16, 18, 20, 23, *24,
  25,* 62, 69, *70,* 135, *138,* 237, 249,
  251, 253, 274, *289,* 301, *312*
Lynch, G.S., 203, *233*

M

McAdam, D.W., 213, *231, 232, 233*
McCallum, W.C., 126, 133, *138,*
  212, 213, *232, 234*
McCarthy, D.A., 204, *232*
Mackworth, N.H., 305, *312*
McMurthry, J.G., 222, *233*
McSherry, J.W., 213, *232*
Magoun, H.W., 167, 174, 179, *183,* 186,
  188, 202, *232, 233, 234*
Mahut, H., 210, *231,* 294, *313*
Malis, L.I., 136, *138*
Malmo, R.B., 227, 228, *232,* 237, *251,*
  272, 287, *289,* 299, *313*
Mancia, M., 187, *232*
Manghi, E., 187, *232*
Marsh, J.T., 204, *232*
Masahashi, K., 186, *230*
Mason, J.W., 258, 270, *289*
Mayorchik, V.E., 35, *51,* 72, 85, *88*
Melby, J.C., 270, *290*
Meshalkin, L.D., 8, *25,* 55, 59, *70*
Milhailovic, L.J., 139, 140, *153*
Miller, G.A., 280, 281, *289,* 306, 311, *313*
Milner, B., 237, *251,* 287, *289*
Minz, B., 180, *183*
Mishkin, M., 140, *153,* 227, 228, *230, 232,*
  237, *251,* 253, 272, 274, 286, 287, *288,
  289,* 294, 295, 297, 301, *313*
Mitchell, S.A., Jr., 188, *231*
Mizuno, N., 169, 170, *176,* 178, 181, *183*
Moisseyeva, N.J., 54, *70*
Monnier, M., 186, 187, *232, 234*

Morison, R.S., 186, 197, 202, *232*
Moruzzi, G., 186, *233*
Müller, J., 185, *233*
Mundy-Castle, A.C., 53, *70,* 72, *88*

N

Nakagawa, T., 186, *230*
Nakamura, I., 186, *230*
Nakamura, Y., 167, *168,* 169, *170, 172,
  177,* 180, 181, *183*
Nauta, W.J.H., 187, 188, *233,* 258,
  270, *289*
Newman, P.P., 178, *183*
Niemer, W.T., 170, 179, *183, 184*
Nissen, H.W., 237, 249, *250*

O

Okabe, K., 186, *230*
O'Leary, J.L., 213, 215, 220, *231*
Olmos, N., 167, *183*
Otto, D.A., 126, 127, 128, 131, 132, 134,
  135, 137, *137, 138,* 309, *312*

P

Paillard, J., 237, 238, *251*
Pappas, N., 128, *137*
Pásztor, E., 240, *250, 251*
Pavlov, I.P., 4, *25*
Peimer, I.A., 91, *107*
Penaloza-Rojas, J.H., 167, *183*
Perret, C., 187, *230*
Petz, B., 92, *106*
Peymer, J.D., 103, *107*
Phillips, D.G., 109, *122*
Piercy, M.F., 269, *288*
Pinsker, H.M., 294, *313*
Pinto-Hamuy, T., 297, *313*
Pohl, W.G., 298, 301, *313*
Poppen, R., 295, *314*
Porter, R.W., 258, 270, *289*
Potter, A., 170, *184*

Pribram, K.H., 83, *88*, 125, 126, 127, 128, 131, 132, 134, 135, 136, 137, *137*, *138*, 152, *153*, 228, *232*, 237, *251*, 253, 254, 255, 271, 272, 274, *275*, *277*, 280, 281, 282, 285, 286, 287, *288*, *289*, *290*, 293, 294, 295, 297, 299, 301, 302, 304, 305, 306, 308, 309, 310, *312*, *313*, *314*
Prockop, E.S., 228, *232*, 287, *289*
Puchinskaya, L.M., 35, *51*
Purpura, D.P., 222, *233*

R

Rebert, C.S., 213, *231*, *233*
Redlich, D., 259, *288*
Rhines, R., 167, 174, 179, *183*
Richter, C.P., 237, 238, *251*
Rinaldi, F., 102, *107*
Ritter, W., 135, *138*, 213, *234*,
Robertson, R.T., 203, *233*
Robinson, J.A., 258, 270, *289*
Robinson, M.F., 162, *165*
Roitbak, A.I., 121, *122*
Roos, L., 152, *153*, 306, 308, *313*
Rosen, S.C., 136, *138*, 150, 151, *153*, 283, 286, *290*, 294, *314*
Rosina, A., 187, *232*
Ross, L., 237, *251*, 281, *290*
Rossi, G.F., 178, 179, *183*
Rosvold, H.E., 162, *165*, 228, *230*, *232*, 253, 274, 286, 287, *288*, *289*, 297, 305, *312*, *313*, *314*
Rothballer, A.B., 102, *107*
Rougeul, A., 187, *230*
Rowland, V., 117, *122*
Rudomin, P., 181, *183*
Rusinov, V.S., 35, 40, *51*, 72, *88*, 117, *122*

S

Sachar, E.J., 258, 270, *289*
Sasaki, K., 203, 214, *233*, *234*
Sauerland, E.K., 167, *168*, 169, 170, *170*, *172*, *176*, *177*, 178, 180, 181, *182*, *183*
Scarff, T., 222, *233*
Scheff, N.M., 213, *234*
Scheibel, A.B., 179, *184*, 187, *233*
Scheibel, M.E., 179, *184*, 187, *233*
Scheid, P., 136, *137*, 213, *231*
Schlag, J.D., 187, 201, 202, 203, 214, 220, *233*
Schmidt, R.F., 181, *184*
Schulman, S., 157, *165*, 227, *233*
Schwartz, H.G., 264, *290*
Seales, D.M., 213, *232*
Semmes, J., 294, *313*
Shakin, M.I., 103, *107*
Sheatz, G. C., 204, *231*, *232*
Shkol'nik-Yapros, E.G., 238, 249, *251*
Shumilina, A.I., 32, *51*, 237, *251*
Shustin, N.A., 237, *251*
Shvets, T.B., 117, *122*
Simernitskaya, E.G., 9, *25*
Simernitskaya, E.T., 238, *251*
Skinner, J.E., 187, 188, 189, 190, 192, 195, 197, 198, 203, 207, 208, 210, 216, 218, 223, 224, 226, *233*, *234*
Small, I., 72, *88*
Smirnov, G.D., 35, *51*
Smith, D.B., 136, *137*
Snider, R.S., 170, *184*
Sokolov, E.N., 6, 19, *25*, 35, *51*, 72, 84, 85, 86, *88*, 92, *107*, 301, *314*
Sokolova, A.A., 72, *88*
Spencer, W.A., 202, *234*
Spilberg, P.I., 72, *88*
Spinelli, D.N., 125, *138*, 272, *290*, 302, 304, *312*, *314*
Spirin, V.G., 238, 249, *251*
Spong, P., 204, *231*, *234*
Sprague, J.M., 180, *184*
Stamm, J.A., 136, *138*
Stamm, J.S., 139, 140, 141, 150, 151, 152, *153*, 162, *165*, 272, 283, 286, *290*, 294, *314*
Starzl, T.E., 186, 202, *234*
Staunton, H.P., 203, 214, *233*, *234*
Stefanis, C., 221, *232*
Stepien, C., 237, *250*
Stepien, I., 140, *153*, 237, *250*
Sterman, M.B., 167, 180, *182*, *184*
Stevenson, J.A.F., 294, *312*
Storey, J., 270, *290*
Storm van Lewen, V., 92, *107*
Stoupel, N., 212, *230*
Szendröy, G., 245, *252*
Szwarcbart, M.K., 305, *314*

T

Talavrinov, V.A., 103, *106*
Tecce, J.J., 213, *234*
Thompson, B., 286, *290*
Tikhomirov, O.K., 19, *25*
Tissot, R., 187, *234*
Tokareva, V.A., 102, *107*
Torri, H., 186, *230*
Torvik, A., 177, 178, 179, *184*
Tsvetkova, L.S., 15, 20, *25, 26*
Tubbs, W.E., 125, *138,* 287, *290,* 310, *313*
Tubbs, W.T., 311, *314*
Tucker, L., 135, *137*
Tunturi, A., 93, *107*

U

Ukhtomski, A.A., 43, *51,* 308, *314*
Unna, K.R., 180, *183*

V

Vanzulli, A., 204, *231*
Vaughan, H.G., Jr., 135, 136, *138,* 213, *231, 234*
Velasco, M., 187, 188, 189, *234*
Velluti, R.A., 181, *183*
Vernier, V.G., 204, *231*
Vest, B., 274, 286, *289*
Viernstein, L.J., 162, *165*
Villablanca, J., 201, 202, 203, 214, 220, *233*
Vinogradova, O.S., 6, *26*
Vishevskaya, A.A., 103, *107*
Voronin, L.G., 102, *107*

W

Waldman, A.V., 102, *107*
Walker, A.E., 157, *165*
Wall, P.D., 294, *312*
Walsch, E., 92, *107*
Walsh, J., 210, *231*
Walshe, F.M.R., 238, *252*
Walter, W.G., 29, *51, 53, 70,* 72, 84, *88,* 91, *107,* 109, 110, 111, 120, *122,* 126, 133, *138,* 212, 213, *232, 234,* 308, *314*
Waxler, M., 274, 286, *289*
Weinberger, N.M., 188, *234*
Weiskrantz, L., 139, 140, *153,* 249, *250,* 257, *289*
Welt, L., 237, *252*
Wendt, R., 180, *183*
Werre, P.E., 53, *70,* 92, *107*
Whitlock, D.G., 187, *233*
Wilson, W.A., Jr., 305, *314*
Winkler, C., 170, *184*
Winter, A.L., 126, 133, *138,* 212, 213, *234*
Wolfe, J.B., 237, 249, *250,* 286, *289*
Wolstencroft, J.H., 178, *183*

Y

Yamaguchi, N., 186, *230*
Yarbuss, A.L., 16, 17, 18, *24, 25, 26*

Z

Zanchetti, A., 179, *183*
Zhirmunskaya, E.A., 84, *88*

SUBJECT INDEX

A

Acceptor of the consequences of action, 20
Accuracy, activation and, 164
Acquisition, nucleus medialis dorsalis and, 162
Activation
  of nucleus medialis dorsalis, 160
    during delayed response tasks, 162
  of prefrontal cortex, 164
  regulation of, 5-10, 38
    evoked potentials and, 29, 34-35, 49-50
    thalamic stimulation and, 186-187
Activation reaction, 84
  electroencephalogram indices for, 71
  types of, 85-86
After-discharge, 147, 148
  effects of, 110
  identification of electrodes using, 110
Alerting, 186
Allassostasis, 271, 281
Alpha rhythm
  asymmetry of, 8, 69, 82
    frontal lobe lesions and, 62-69
    in normal subjects, 54-59
    relation to orienting reaction, 59-62
  effects of attention activation on, 53
  effects of mental diseases on, 102-103
  frequency composition of, 84
  lesion site and, 39, 40, 43, 48
  in nucleus medialis dorsalis, 159
  suppression of
    meaningfulness and, 8, 71-86

  mental work and, 91-92
Alternation tasks, *see also* Delayed alternation tasks
  distraction in, spatial cues and, 299
  frontolimbic lesions and, 295
  prefrontal lesions and, 237, 249
  thalamocortical blockade and, 223-225
Aminasine, effect on biopotential correlations, 98-102, 106
Aminia, 63
Amplification, contingent, 116-117
Amygdalectomy, distractibility and, 272
Amygdaloid nucleus, corticosteroid response and, 270
Anamnesis, 64
Anosmia, 63
Anterior central gyrus, biopotential correlations and, 98, 102, 103
Anterior frontal cortex, 126
  biopotential correlations and, 102
  negative shifts in, 136
Anterior insula, 294
Anterior lateral cortex, 218-219, 221
Anterior sigmoid gyrus, 189-191, 198-200, 204-205, 214, 216-219, 221, 238
  augmenting response in, 192-193, 220
  epileptogenic discharges in, 200
  lesion of, 264
  recruiting response in, 192-194, 200, 207-209, 221
  spontaneous spindle bursts in, 192, 194, 195
Anterior suprasylvian gyrus, 192-193, 198-199, 214, 216-217

epileptogenic discharges in, 201
spontaneous spindle bursts in, 192, 194, 195
Antilocalizationism, 3
Anxiety, 237
Apraxia
  ideational, 237
  motor, 238
Arcuate sulcus, 141, 254
Arousal, 186
  cortical potential changes during, 117
  electroencephalogram desynchronization and, 253
Association, probability of, 118-119
Association potentials, 188
Associative fibers, biopotential correlations and, 98
Atropine, thalamocortical blockade and, 196-197, 198-199
Attention
  alpha wave oscillation and, 9, 53, 55, 57-59
  in delayed response tasks, 163
  effect of lesions on, 40, 63, 64, 66, 72
    convexal, 8
    frontal granular cortical, 227
    frontoparasagittal, 38
    posterior, 7
  electroencephalogram desynchronization and, 253
  evoked potential enhancement and, 204, 212
  input control and, 304
  negative steady potential shift and, 212
  registration in, 304, 306
  regulation of, 9
  relationship to reinforcement, 305
  reticulothalamocortical system in, 186, 187
Attenuation, contingent, 116
Auditory defects, 7
Auditory evoked potentials, enhancement of, thalamocortical blockade and, 208-209
Auditory middle ectosylvian cortex, evoked responses in, 208-209
Augmenting response, 200-201
  negative slow potential shift and, 213-214, 220-221
  reduction of slow potentials associated with, 214-217

thalamocortical blockade and, 189-193, 196, 202-203, 215-217
waveform of, 201-202
Automatism, oral, 63

B

Babinski's reflex, 63
Barbiturates
  effect on parietal responses, 203
  thalamocortical blockade and, 196-199
Basal ganglia, 295
Behavior, see also specific behaviors
  context-dependent, 308, 310-312
  expedient, 4
  thalamocortical blockade and, 222-228
Beta activity, 64
  effects of attention activation on, 53
Biopotentials
  correlations of, 93
    in normal subjects, 93-102
    in schizophrenics, 102-106
  method of recording, 92-93
  slow voltage fluctuations confused with, 110
Blockade, types of, 188
  cryogenic, 188, 222
Brain stem, 63, 188, 281
  interrelations with frontal cortex, 5
  reflex facilitation and, 179
  reflex inhibition and, 167, 180
  sites of origin for reticulospinal tract in, 179

C

Cat
  orbital-cortically induced inhibition in, 169-182
  thalamocortical blockade in, see Inferior thalamic peduncle
Caudate nucleus, 163, 286
  stimulation of, 192-193, 195
Caudate spindle, 195
  inferior thalamic peduncle blockade and, 192-193, 195
Cella turcica, changes in bones of, 63
Centrum semiovali, 41
Children, contingent negative variation in, 121

SUBJECT INDEX

Choice, goal-directed, ability to make, 20-21
Claustrum, 188
Cochlear nucleus, stimulation of, 208-209
Coding, role of frontal lobes in, 125
Cold, *see* Blockade, cryogenic
Commissural fibers, biopotential
  correlations and, 98
Conditioning
  contingent negative variation and, 121
  cortical, 169
  of galvanic skin response, 264-266,
    269-270
  of negative slow potentials, 215-216,
    218-220
  potential changes during, 117
  prefrontal and premotor ablation and,
    239-249
  reinforcement processing in, 305
Consciousness
  confused, 23
  loss of, 40
Context, 308, 310-312
Contingency, 113-114
  in delayed response and alternation tasks,
    308
  effect on response, 116
Contingent interaction, 116-122
Contingent negative variation, 117-122,
  125-126, 213, 222, 308-309, 310
  delayed response tasks and, 136
  extinction of, 118, 120
  method of measurement, 126-127
  motor response and, 135
Convulsions elicited by electrocortical
  stimulation, 144, 149
  threshold of, 150
Corneal blink reflex, 117, 168
Coronal cortex, 190, 214, 216-219, 221
  augmenting response in, 220
Corpus callosum, 39, 75, 293
Correction, delayed, 63
Cortex
  depolarization of, 120
  elective activation of, 34-35, 49, 50
  irritation of, influence on evoked potentials, 35, 41
  lability of, 35, 50
  reactivity of, 35
Cortical tone, 5
Corticosteroid response, 270-271, 284-285,
  287

galvanic skin response and, 258-263
Counting, 64
Criticism, defects of, 22, 38, 40, 63, 66
Cruciate sulcus, 238
Cue(s)
  in delayed response and alternation tasks,
    301
  enhancement of, 286
  presentation of
    effect on thalamic cell discharge,
      159-160
    electrical stimulation during, 162
  spatial
    distraction and, 299
    internal and external, 298-299

D

Delayed alternation task(s)
  context-dependent behavior in, 308
  cue variables in, 301
  effect of electrocortical stimulation on,
    139-140, 283
  locus of, 148-150
  method of investigation, 140-144
  time of, 144-148
  lesion site and, 286
  reinforcement processing in, 305-307
Delayed response task(s), 125, 127-128, 237
  behavioral deficits in, 227-228
  context-dependent behavior in, 308
  cue variables in, 301
  double
    signaled, 128-130
    unsignaled, 131, 132
  galvanic skin response and, 253
  nucleus medialis dorsalis and, 159-163
    lesions of, 157
    method of investigation, 158
    prefrontal cortex and, 163-165
  psychological processes involved in, 139
  retroactive and proactive inhibition in,
    299
  single, 131-133, 134
  transcortical negative variations and,
    129-137
Delta rhythm, 48, 64, 66
  intensification of, 77, 82
  pathological focus and, 72

SUBJECT INDEX

Delusions, biopotential correlations and, 103-106
Dendrites, apical, 117
  depolarization of, 121
Desynchronization, *see* Activation
Diencephalic region, 43
  pathological influence on, 40
  regulation of evoked potentials and, 49
Discrimination task(s), 30, 308
  auditory, 257
  behavioral deficits in, 227-228
  delayed, electrical stimulation and, 151
  influence on evoked potentials, 32-33, 36, 41, 42, 45-46, 48-49, 50
  reinforcement processing in, 305-307
Disinhibition, 237
  prefrontal and premotor ablations and, 240
  model for, 245
Dispersive convergence, 114
Distractibility, 228, 249, 272-276, 279-280, 285-286, 287
  behavioral habituation and, 276-278
  inhibition and
    deficit in, 278-279
    retroactive and proactive, 299
  spatial deficit and, 277-278, 287
  thalamocortical blockade and, 225-226
Dog, behavioral effects of prefrontal and premotor ablations in, 238-244
Dominant focus, 308-309
Dorsolateral frontal cortex, nucleus medialis dorsalis responses to stimulation of, 163
Drawings, inert stereotypes in, 13
Dreams, phasic states and, 5
Drugs, *see also specific drugs*
  electroencephalogram synchronization and, 196
Dura, 38
Dyskinesia, treatment of, 109

E

Echinococcus, 81
Echolalia, 10, 64
Echopraxia, 64
Electrocerebrogram, analysis of, 72
Electrocorticogram, relation to electroencephalogram, 114

Electrodes
  effective impedance of, 110
  effects of, 109-111
  expectancy waves and, 117-118
  intrinsic time constants of, 118
  method of implantation, 140-141
Electroencephalogram, *see also* Alpha rhythm, Evoked potentials
  desynchronization of, 269, 285
    galvanic skin response and, 253-258, 271, 287
  frequency analysis of, 72
  lesion site and, 39, 40, 43, 44, 48, 72
  recording method, 54-55, 59, 73
  relation to electrocorticogram, 114
  in schizophrenics, 103
  in sleep, 195-196
  synchronization of, effects of drugs on, 196
  thalamic stimulation and, 186-187
Electromyogram, 120
Electrooculogram, 133
Emotions, disturbance of, 23, 63
  treatment for, 109
Encephalitis, periventricular, 81
Eosinopenia, 258, 270
Epilepsy, 63
  treatment of, 109
Epileptic activity, 40, 43, 48, 66
  thalamocortical blockade in reduction of, 197, 200, 201, 203
Equalization phase, 5
Evoked potential(s), 29-30
  elicitation of, 30-31
  enhancement of, 9
    association and nonprimary, 210-212
    primary, 204-209
    at thalamic and cortical levels, 208-211
  influence of verbal instructions on, 9, 49-50
  in normal subjects, 31-35
  in patients with frontal lobe lesions, 35-49
  record analysis of, 111
  somatosensory, 136
  thalamocortical blockade and, 188
Excitation
  alpha wave oscillation during, 58-59
  conversion into inhibition, 41
Exophthalmia, 63
Expectancy

SUBJECT INDEX 325

influence on evoked potentials, 9, 31-35,
    36, 41-45, 48-49
  negative steady potential shift and, 212
Expectancy wave, 6, 117-122, *see also* Contingent negative variation
Extinction of orienting response, 115
Eye movement(s)
  preliminary orientation and, 16-19
  steady potential shifts and, 213

F

Facial nerve, paresis of, 63
Facilitation, 280
  information storage and, 244-249
  of reflexes, orbital-cortically induced,
    174-176, 178, 179, 182
Fatigue
  alpha wave oscillation during, 58-59
  influence on evoked potentials, 39
Feedback, transcortical negative variation
    and, 135
Foreperiod, 308, 309
Frontal granular cortex, 163
  behavioral effects of blockade of, 227
  distractibility and, 277-280, 285-286, 287
  efferent connections of, 187-188
  galvanic skin response and
    conditioning, 264-266, 269-270
  corticosteroid response and, 258-263
  electroencephalogram and, 253-258
  recruiting response and, 189, 190,
    192-194, 196, 202-203
  thermoregulation and, 266-271, 285, 287
Frontal lobe, anatomical considerations,
    293-294
Frontal lobe lesions
  behavioral effects of, 4-23
  extent of, 111
  influence on evoked potentials, 41-43
  localization of, reactivity of alpha
    frequencies and, 80-81
  symptoms of, 63-64, 75, 77
Frontal sigmoid gyrus, 238
Frontolimbic system, allassostatic function
    of, 281-282
Frontoparasagittal region, sarcoma of, 38-39
  influence on evoked potentials, 39-40
Frontopremotor lesion, 43

influence on evoked potentials, 43-46
Frontotemporal lesion, 63, 64
Fundus, varicose veins in, 46

G

Gait, astasia-abasia types, 63
Gallamine triethiodide, 169
Galvanic skin response, 62, 71, 284, 301
  conditioning, 264-266, 269-270
  corticosteroid response and, 258-263
  electroencephalogram desynchronization
    and, 253-258, 271, 287
Glia, 222
  activation of, 121
Gnosis, disturbance of, 7
Grasping reflex, 63
Gyrus proreus, 189, 204-205
  lesion of, 264

H

Habituation, 115-116, 137, 215, 309
  behavioral, 271-272, 286, 301-304
    distractibility and, 225, 276-280
    galvanic skin response and electroencephalogram and, 254, 287
    registration and, 306
  effect on orienting reaction, 6
  of frontal negativity, 135
  inhibitory information and, 249
  of transcortical negative variations, 135,
    136
Headaches, 38, 63
Heat lesions, 189, 192-193, 196
  behavioral effects of, 228
Hemiparesis
  left-side, 38
  right-side, 40
Hippocampectomy, distractibility and, 272
Homeostasis, 271, 280, 281
Hydrocephalus, 63
Hyperactivity, 228, 237, 270
Hyperpolarization, 284
Hyperreactivity, 270, 301
Hyperreflexia, 63
Hypertension, 63
  intracranial, 38

Hypnosis, effect on contingent negative
   variation, 118, 120
Hyposmia, 63
Hypothalamus, 188, 266
   galvanic skin response and, 258
Hypothesis formation, disturbance of, 15,
   19-21

I

Idiodromic projection, 114-115
Impulsiveness, 15, 63, 66
Inferior horn, 47, 81
Inferior thalamic peduncle, 187, 188
   augmenting response and, 189-193, 196,
      202-203, 215-217, 220
   conditioned cortical responses and,
      218-220
   cryogenic blockade of, 229-230
      behavioral effects of, 222-229
      compared with ablation and lesions, 222
   enhancement of evoked potentials and
      association and nonprimary, 210-212
      primary, 204-209
      at thalamic and cortical levels, 208-211
   lesions of, 226-227
   recruiting response and, 189-194, 196,
      202-203, 215-217, 220
   spontaneous spindle bursts and, 192-196
Inferotemporal cortex, 141
   electrical stimulation of, 151-152
Information
   distinctive, inability to single out, 19-20,
      21
   storage of, facilitory and inhibitory,
      244-249
Inhibition, 284
   of ascending activating system, 40
   behavioral, 167
   conditional, prefrontal and premotor
      ablations and, 240-249
   contingent negative variation and,
      121-122
   deficit in, 277-279
   differential, 240
   external, 272, 309
   information storage and, 244-249
   intensification by motor task, 41
   internal, 186-187, 222, 272, 309
   of nucleus medialis dorsalis, 159

   parietooccipital lesion and, 50
   postsynaptic, of masseteric reflex,
      180-181, 182
   presynaptic, 181, 182
   of pyramidal neurons, 121
   reciprocal orbital gyrus and, 180
   of reflexes, orbital-cortically induced,
      167-174, 178-182
   retroactive and proactive, in delayed
      response tasks, 299
   of selective traces, 23
   transmarginal, 45-46
      verbal instructions and, 9
Initiative, derangement of, 22
Input control, 300-304
Instructions
   effect on contingent negative variation,
      118
   influence on evoked potentials, 49-50
      in normal subjects, 31-35
      in patients with frontal lobe lesions,
         35-49
   meaningfulness imparted by, 71
   regulation of activation by, 6-10
Intellectual activity, 6, 7
   alpha rhythm during, 8-9, 55, 57, 67,
      91-92
   biopotential correlations during
      in normal subjects, 95, 96-98
      in schizophrenics, 104
   in diagnosis of frontal lesions, 21
   lack of orienting basis of action in, 18-19
Intellectual disorders, 40, 64
Intention, 5, 21
Interface, 281-282
Interference, inhibition of, 299-300
   behavioral disturbances and, 305-307
   input control and, 300-304
   model for, 306-312
Internal capsule, 81
Intralaminar medial thalamus, stimulation
   of, 214
   steady potential shift and, 213
Island of Reil, 41

K

Kinesthetic defects, 7
Kohs block, 15

SUBJECT INDEX

L

Latency
 development of contingent negative variation and, 120
 distractibility and, 274-275, 285-286
 of frontal responses, 114
Lateral geniculate body, enhancement of evoked potentials and, 206, 210-212
Lateral ventricle, 63, 81
Lenticular nucleus, 41
Leucomeningitis, 81
Leukotomy, galvanic skin response conditioning and, 269
Limbic system, 83, 294
 corticosteroid release and, 258
Link cube, 15
Logical codes, 20

M

Masseteric nerve, 168
Masseteric reflex
 elicitation of, 169
 facilitation of, 175, 176
 inhibition of, 168-174, 180-182
Meaningfulness, *see also* Instructions
 alpha rhythm and, 60-62, 67-69, 74-76, 78-83, 85-86
 orienting reaction and, 6-9, 71
 probability and, 116
Medial gyrus, 40-41
Medial internal capsule, cryogenic blockade of, 216-217, 220
Medial thalamus, difference from mesencephalic reticular formation, 187
Mediobasal lesions, 63, 66
 regulation of evoked potentials and, 49
Mediobasal motor cortex, 293
Medulla
 bilateral projections to, 178
 effect on inhibition of masseteric reflex, 170, 172
 site of origin of reticulospinal fibers in, 179
Medullary inhibitory area, 180
 electrical stimulation of, 172-173, 174-175
 evoked potentials in, 177, 178, 179, 181-182

Memory, 249
 short-term, 125, 139, 151-152, 286
 nucleus medialis dorsalis and, 162, 165
 spatiokinetic, 272, 280
 verbal-auditory, 64
 working, 282-283, 285
Memory trace, maintenance of, 125
Mental disorder, treatment by electrodes, 109
Mesencephalic reticular formation
 difference from medial thalamic system, 187
 high-frequency stimulation of, 186, 187
 lesions of, 186, 187, 195, 202
  effect on parietal responses, 203
  epileptic focus produced by, 197, 200, 201, 203
  spontaneous spindle bursts and, 192-194, 196
Midcoronal gyrus, epileptogenic discharges in, 200, 201
Mirror symptom, 64
Mnemonic processes, 163-164
 principalis cortex in, 151
Modality signature, 114
Monkey
 contingent negative variation in, 126
 delayed response tasks and, 128-137
 delayed response task and
  behavioral deficits in, 227-228
  electrical stimulation and, 139-152
  galvanic skin response and, 253
  nucleus medialis dorsalis and, 158-165
 distractibility in, 277-280, 285-286, 287
 galvanic skin response in
  conditioning, 264-266, 269-270
  corticosteroid response and, 258-263
  electroencephalogram and, 253-258
 thermoregulation in, 266-271, 285, 287
Motivation, contingent negative variation and, 213
Motor activity, 30, 228
 inert stereotypes and, 12-14
 influence on evoked potentials, 33-34, 37, 41-42, 45, 47
 nucleus medialis dorsalis and, 160
 prefrontal and premotor lesions and, 226, 238-242
 slow negative potentials and, 213
 thalamocortical blockade and, 226
 transcortical negative variations and, 133, 135

SUBJECT INDEX

Motor analyzer
  biopotential correlations of, 94, 98, 103
  lesion in, 23
Motor nerve, reflex inhibition and, 168
Motor perseverence, 38, 63, 64
Motor zones, 293
  biopotential correlations of, 95, 96, 104
  negative shifts in, 136
  negative steady potential in, 213
  waveforms in during waiting period, 129, 131, 135, 136
Multiple-choice tasks, reinforcement processing in, 305-307
Muscle tone, 63
  inhibition of, 167

N

Nausea, 63
Neocortex, 281
Nerve VII, paresis of, 40, 46
Nerve XII, paresis of, 40
Neuropil, 281, 284
  graded response transmission and, 281-282
Nocioceptive reflex, 264
Nonspecific brain systems, terminology for, 185-186
Noticing order, 308
Novelty response, *see* Orienting response
Nucleus centralis medialis, 190-191, 206
  spontaneous spindle bursts in, 192-194
  stimulation of, 192-194, 208-209, 221, 223
    recruiting responses and, 200, 216-217
    reduction of slow potentials and, 214, 216-217
Nucleus medialis dorsalis, 254, 294
  cell discharge patterns of, 159
    during delayed response tasks, 159-163
  lesions in, 227, 228
  parvocellularis region of, 159
  relationship to prefrontal cortex, 157, 163-165
Nucleus paracentralis, 220
Nucleus reticularis gigantocellularis, 179
  corticoreticular fibers to, 178
  reflex inhbition and, 172, 174, 181
Nucleus reticularis lateralis, 174, 179

Nucleus reticularis pontis caudalis, 175, 179
  evoked potentials in, 178
Nucleus reticularis pontis oralis, 175, 179, 180, 181
  corticoreticular fibers to, 178
Nucleus reticularis ventralis, 174, 179
Nucleus ventralis anterior
  cryogenic blockade of, enhancement of primary evoked potentials and, 204-206
  stimulation of, 215
Nucleus ventralis lateralis, 192-193
  augmenting responses and, 189-191, 200-201, 203, 216-217, 221
  reduction of slow potentials and, 214, 216-217
Nucleus ventralis posteromedialis, 220
Nystagmus
  caloric, hyperreflexia of, 63
  spontaneous, 63

O

Obstinate progression, 226-227, 228
Occipital zones, 126
  biopotential correlations of, 95, 96, 97, 104, 105
  negative shifts in, 136
Occipitotemporal sulcus, 141
Ocular fundus, stagnant phenomena in, 63
Optic nerve, 212
Optic tract, 280
Orbital gyrus, 168, 214
  divisions of, 294
  electrical stimulation of, 167, 177, 180, 181
  reflex facilitation and, 174-176, 178, 179, 182
  reflex inhibition and, 172-174, 180-181
Orbital sulcus, 168
  amplitude changes of masseteric reflex and, 175-176
Orbitofrontal cortex, ablation of, 204-205
  behavioral effects of, 226-227
  enhancement of primary evoked potentials and, 204-205
Orbitoinsulotemporal cortex, 293
Orienting reaction, 15-21, 22, 29
  alpha waves and, 59-62, 69, 92

development of, 112
extinction of, 115
firing patterns of thalamic neurons
  during, 162
galvanic skin response component of, *see*
  Galvanic skin response
inhibition in, 309
pathological changes in, 62
psychophysiological components of, 9,
  301
stability of, 6-9
symptoms of, 6
vegetative components of, 71
Overtraining, 216
  negative steady potential and, 213,
    218-219, 221
Oxygen availability, 109-110

P

Papilloedema, 63
Parabiosis, 43
Paradoxical phase, 5
Paradoxical reaction, parietooccipital
  lesion and, 50
Parietal cortex
  frontal lobe ablations and responses in,
    203
  habituation in, 116
  transcortical negative variation in, 131,
    133
Parietooccipital regions
  evoked potentials in, 31-33, 36-37, 41-42,
    43-46, 47, 48-49
  lesions in, 21, 81
    orienting reaction and, 7
    paradoxical reaction and, 50
    problem solving and, 19
    visual perception and, 19
Parietotemporal lesions, 81
Periamygdaloid cortex, 294
Perifocal reaction, 38
Perseveration, 237
Picrotoxin, effect on masseteric reflex
  inhibition, 175, 176, 181
Pontine facilitatory area
  electrical stimulation of, 172-173, 174-175
  evoked potentials in, 177, 178, 179,
    181-182

Pontine tegmental region, 81
  bilateral projections to, 178
Postcentral region
  orienting reaction and lesions of, 7
  transcortical negative variation in, 135
  waveforms in, during waiting period, 129,
    131, 133, 135, 136
Posterior cranial fossa, 81
Posterior lesions, reactivity of alpha
  frequencies and, 82
Posterior orbital cortex, 294
Posterior-parietal zones, biopotential correlations of, 95, 96, 102, 103, 104
Posterior sigmoid gyrus, 190, 200, 223
  recruiting responses in, 192, 194
  spontaneous spindle bursts in, 192, 194
Posterior suprasylvian gyrus, spontaneous
  spindle bursts in, 192, 194
Posterofrontal lesion, 46-48, 67
  influence on evoked potentials, 48-49
Posture, praxis of, 64
Potentials, *see also* Evoked potentials, Slow
  potentials
  slow changes of, 110, 125
Praxis, disturbance of, 7
Precentral motor cortex, waveforms in, during waiting period, 129, 131, 135, 136
Precruciate sulcus, 238
Prefrontal zones
  biopotential correlations of, 95, 96, 97,
    98
  lesions in, 237, 238
    behavioral effects of, 238-249
  relationship to thalamus, 157, 164-165
  role in delayed alternation tasks, 139
  spontaneous activity of, 164
Premotor zone
  expectancy waves and, 118
  lesions of, 40, 43, 238
    behavioral effects of, 237-249
    influence on evoked potentials, 43-46
Pressure
  intracranial, 38, 40, 46, 63, 75, 80, 81, 83
  of spinal fluid, 63
Primary signaling system, 112
Principalis cortex, delayed response tasks
  and, 150-152
Principal sulcus, 141
  role in delayed response tasks, 140
  electrical stimulation and, 152

Probability, *see* Contingency
Problem solving, disturbances of, 14-21
Procaine hydrochloride, 169
Progressive equivocation, 119
Proreus gyrus, 238
Pseudomotor neurons, 258
Psychoregulation, 4
Pyramidal insufficiency, 63
Pyramidal neurons, presynaptic inhibition of, 121
Pyramidal tract, 258

R

Reaction time, *see* Latency
Recoding, inability for, 11
Recovery function, 308-309
Recruiting response, 200
    frontal granular cortex and, 189, 190, 192-194, 196, 202-203
    inferior thalamic peduncle and, 189-193, 196, 202-203
    negative steady potential shift and, 213, 220-221
    reduction of slow potentials and, 214-215, 216-219
    thalamocortical blockade and, 204-209, 215, 216-217, 222-224
    waveform of, 201-202
Reflex
    facilitation of, 174-176, 178, 179, 182
    inhibition of, 167-174, 178-182
    masseteric, *see* Masseteric reflex
    nocioceptive, 264
    segmental, 264
    soleus, *see* Soleus reflex
    spinal, 167
    tendon, 46, 63
Reflex arc, 3, 4-5
Regulation, 4-5
    of activation process, 5-10, 22-23
    of evoked potentials, 29-30, 49-50
        methodology, 30-31
        in normal subjects, 31-35
        in patients with frontal lobe lesions, 35-49
    symptoms of derangement of, 5-10
        in complex behavior, 10-14
Reinforcement
    partial, effect on contingent negative variation, 119-120

processing, 305-307
    social, 121
    steady potential shifts and, 150-151
    symbolic, 121
Repetition
    of instructions, 10-12
    of movements, 11
Rest, biopotential correlations during
    in normal subects, 94-95, 98
    in schizophrenics, 104, 105
Reticular activating system
    ascending, 186
        Aminasine blockage of, 102
        inhibitory influences on, 40
        relay of signals through, 114
    enhancement of evoked potentials and, 212
Reticular formation
    descending influences on, 9
    difference from medial thalamic system, 187
    effect of Aminasine on, 98-99, 106
    high-frequency stimulation of, 186, 187
    lesions of, 186, 187, 195, 202
        effect on parietal responses, 203
        epileptic focus produced by, 197, 200, 201, 203
        spontaneous spindle bursts and, 192-194, 196
    reflex inhibition and, 167, 174, 181-182
Reticulospinal tract, 177-179
    orbitomedullary fibers and, 178, 182
    orbitopontine fibers and, 179, 182
Reticulothalamocortical system, functions of, 186
Rolandic cortex, slow negative potential in, 213

S

Schizophrenics, biopotential correlations in, 102-106
Secondary signaling system, 7, 112, 113
Sedatives, effects on synchronous waveforms, 6
Segmental reflex, 264
Selectivity, derangement of, 23
Sensory-motor cortex, 258
Sensory nerves, peripheral, 281
Septum pellucidum, 63

SUBJECT INDEX

Servomechanism, 3
Sleep
  electroencephalogram in, 53, 186, 195-196
  inducement of, 167
  spindle, inferior thalamic peduncle and, 195, 196
  violation of, 40
Slow potentials, 212-214, 220-222
  reduction of
    associated with augmenting responses, 214-215, 216-217
    conditioned negative slow potentials, 215-216, 218-220
Sluggishness, 64
Social influence, 113
  effect on contingent negative variation, 118, 120-121
Sodium methohexital, 169
Soleus reflex
  elicitation of, 169
  facilitation of, 175
  inhibition of, 172-173, 180
    bulbar reticular formation and, 174-175
Spatial deficit, 7, 19, 277-278, 286, 287
  cue processing and, 151, 272, 297-299
Speech, 7, 238
Spinal fluid, pressure of, 63
Spinal reflexes, inhibition of, 167
Spindle bursts, 186
  thalamocortical blockade and, 188, 190-196
Spontaneity, lack of, 64
Startle reactions, lack of activation of nucleus medialis dorsalis and, 160
Stereotype(s)
  dynamic, 237, 249
  inert reproduction of, 12-14
  logical, 20
  motor, 23
Stereotypy, 63
Stimulation
  acoustic
    alpha rhythm and, 59-62, 67-69, 74-85
    effects on electroencephalogram and galvanic skin response, 253-256, 269-270, 287
  electrical, 30, 110, 293
    for cortical conditioning, 169
    delayed response tasks and, 139-152, 158-165

  evoked potentials and expectation of, 31-34, 36, 41-45, 48-49
  galvanic skin response and corticosteroid response to, 258-263, 270-271
  masseteric reflex inhibition and, 170-174, 180-181
  for reflex elicitation, 169
  therapeutic, 109-110
  modality of
    response attenuation and, 115
    response latency and, 114
  optic, 30, 223
    electroencephalogram and galvanic skin response and, 256-258, 269-270
    evoked potentials in frontal lobe lesion patients and, 35-49
    evoked potential in normal subjects and, 31-35
    thalamocortical blockade and, 204-211
Stimulus
  conditional, 112-113
    contingent negative variation and, 121
  imperative, 112-113, 128, 237
  indicative, 112
  intensity of
    cortical phasic states and, 5
    frontal response and, 115-116
  meaningfulness of, see Meaningfulness
  novel, responsiveness to, 253, 264, 271, 285, 301
  relation to separate points of cortex, 9
  signaling, influence on evoked potentials, 40, 49
  terminology for, 112
  unconditional, 112
    contingent negative variation and, 121
Striatal system, 81
Strychnine, effect on masseteric reflex inhibition, 175, 176, 181
Subcortical ganglia, 43, 75
Subthalamus, 188
Sulcus principalis, 254, 283
  electrical stimulation of cortex around nucleus medialis dorsalis activity and, 162
Superior gyrus, 40-41
Superior longitudinal sinus, 43
  evoked potentials and, 43-46
Superior temporal sulcus, 141
Suppressive reaction, see under Alpha rhythm

Switch-overs, inability to perform, 11-12, 14
Sylvian fissure, 63, 66

T

Task difficulty, biopotential correlations and, 98
  inert stereotypes and, 12-14
Temporal organization, frontolimbic formations in, 295, 297
Temporal zones, 63, 64, 188, 294
  biopotential correlations of, 94
  orienting reaction and, 7
Tendon reflexes, 46
  asymmetry of, 63
Δ-1-Tetrahydrocannabinol, thalamocortical blockade and, 196-197, 198-199
Thalamic neurons, spontaneous activity of, 158-159
Thalamic nuclei, 293
  relay of signals from, 114, 115
  stimulation of, 201
    augmenting responses and, 189, 192-193, 201-202
    high-frequency, 186-187
    low-frequency, 186-187, 197, 200
    recruiting responses and, 189, 192-193, 201-202
Thalamocortical system, projections of, 187
Thalamus, 136, 157
  recruiting responses in, 192, 194
  spontaneous spindle bursts in, 192, 194
Thermoregulation, 266-269, 270-271, 285, 287
Theta rhythm, 64, 67, 69
  influence of instructions on, 43, 45
  intensification of, 77, 82
  site of lesion and, 72
Third ventricle, anterior division of, 47
Thought
  effect of phasic states on, 5
  fragmentation of, 63
  mode of, alpha rhythm and, 91
  verbal, disturbance of, 20, 23

Tibialis anterior, inhibition of reflex to, 180
Transcortical negative variation, 129
  topography of, 131, 136
Transmission, speed and detail of, 281-282
Transparent septum, 66
Trigeminal motor nuclei, 177
  mediation of inhibitory influences to, 178-182
Trigeminal nerve, 169
  reflex inhibition and, 168

U

Ultraparadoxical phase, 45

V

Ventrolateral thalamus, 136
Ventromedial bulbar reticular formation
  electrical stimulation of, reflex inhibition and, 167, 174, 181-182
Verbal tasks, inert stereotypes in, 14
Visual cortex, 223
  activation of, 30
  evoked potentials in, 204-209, 212, 222-224
  habituation in, 116
Visual defects, 7, 16-19, 63
Visual tasks, *see also* Discrimination tasks
  constructive, derangements in, 15-19
Vomiting, 63

W

Waiting period, elicitation of contingent negative variation during, 128-129, 133, 135
Wakefulness, alpha rhythm during transition from, 53, 58-59
Waveform
  definition of cortical responses by, 203
  of recruiting and augmenting responses, 201-202, 214